Geography and History

3

Geography and History 3 is a collective work, conceived, designed and created by the Secondary Education department at Santillana, under the supervision of **Teresa Grence Ruiz**.

WRITERS

Martin Minchom
Lesley Thompson
Pedro Adiego Sancho
Olga Cabeza García
Rosa López Pérez
Ana Redondo del Río
María Luisa Vázquez Sánchez
Francisco Javier Zabaleta Estévez
Cynthia Donson
Fernando Abascal Altuzarra
José M.ª Caballero Martínez
María Ángeles Fernández de Bartolomé
Juan Martín Martín
Raquel Rubalcaba Bermejo
Javier Velilla Gil

PROOFREADING
Ricardo Gómez Tovar

MANAGING EDITOR
Sheila Tourle

Do not write in this book. Do all the activities in your notebook.

SANTILLANA
Richmond

Geography index

History index

The book is organised into 12 units. Every section of a unit helps the students to develop their key competences.

Opening double page: presentation of the unit

The **contents** of the unit are presented at the beginning.

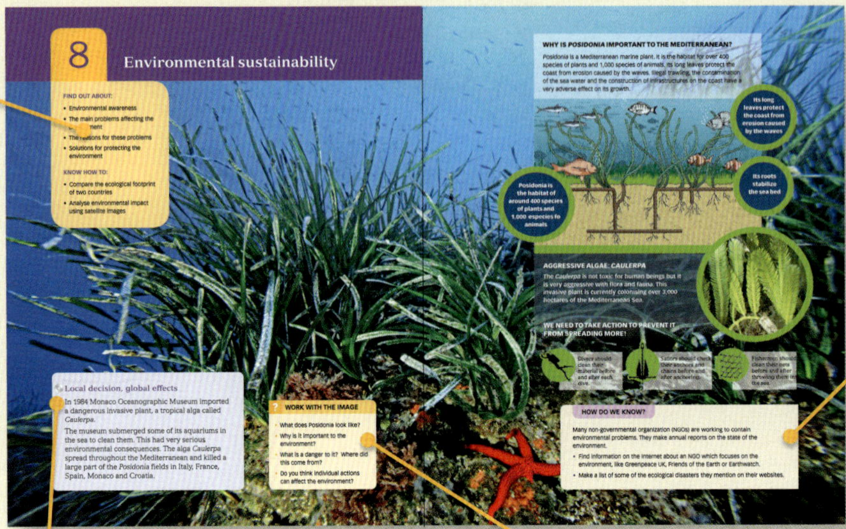

How do we know?: the student is introduced to the techniques and sources of geographers and historians.

A **CD symbol** shows that the text is recorded.

The **Work with the image** section checks the student's understanding of the visual material and the text.

The content pages

The **main contents** are clearly presented.

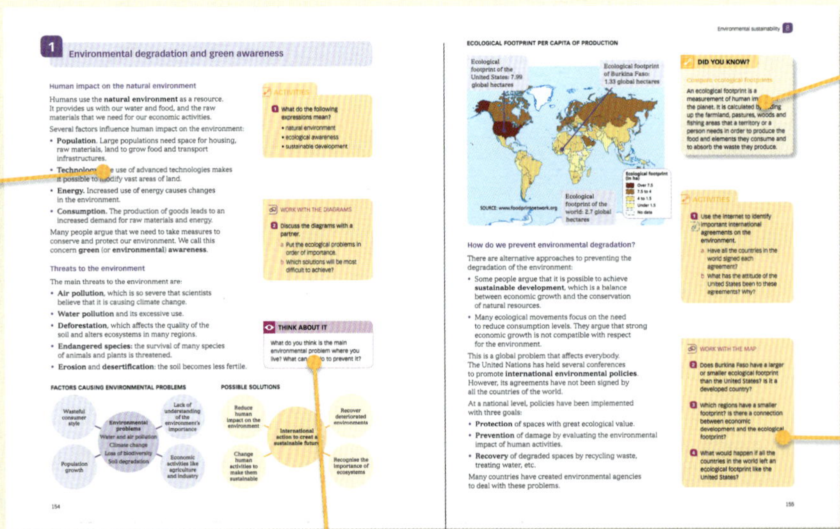

The **Did you know?** section widens the students' range of knowledge.

A series of **activities** cover all the main contents of the page.

Think about it invites the students to reflect about challenging concepts.

Activity Round-up: learning to learn

This section always begins with a **summary** of the key ideas.

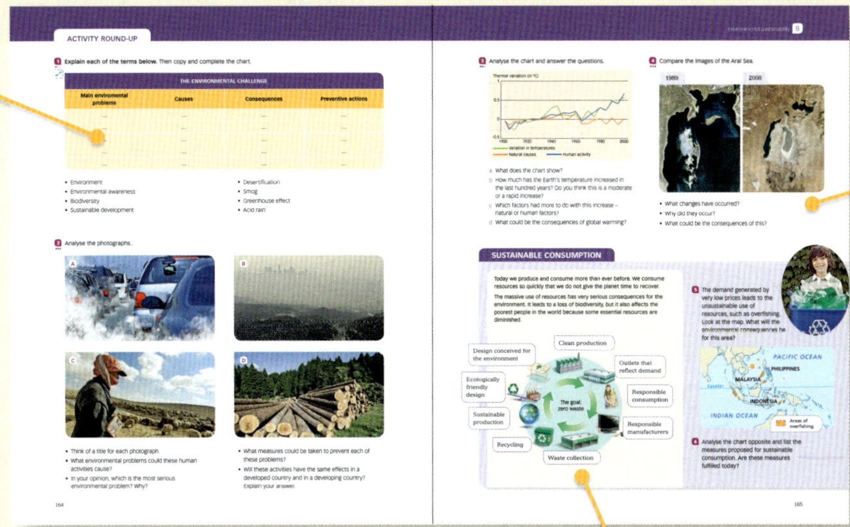

Numerous activities provide a full **round-up** of the unit, and encourage learner autonomy.

The **final section** analyses environmental problems in the Geography units, and our inheritance from the past in the History units.

Know how to...: key competences

The students develop **key skills** in Geography and History.

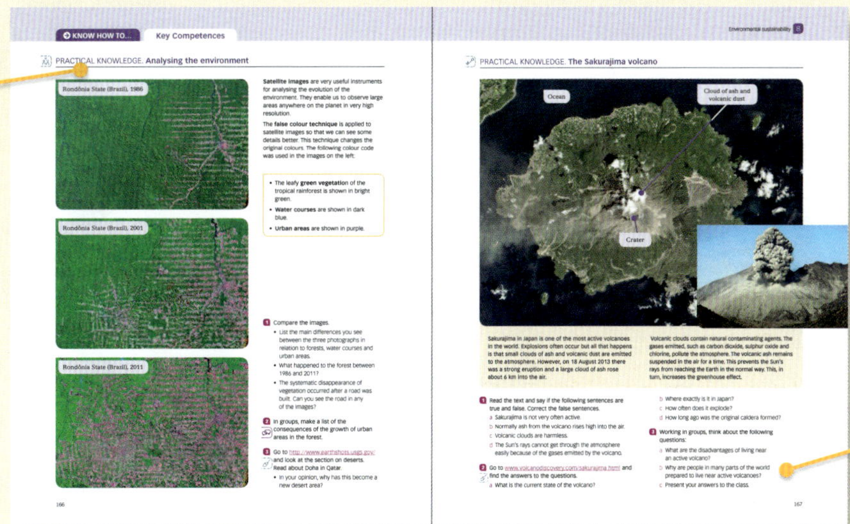

The students do individual activities, pairwork and **group projects**, making full use of internet resources.

Key Competences

- Linguistic competence
- Competence in mathematics, science and technology
- Digital competence
- Social and civic competence
- Cultural awareness and expression
- Learning to learn
- Initiative and entrepreneurship

1 Physical geography

FIND OUT ABOUT:

- How relief is formed
- Types of relief
- The relief and water of the continents
- The Earth's climates and landscapes
- The physical relief of Spain

KNOW HOW TO:

- Use maps to link geographical features to each other
- Analyse the effects of cyclones

Base camp

? WORK WITH THE IMAGE

- How do you imagine the view from the top of Everest?
- Who first climbed Everest? Look for information about other climbers who have done it.
- Find out about Reinhold Messner. What did he achieve on Everest?
- What is a sherpa? What is a porter?

The highest peak in the world

Everest, at 8,850 metres, is the highest mountain in the world. On 29th May 1953, the New Zealander Edmund Hillary and the sherpa Tenzing Norgay, both members of a British expedition, reached the top for the first time. Other climbers had tried to do the same unsuccessfully since the first expedition to Everest in 1920.

The 1953 Everest expedition

Total members: **406**

The 14 mountains over 8,000 metres high

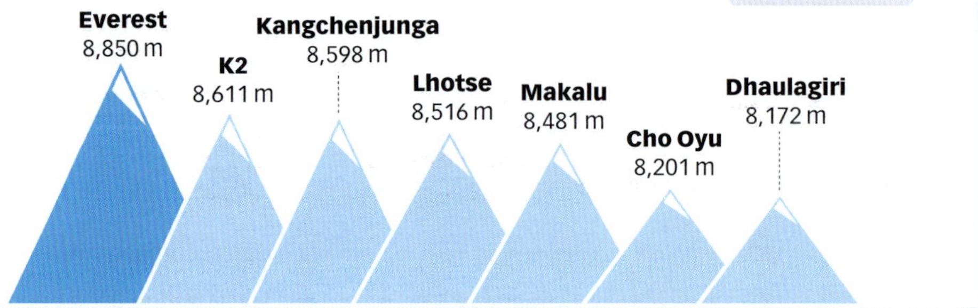

Theodolite: *the instrument that was first used to calculate the height of Everest.*

DID YOU KNOW?

Mount Everest owes its name to the topographer Sir George Everest (1790–1866). He calculated the height of the mountain – but he never climbed to the top.

The South Col route

Edmund Hillary

Tenzing Norgay

Route
Camps
Frontier

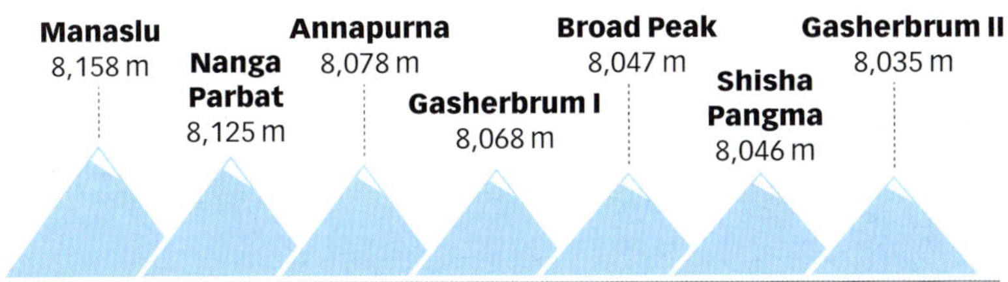

HOW DO WE KNOW?

Geodesy is the scientific study of the shape and dimensions of the Earth, and its orientation in space. It gives us the geographic coordinates of a place. This enables other fields like topography and cartography to make maps and plans.

- What are topography and cartography?
- Which mountains on Earth are higher than 8,000 metres? In which mountain ranges are they situated? Investigate.

1 Relief formation

What is relief?

The variations in elevation of the Earth's surface are called **relief**. These variations are found on the surface of our continents and on the ocean bed.

Many changes in relief take place over an extremely long period of time.

How is relief formed?

The theory of ~~plate tectonics~~ explains relief formation. According to this theory, the Earth's crust is made up of **plates** which float on the top layer of the mantle. They separate, slide against each other or collide. New types of relief are then formed.

- When the plates separate, **rifts** (cracks) are formed in the ocean's crust. The magma inside the earth pushes up through the cracks. When it cools, it forms a new crust. This separation also takes place on the continents. The Great Rift Valley is a huge fracture crossing a large area of eastern Africa.
- A plate on the ocean's crust sometimes collides with the edge of a continental plate, and is pushed down into the Earth's interior. High temperatures melt the crust, and turn it into magma. The area where the crust is destroyed is called a **subduction zone**.
- Magma pushes up through the continental crust, creating **volcanoes** on the Earth's surface. This was the origin of the Andes mountain chain.
- Two continental plates sometimes collide, and the pressure forms **folds** (undulations) in the Earth's surface. **Faults** occur when the crust is rigid and fractures. In this case, some blocks sink while others rise and form mountains.

The biggest changes in the Earth's crust take place at the boundaries between plates. They are areas of intense seismic and volcanic activity.

THE MOVEMENT OF TECTONIC PLATES

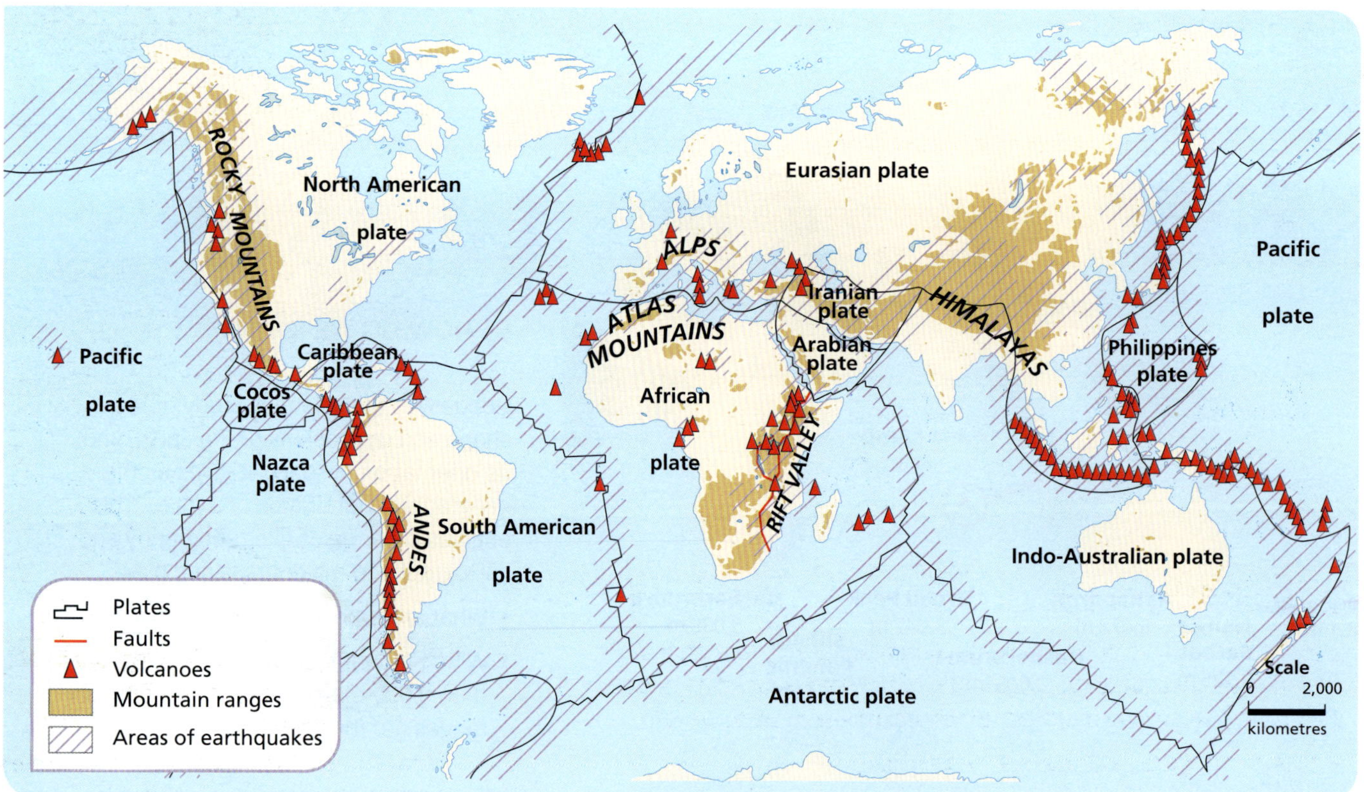

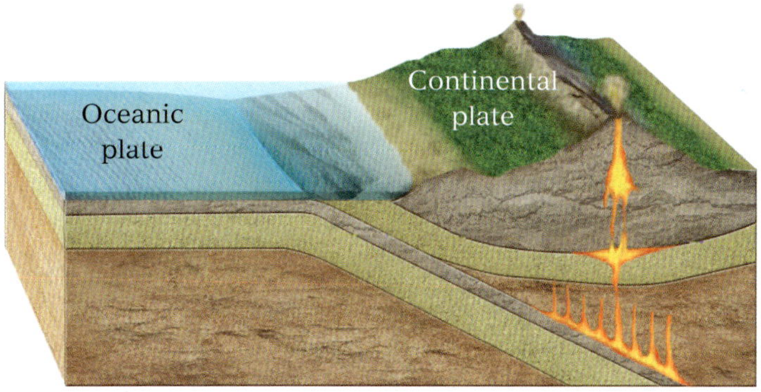

Subduction.

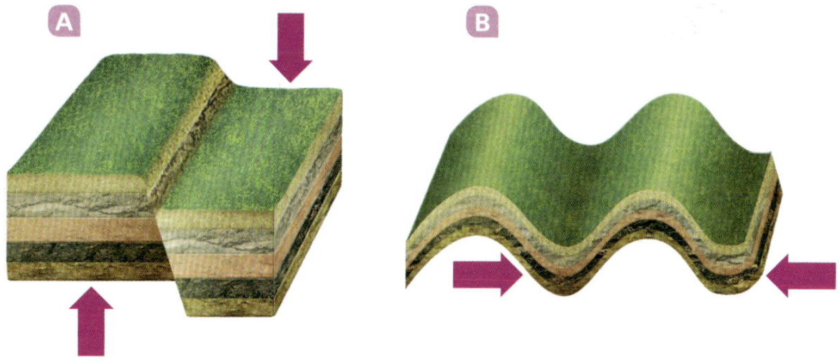

The formation of faults (A) and folds (B).

Exogenous processes

In addition to forces beneath the Earth's surface, relief is also formed by **exogenous** (or external) processes:

- **Erosion** of rocks, which are worn away, fragmented or dissolved.
- **Transport** of the eroded materials.
- **Sedimentation**, or the deposition of materials in another place.

The erosion of rocks is caused by the interaction of **external agents** over an extended period of time. These forces are mainly **water**, **wind** and **temperature**:

- Rivers carve out deep valleys and gorges. Glaciers, or large masses of ice, also cause erosion when they move.
- The force of the sea erodes cliffs, changing the coastline. Sea water transports the material from the cliffs and deposits it in other areas, forming new beaches.
- Sharp changes in temperature can break up rocks. Water sometimes filters down through the cracks in hard rocks. If the water freezes, it then exerts pressure on the rocks and may break them into pieces. The **scree** on mountains are accumulations of rock fragments which originated in this way.
- Limestone and gypsum do not fragment. The action of water dissolves them and they form **karstic relief**. The karst rock formations in caves and subterranean galleries were formed in this way.

THINK ABOUT IT

Think about your area. Are there any interesting rock formations? Which external agents (like water and wind) have shaped its relief? And have people also modified its natural relief? In what ways?

- Share your ideas with the rest of the class.

WORK WITH THE PHOTOS

1. What processes do the photos show? How do they create, destroy or modify relief?

The Grand Canyon was formed by the Colorado River in Arizona (USA).

The rocks of Ciudad Encantada (Cuenca) are karst formations, which have been eroded by water.

2 The Earth's relief

Continental relief

The main forms of continental relief are the following:

- **Mountains**, which may be isolated or grouped together in **mountain ridges**. **Valleys** are areas of low land between mountains.
- **Plains** are flat areas of land near the coast or in the basins of large rivers. **Plateaus** are high raised plains. **Depressions** are very low areas that may even be below sea level.
- On the **coast** we find other relief forms such as **cliffs**, which are steep rock formations, and **beaches**, which are found in flat coastal areas.
- **Gulfs** and **bays** (small gulfs) are where the sea extends inland. **Capes** and **peninsulas** are areas of land that extend into the sea.

Oceanic relief

Relief on the ocean floor is as varied as continental relief:

- A **continental shelf** is a large plateau that surrounds a continent, reaching a depth of 200 m. A **continental slope** leads down from the continental shelf into the deeper part of the ocean.
- **Abyssal plains** are vast plateaus at a depth of 3,000-7,000 m. **Mid-ocean ridges** are long mountain ridges that are found on the abyssal plains. Some of the peaks of underwater mountains rise above the **ocean's** surface and form islands.
- **Ocean trenches** are long, deep cracks in the ocean floor. The deepest ocean trench is Challenger Deep (11,000 m) in the Pacific Ocean.

THE EARTH'S RELIEF

The influence of relief on our lives

The nature and altitude of relief influences climate, vegetation and land use. It affects both population and economic activity:

- Most of the world's population lives in **flat areas** below an altitude of 500 metres. These areas are mainly near the coast and in river valleys. The climate is mild, conditions are favourable for agriculture and there are good communications.
- **Mountainous areas** are usually sparsely populated. The climate is cold and wet, there is little agriculture and communications are expensive. The economy is often based on livestock or mining and, more recently, tourism.

However, **technological advances** have reduced our dependence on the environment. For example, modern communications reduce the isolation of mountain communities.

ACTIVITIES

1 Define the following terms:

- Mountain range
- Continental shelf
- Abyssal plain
- Depression
- Ocean ridge
- Gulf

2 Look for three photos on the Internet showing relief that has been modified by water, wind and people. In groups, show the photos to your classmates and explain the processes that have taken place.

3 Compare continental and oceanic relief. Are there any similarities?

WORK WITH THE IMAGE

4 What are the main forms of continental relief?

5 Which types of relief are found in a coastal landscape?

1. Peak. 2. Mountain range. 3. Valley. 4. Plateau. 5. Plain. 6. Depression. 7. Cliff. 8. Cape. 9. Gulf. 10. Beach. 11. Island. 12. Peninsula. 13. Continental shelf. 14. Continental slope. 15. Abyssal plain. 16. Oceanic ridge. 17. Marine trench.

3 The continents

- **Africa.** The continent has a high average elevation of 750 metres because of its many plateaus. There are great deserts, like the Sahara and Kalahari. The main mountain ranges are the Atlas Mountains and the Drakensberg Mountains. The Great Rift Valley is in the east. The coastline is regular.
- **Asia.** Mountain ranges are mainly in the centre. The Himalayan mountain range contains Mount Everest, the highest peak in the world. Plateaus include the Plateau of Tibet, the highest in the world. Vast plains are crossed by long rivers. The main basins contain inland seas: the Caspian, Aral and Dead Seas. The coastline is irregular. There are many peninsulas, gulfs, islands and archipelagos: Japan, the Philippines, Indonesia…
- **America.** The great mountain ranges (the Rocky Mountains, Sierra Madre and the Andes) run parallel to the Pacific Coast. In the interior, plateaus include the Matto Grosso and the Bolivian Altiplano. The Great Plains are in North America. The coastline of North America is more irregular than that of South America.
- **Europe.** Great plains occupy the centre and east of the continent. The oldest mountains are in the north (Scandinavian mountains) and east (Urals), but they are not high because of erosion. High, young mountain ranges form an arc around the Mediterranean Sea: the Pyrenees, Alps, Carpathians, Balkans and Caucasus. The coastline is irregular. It has many peninsulas and islands like Iceland and Sicily.
- **Oceania.** Relief varies from one island to another. Australia has coastal plains, low desert in the interior and the Great Dividing Range in the east. New Guinea and New Zealand are mountainous. The Pacific islands are mainly low.
- **Antarctica.** It has the highest average altitude (2,000 metres) and is mainly covered by ice.

THE EARTH'S RELIEF

WORK WITH THE MAP

1 Prepare a table with information about Africa, Asia, America, Europe and Oceania. Include at least one example of each of the following:

- Mountain ranges. (Name the highest peak on each continent.)
- Plateaus and plains.
- Peninsulas and islands.

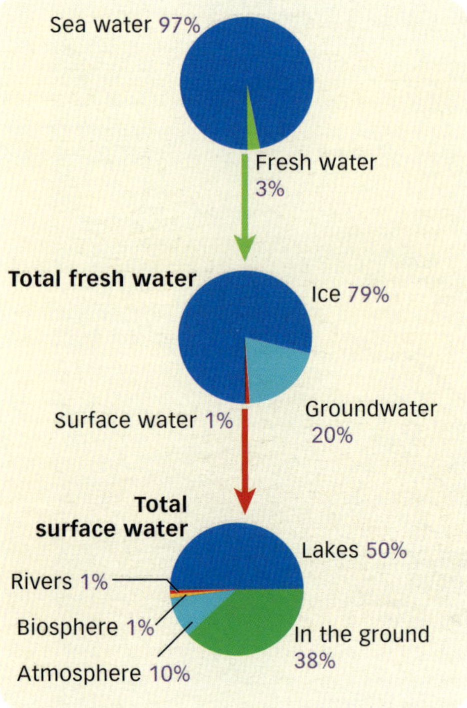

The distribution of water on Earth.

4 Water

Water covers more than 70% of the Earth's surface. Some water is found on the continents, but most of it is in the oceans and seas.

Continental water

Rivers, lakes, groundwater and ice contain **fresh water.** They make up about 3% of all the Earth's water.

- **Rivers** are continuous flows of water that originate in rain, melting ice or groundwater. They flow into a sea, a lake or another river. Rivers are important because they provide water for human consumption, agriculture, hydroelectric energy and industry. They are also important communication routes.
- **Lakes** are enclosed masses of water which accumulate in depressions. Salt water lakes are called **inland seas**.
- **Groundwater** accumulates when rain and river water infiltrates porous ground. This water flows downwards until it reaches a layer of impermeable rock and forms an underground deposit called an **aquifer**. Groundwater feeds rivers and lakes, and it is our main reserve of drinking water.
- Ice is found at the **polar ice caps** and in mountain **glaciers**. Antarctica holds most of the Earth's fresh water in the form of ice.

Marine water

Oceans and seas hold 97% of the Earth's water. They are in constant movement:

- **Marine currents** are masses of water which move like rivers through the oceans. A **warm current** has a higher temperature than the surrounding water. A **cold current** has a lower temperature. Currents have a considerable influence on the climate of coastal regions.
- **Tides** are the daily rise and fall in the sea level caused by the gravitational pull of the Moon and Sun. **High tide** is when the sea level rises, and **low tide** is when it falls.
- **Waves** are the undulating movements of water on the surface of the sea. They are caused by the wind.

WORK WITH THE GRAPH

1. What percentage of the Earth's water is saltwater? What percentage is fresh water?
2. Which types of water are continental? Which one has the greatest volume?

THINK ABOUT IT

Scientists say that rising temperatures are causing ice to melt in Antarctica.

- What is the cause of these rising temperatures?
- If this trend continues, what effect will it have on our planet?

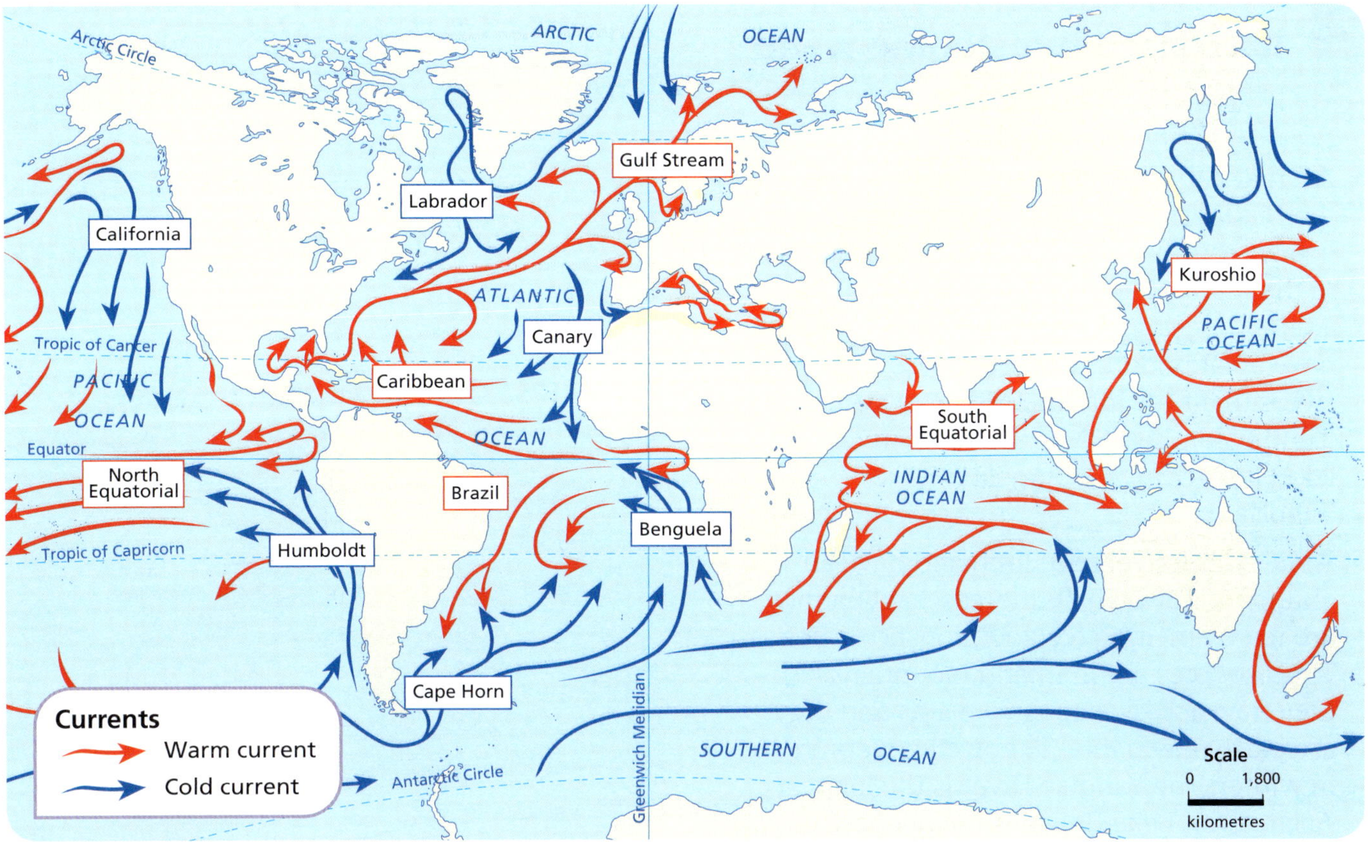

OCEAN CIRCULATION

WORK WITH THE MAP

3 Why are European coastal regions warmer than most places at the same latitude?

Water as a scarce resource

Water is a precious resource:

- Most water is not **suitable for consumption**. Nearly all the Earth's water is salt water and cannot be used. Most fresh water is found in the form of ice or deep under the ground.
- **Overexploitation** of the Earth's water has led to the depletion of aquifers, and a sharp reduction in the size of many lakes.
- The **distribution** of water on Earth is uneven. Its availability can vary between seasons, and from one year to the next.

Hydraulic engineering (for example reservoirs, canals, desalination plants and wells) helps to make water available all year round. These infrastructures have a considerable impact on landscapes and the environment.

THINK ABOUT IT

The importance of water in our lives

Water is essential for life. We need it for drinking and we use it for washing, irrigation, transport, energy production, and in industrial processes. It is also an important tourist resource. For thousands of years, people have made their homes by the water in coastal areas and near rivers, lakes and springs.

- In small groups, exchange ideas about all the ways that water is used. Then choose three examples to illustrate the influence of water in our lives, and discuss them with the class.

ACTIVITIES

4 Define the following terms:

- Inland sea
- Reservoir
- Aquifer
- Glacier
- Tide
- Wave

5 Where does most of the water that we use come from?

5 Rivers and lakes of the world

- **Africa.** Among the rivers that flow into the Mediterranean Sea, only the Nile River has an abundant flow. It is the longest river in the world. The rivers that flow into the Atlantic Ocean, like the Niger and Congo, have an abundant flow. So too do the rivers that flow into the Indian Ocean, like the Zambezi River. There are many lakes in Africa. The biggest is Lake Victoria.
- **Asia.** Many rivers are long and have an abundant flow, which increases when they are fed by melting ice and monsoon rains. Their sources are in inland mountains, they flow across extensive plains and they are partly navigable. The Arctic rivers freeze in winter. The Yangtse River is the longest Asian river. Lake Baikal is the deepest lake in the world, and the Caspian Sea is the biggest lake.
- **America.** On the Atlantic side, rivers like the Mississippi and the Amazon are the longest and have the greatest flow. On the Pacific side, rivers are shorter because their sources are in the mountains near the coast. The five Great Lakes are in North America.
- **Europe.** The flow of European rivers is determined by climate. The Atlantic rivers have the greatest flow because they receive abundant rainfall all year. As a result, they are navigable. In general, the Mediterranean rivers have a low, irregular flow and their water levels decrease in the summer. Important European rivers include the Danube, which flows into the Black Sea, and the Volga, which is the longest on the continent. The biggest lakes are Ladoga and Onega in north-west Russia.
- **Oceania.** The longest rivers are the Murray and its tributary, the Darling, in Australia.

THE WORLD'S RIVERS AND LAKES

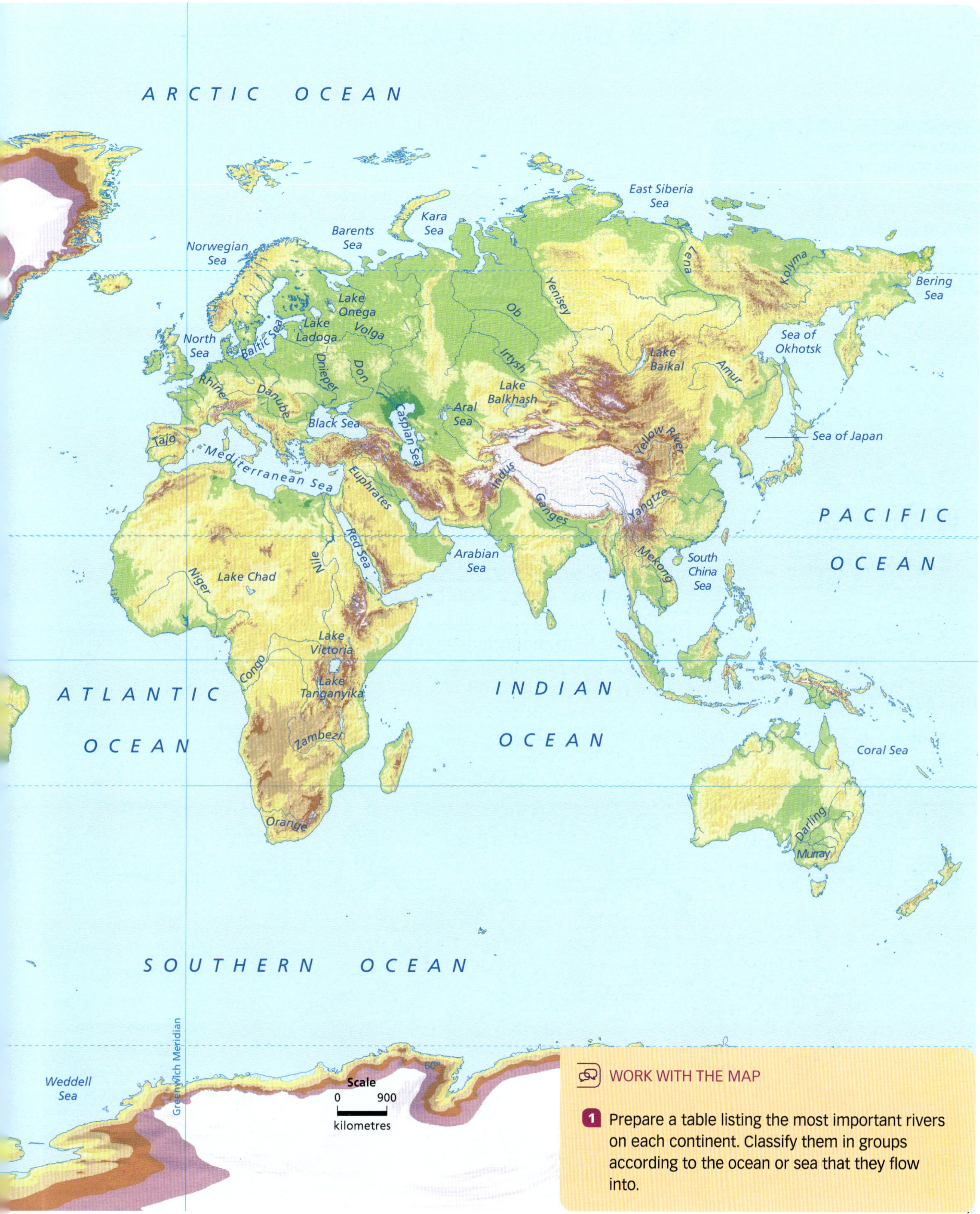

WORK WITH THE MAP

1. Prepare a table listing the most important rivers on each continent. Classify them in groups according to the ocean or sea that they flow into.

6 Climates and landscapes

The angle of the Sun's rays and climate zones.

The sun does not warm all the Earth's surface to the same extent. When the Sun's rays are perpendicular (as in the tropical zone), the amount of solar energy is considerable. When the Sun's rays fall at an oblique angle (as in the polar regions), the amount of energy is very low. In terms of latitude, there are five main climate zones:

- A **hot zone**, between the tropics. The average annual temperature is over 20 °C.
- Two **temperate zones**, between 30° and 60° latitude in both hemispheres. Solar radiation is not as high, and there is a clear seasonal difference between summer and winter.
- Two **cold zones**, from 60° to 90° latitude in both hemispheres. Solar radiation is weak and temperatures are low all year round.

Altitude, **relief** and proximity to the **sea** also influence climate. There are different types of climates within each main climate zone: **hot** (equatorial, tropical and desert), **warm** (Mediterranean, oceanic and continental) and **cold** (polar and mountain).

Climate has a great influence on human societies. Extreme temperatures and humidity make life more difficult: deserts, forests, high mountains and the polar regions are sparsely populated. More than 70% of the world's population lives in temperate zones.

WORK WITH THE IMAGE

1. Explain the influence of the angle of the Sun's rays on the Earth's climates.

THE EARTH'S CLIMATES

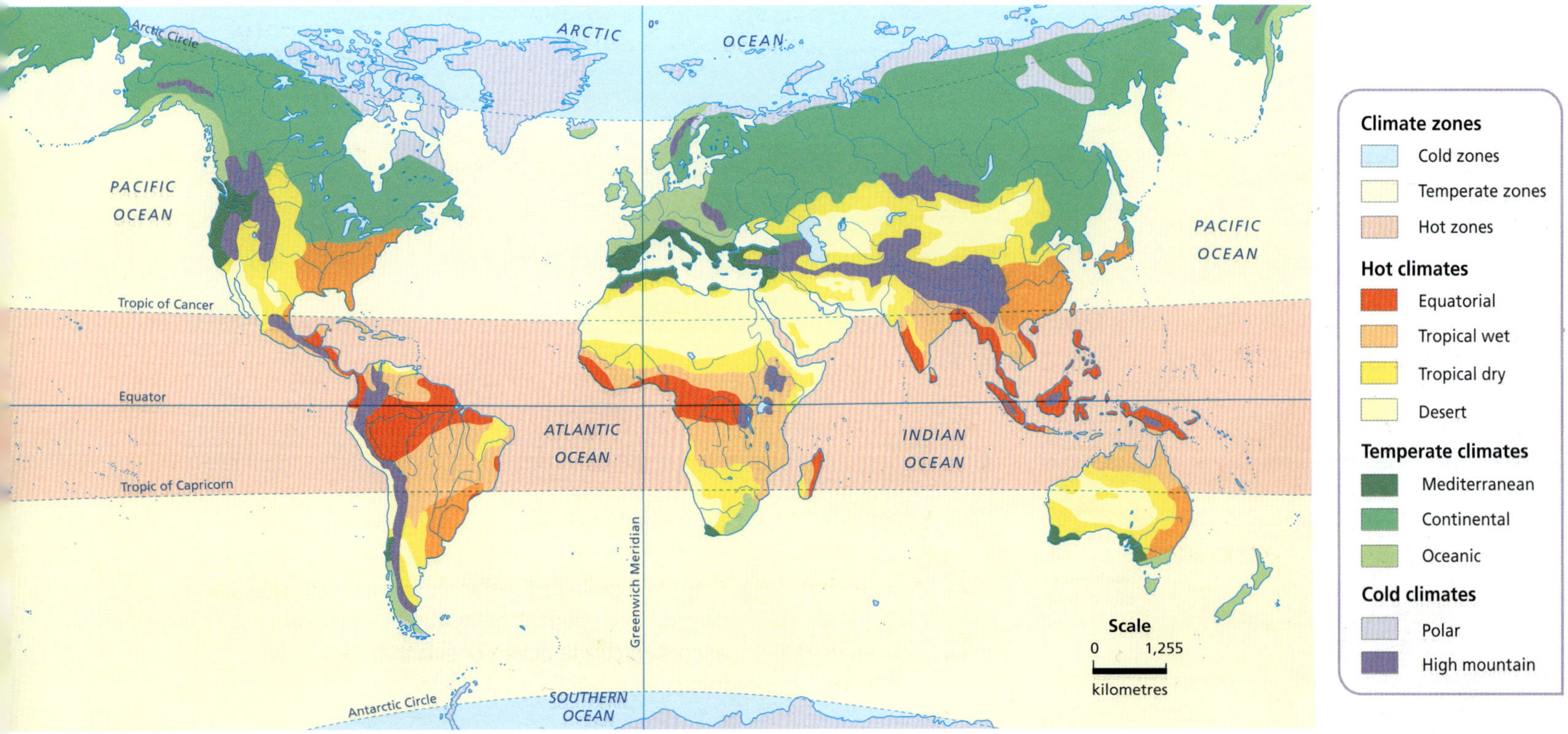

	CLIMATES AND LANDSCAPES				
	CLIMATE	**LOCATION**	**TEMPERATURES**	**PRECIPITATION**	**LANDSCAPE**
HOT CLIMATES	**Equatorial**	Near the Equator, between 10° N and 10° S.	Average annual temperatures are around 25 °C, with little variation (less than 3 °C). It is hot all year round.	Total annual precipitation is higher than 2,000 mm. It rains every month.	Rainforest (dense green vegetation).
HOT CLIMATES	**Tropical**	Between 10° and 30° N and 10° and 30° S, at the same latitude as the desert climate.	Average annual temperature is over 20 °C. Variation is between 3 °C and 10 °C, increasing further from the Equator.	The annual total ranges between 2,000 mm (south-east Asia) and less than 400 mm (near deserts). Seasons alternate between dry and wet.	Tropical rainforest and savannah (a grassy plain with few trees).
HOT CLIMATES	**Desert**	It shares latitudes with the tropical climate.	Average annual temperatures are high, over 20 °C. There are big differences between day and night.	Scarce, under 250 mm, and irregularly distributed.	Desert. Vegetation is only found at oases.
TEMPERATE CLIMATES	**Mediterranean**	Between latitudes 30° and 40°.	Average annual temperature ranges between 10 °C and 18 °C. Temperature variations are between 12 °C and 16 °C. Summers are hot and winters are mild.	This ranges between 900 mm and 300 mm a year. Distribution is irregular and summers are very dry.	Mediterranean woodland (holm oak and cork trees) and scrub.
TEMPERATE CLIMATES	**Oceanic or Atlantic**	Between 40° and 50° latitude (and even up to 60°).	Annual mean temperature ranges from 10 °C to 15 °C. Variation is low (from 10 °C to 12 °C) Winters are mild and summers are cool.	Annual total is around 1,000 mm. It is regular throughout the year.	Deciduous forest (oak and beech) and moorland (scrub land with dense vegetation).
TEMPERATE CLIMATES	**Continental**	Inland, in North America, Europe and Asia.	Average annual temperatures are between 0 °C and 10 °C. Variation is very high. Winters are long and cold, and summers are short and hot.	This ranges between 750 mm and 300 mm. It is irregular, with maximum precipitation in summer.	Taiga or conifer forests (evergreen species like fir and pine trees). Prairies and steppes.
COLD CLIMATES	**Polar**	Near the poles.	Average annual temperatures are below 0 °C. It is never higher than 10 °C.	This is scarce, less than 250 mm annually. It mainly falls as snow.	Tundra (an area of mosses and lichens).
COLD CLIMATES	**High mountain**	High mountainous areas.	These decrease with altitude (0.6 °C every 100 metres). Summers are short and cool. Winters are long and very cold.	This increases with altitude, especially on the side where the wind blows.	Vegetation varies according to altitude: trees, bushes, scrub and prairie land.

WORK WITH THE MAP

2 Which climates are found in Europe? How are they distributed?

3 If two places are at the same latitude, will they always have the same climate?

ACTIVITIES

4 How many climate zones does the Earth have?

5 What is precipitation like in a desert climate? And in a polar climate? How do these climates differ?

6 Describe the equatorial and continental landscapes.

7 Does climate influence the distribution of world population? Explain your answer.

A granite landscape in Galicia.

Eroded limestone in Torcal de Antequera (Málaga).

A clay landscape near the Najerilla River in La Rioja.

A volcanic landscape in Gata Cape (Almería).

WORK WITH THE PHOTOS

1 Can landscapes similar to the ones in the photos be found in other parts of Spain?

7 The relief of Spain

Spain consists of most of the Iberian Peninsula, the Balearic and Canary archipelagos and the cities of Ceuta and Melilla.

The formation of Spain's relief

A long mountain range was formed in the west of the Iberian Peninsula about 300 million years ago. Over time, it was completely eroded and formed a large raised plateau. During the **Tertiary Era** (65 million to 2 million years ago), the internal forces of the Earth caused great changes:

- The plateau fractured. Some blocks sank down, filled with sedimentary material and became the Inner Plateau.
- Other blocks rose and became the Galician Massif, the western part of the Cantabrian Range, the Mountains of Toledo and Sierra Morena.
- The materials deposited at the edge of the plateau twisted, and became the eastern part of the Cantabrian Range and the Iberian Mountain Chain.
- The Ebro and Guadalquivir depressions were formed at the same time as the Pyrenees and the Baetic Chain.
- Volcanic activity on the bed of the Atlantic Ocean formed the Canary Islands.

The rocky substrate

The **substrate** of Spain is very varied. There are four main types of rocks:

- **Siliceous rocks** like granite or slate. These are very old, rigid rocks from the Primary or Palaeozoic era which began 570 million years ago. They are found mostly in the west of the Peninsula.
- **Limestone rocks** come from marine sediments deposited during the Secondary or Mesozoic era which began 230 million years ago. They are abundant in the Pyrenees, the eastern half of the Cantabrian Range, the Iberian Mountain Chain, the Catalan Coastal Chain and the Baetic Chain.
- **Clays** are fine rocks deposited in low areas in the late Tertiary era and during the Quaternary era, 1.6 million years ago. They predominate in the Inner Plateau, Ebro and Guadalquivir basins, and the coastal plains.
- **Volcanic rocks** were formed in the Tertiary era and are common in the Canary Islands. They can also be found on the Peninsula in the provinces of Ciudad Real, Almería and Gerona.

PHYSICAL RELIEF OF SPAIN

The relief of Spain

Spain has an average altitude of 660 metres and is the second most mountainous country in Europe after Switzerland. **Peninsular relief** has three main features:

- The **Inner Plateau** forms the central part of the Iberian Peninsula. It is divided into two sub-plateaus by the Central Mountain Chain. The Northern Sub-plateau is made up of the sedimentary basin of the Duero and the Castilian low-relief plains. The Southern Sub-plateau is crossed by the Mountains of Toledo and is formed by the sedimentary basins of the Tajo and the Guadiana, the plain of La Mancha, and the low-relief plains of Extremadura.
- The Galician Massif, the Cantabrian Range, the Iberian Mountain Chain and the Sierra Morena **surround the Inner Plateau**.
- The following features are **beyond the Inner Plateau**: the Pyrenees, the Baetic Chain, the Catalan Coastal Chain and the Guadalquivir and Ebro river basins.

The relief of the coasts and islands

Spain has more than 4,000 kilometres of **coastline**.

- The coasts of **Cantabria** and **Galicia** have rocky cliffs and estuaries (*rías*).
- The coasts of **western Andalusia** receive sediment transported by the Guadalquivir and are very sandy.
- On the **Mediterranean coast**, long beaches alternate with cliffs in places where the mountains reach the coast.

The **relief of the islands**: the Balearic Islands are an extension of the Baetic Chain. Mount Teide (3,718 metres), the highest mountain in Spain, is on Tenerife in the Canary Islands.

WORK WITH THE MAP

2 Which mountains surround the Inner Plateau? Which ones cross it?

3 Identify the main relief features beyond the Inner Plateau.

The mouth of the Nalón River in the estuary (or *ría*) near Pravia.

The Guadiana River as it flows through Medellín (Badajoz).

Ibón de Acherito, in the Pyrenees of Aragón.

WORK WITH THE PHOTOS

1 What type of surface water does each photo show?

8 Water in Spain

Rivers

The Iberian Peninsula has three **watersheds**, which are the dividing lines between drainage basins: they are the Cantabrian, the Atlantic and the Mediterranean. They have very different characteristics.

- **The Cantabrian watershed**: the rivers which flow into the Cantabrian Sea.
  - These rivers are short because their source is in mountains near the sea. Their flow is abundant and regular throughout the year.
  - From east to west, the main rivers are the Bidasoa, Nervión, Nalón, Narcea, Navia and Eo.
- **The Atlantic watershed**: the Galician rivers and those which cross the Inner Plateau and the Guadalquivir depression.
  - Except for those in Galicia, these rivers are long. They cross wide plains and their flow is abundant thanks to their tributaries. However, their flow is irregular and diminishes in summer.
  - From north to south, the main rivers are: the Miño, with its tributary the Sil; the Duero, whose basin is the largest in Spain; the Tajo, which is the longest river on the Peninsula; the Guadiana, which in some places runs underground through limestone terrain; the Guadalquivir, which forms an extensive landscape of marshes where it flows into the Atlantic Ocean.
- **The Mediterranean watershed** is smaller than the Atlantic watershed. This is because the Iberian Mountain Chain and Baetic Chain (which divide the two watersheds) are close to the Mediterranean Sea.
  - The Ebro is long and has an abundant flow because it rains a lot near its source and it is fed by tributaries in the Pyrenees. At its mouth, it forms an extensive **delta**.
  - The other rivers in the Mediterranean watershed are shorter, with a weak and very irregular flow. They include the Ter, Llobregat, Turia, Júcar and Segura.
  - This watershed also has ***ramblas***, which are dry river beds that only carry water when it rains heavily, usually in autumn.

RIVERS AND LAKES OF SPAIN

Lakes

There are many **lakes** and **wetlands** in Spain (over 2,000), but they are not very large. They are of great ecological value and form a natural habitat for many species of plants and animals. They also provide stopovers for migratory birds.

Mountain lakes like Sanabria, Peñalara and Laguna Negra, as well as a number of small Pyrenean lakes (called *ibones*), originated in glaciers. Other lakes, such as Lagunas de Riduera, were formed when limestone rock dissolved in water.

Groundwater

An **aquifer** is formed from surface water which filters into the ground. The water runs down through porous materials until it reaches a layer of impermeable rock where it accumulates.

In Spain, this natural resource is very important during droughts, and in regions with few rivers and lakes. **Groundwater** is mainly used for human consumption and agricultural irrigation.

ACTIVITIES

2 Which river watersheds are found in Spain? What are the rivers like in each watershed?

3 Explain why aquifers are so important.

WORK WITH THE MAP

4 Name three important rivers in each watershed.

5 What is the longest river on the Iberian Peninsula? Does it only flow through Spain?

6 Are Atlantic rivers longer or shorter than Cantabrian rivers? Explain your answer.

7 Find the mountain lakes mentioned in the text. In which mountain ranges are they located?

ACTIVITIES

1. Does relief influence climate on the Peninsula?
2. Is the Mediterranean climate the same everywhere?
3. Why is a mountain climate found in many parts of northern Spain?

WORK WITH THE MAPS

4. Where do we find the highest average temperatures? Where do we find the lowest?
5. Where does it rain most?
6. Is there a relation between areas with high temperatures and areas with low rainfall?

9 Climates and landscapes of Spain

Factors influencing climate

Climate is influenced by the following factors:

- **Latitude**. The Iberian Peninsula and the Balearic Islands are in the southern half of the temperate zone. They therefore have seasons and receive a lot of sunlight. The Canary Islands, situated at a subtropical latitude, have mild, steady temperatures all year round.
- **Location**. Spain is on one continent (Europe), but close to another (Africa). It is between two great masses of water, the Atlantic Ocean and the Mediterranean Sea. As a result, it receives hot and cold air masses from both land and sea.
- **Influence of the sea**. This regulates temperatures along the coasts and on the archipelagos throughout the year. However, inland there are much greater variations between high and low temperatures.
- **Relief**. At higher altitudes, temperatures fall and precipitation increases. The orientation of the mountains determines whether a place receives a lot of wind and sunlight. Mountain ranges act as barriers which block air masses.

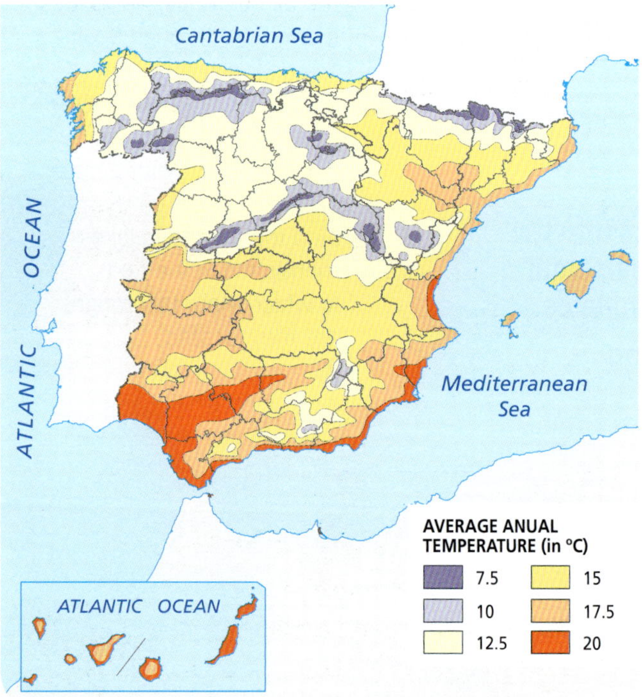

Map of isotherms. The lines connect places with the same annual mean temperature.

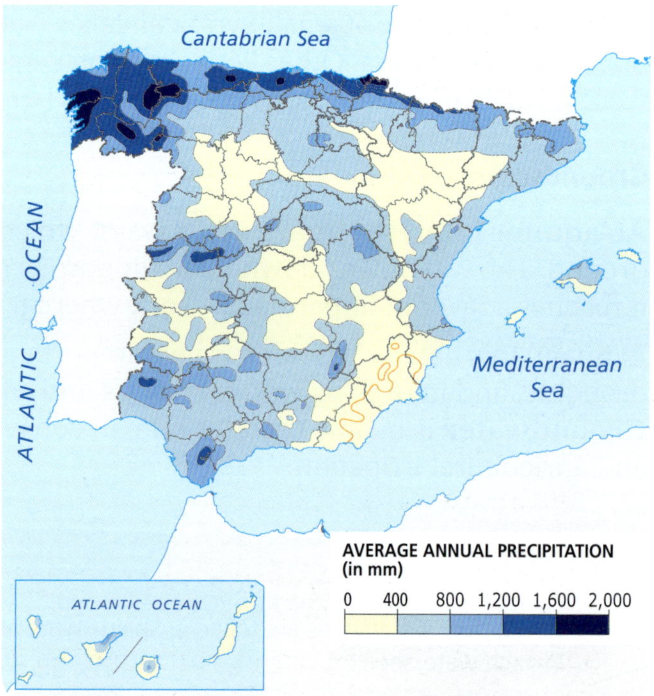

Map of isohyets. The lines connect places with the same amount of precipitation.

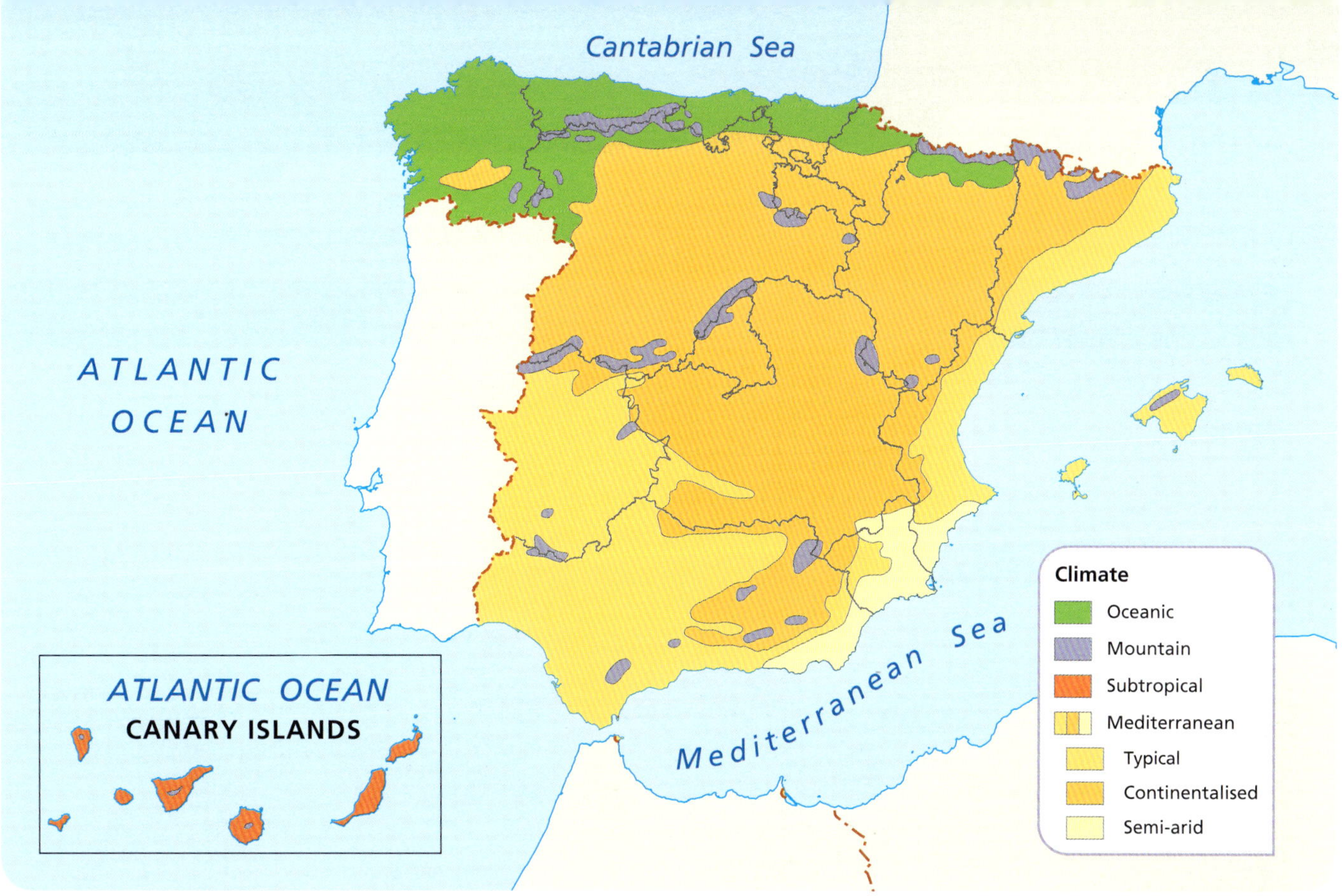

CLIMATES OF SPAIN

Climates and landscapes

- **The oceanic climate** extends across northern Spain. Temperatures are mild, with cool winters and warm summers. Precipitation is abundant and regular (more than 800 mm annually).

  The landscape has been humanised. In many areas, deciduous oak and beech trees have been replaced by pine and eucalyptus trees, as well as pastures for cattle.

- **The Mediterranean climate** is characteristic of most of the Peninsula, the Balearic Islands, Ceuta and Melilla. It has three sub-climates, which all have long dry summers.
  - **Typical**. Winters are cool and summers are hot in most coastal areas. Precipitation is between 300 and 700 mm annually, and it mainly falls in autumn and spring.
  - **Continentalised**. In inland areas, temperatures are very low in winter and high in summer. Precipitation is between 400 and 600 mm annually.
  - **Semi-arid**. Winters are mild and summers are hot in the south-east. Precipitation is less than 300 mm annually. There are long periods of drought.

  Mediterranean woodland (holm oak and cork trees) is no longer very extensive. Intensive land use has transformed these landscapes.

- **The subtropical climate** is characteristic of the Canary Islands. Temperatures are mild all year round. Precipitation is scarce and mainly occurs in winter.

  Vegetation includes conifers and laurel trees (laurisilva), and endemic species like the dragon tree and the Canary palm.

- **The mountain climate** is found in the highest mountain ranges. Winters are long and cold, and summers are short and cool. Precipitation is abundant. Vegetation changes with altitude.

WORK WITH THE MAP

7 Study the map and answer the questions.

a What type of climate predominates in Spain?
b Where is an oceanic climate found?
c What is the climate like in the Canary Islands?
d What is the climate like in the south-east of the peninsula?

ACTIVITY ROUND-UP

1 Look for extra information about the continents. Then copy and complete the table in your notebook.

CONTINENT	RELIEF	WATER	CLIMATE	LANDSCAPES
Europe	...	Longest rivers: Volga, Danube and Ural. Many small lakes.	...	...
Asia	The highest mountain ranges and plateaus in the world.	...	...	...
Africa	...	...	...	Rainforest, Mediterranean and high mountain vegetation.
America	...	...	Hot, temperate and cold.	...
Oceania	Many islands.	...	...	...
Antarctica	...	...	Polar.	...

2 Make a similar table for Spain.

3 Define the following terms:

- Relief
- Fault
- Tectonic plate
- Subduction zone

4 List the main forms of continental relief (inland and coastal). Describe three of them.

5 Describe the main forms of oceanic relief.

6 Why do mountain ranges form when two continental plates collide?

7 Look at the graph and write the words corresponding to each letter. Then answer the question.

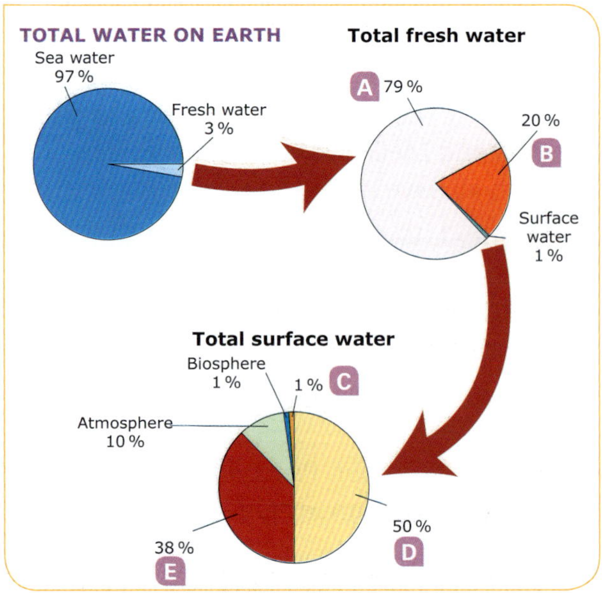

- If water covers more than 70% of the Earth's surface, why do we think of it as scarce?

8 Analyse the climograph and answer the questions.

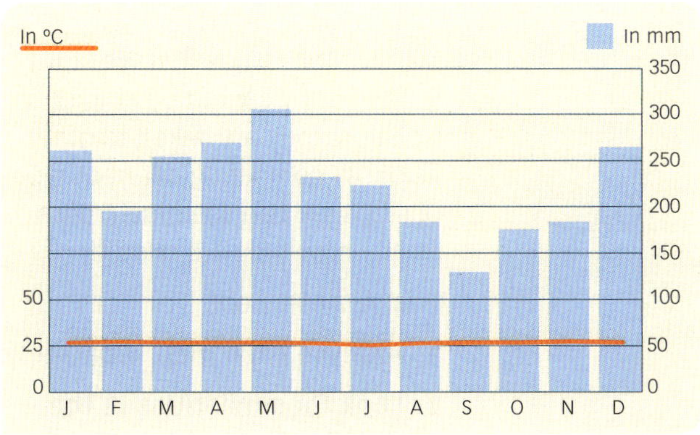

a Is precipitation heavy or light? Is it regular or irregular?

b What are temperatures like? Is there a wide or narrow range of temperatures?

c Are there clearly defined seasons throughout the year?

d What type of climate do you think it is?

e In which regions is it found?

9 Look at this photo of a *rambla* in Murcia. During which period of the year was this photo probably taken? Do you think the landscape looks like this all the year round?

10 Use an atlas to locate the most important mountains and rivers in the following countries:

- United States
- Russian Federation
- South Africa
- China

MANAGING WATER RESOURCES

Spain depends on rainfall to feed its rivers and refill its aquifers. However, most of the Peninsula has a Mediterranean climate so rainfall is irregular. River flows vary throughout the year and reservoirs are required to ensure a regular supply of water.

On the other hand, most rainfall is lost due to evaporation, inadequate irrigation systems and leaking water pipes. In some places, up to half of the available water is lost in this way.

Hydraulic works are expensive to build and have an adverse effect on natural ecosystems. Water transfers (which move excess water from river basins to areas where there is a shortage) are highly controversial.

Consumption in numbers

- Water consumption in Spain is among the highest in the world. In 2012, each person consumed 122 litres on average.
- For less than two euros, each Spaniard receives one thousand litres of drinking water at home.
- 80% of the water in our country is used for irrigation. A lot of this water is wasted.
- Spain is the country with the largest number of reservoirs per person. There are around 1,300.

11 What measures would you suggest in order to make better use of our water resources?

12 Think about how you use water. How could you make better use of it?

13 Look on the Internet for information about three reservoirs in different Spanish watersheds. Which watersheds are they in? How much water do they hold?

TECHNIQUES. Use maps to link geographical features

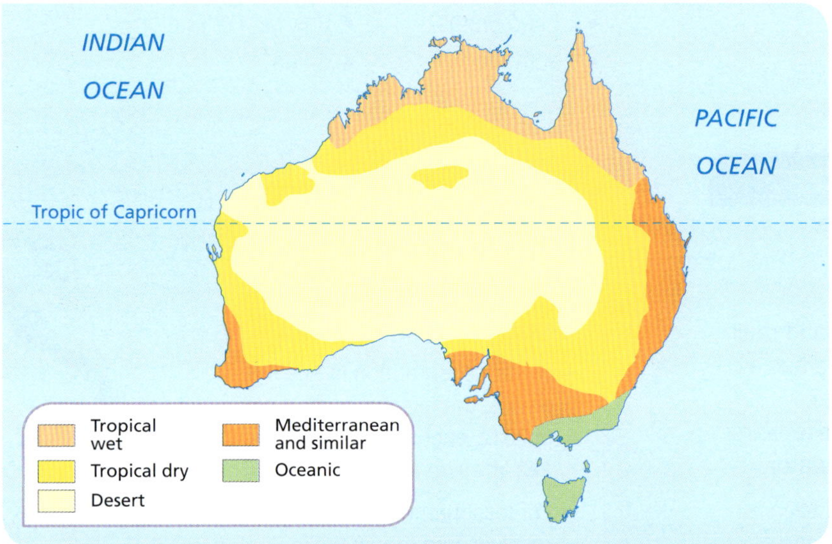

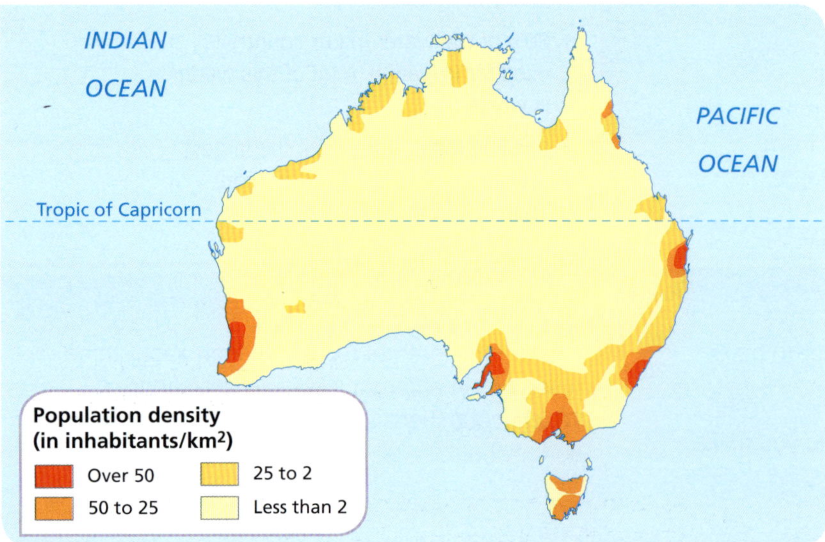

Many of the issues that we study in geography cannot be studied in isolation: they are linked to other fields. By comparing different types of maps we can see much more clearly how they are related.

Follow these steps:

1. **Analyse each map separately.**
2. **Look for similarities in the distribution.** For example, do the least populated areas match the desert climate? It may help if you trace one map and place it on top of the others.
3. **Make a list of the connections that you find.**
4. **Explain each connection.**

1 Analyse the maps of Australia.

- **Physical map**. Identify the main features of physical relief and the course of rivers.
- **Climate map**. Identify the climates. In which regions is each type of climate located?
- **Population map**. Which areas are densely populated? Which are sparsely populated?

2 Trace the population map and place it over the other maps.

- Make a list of the main connections between population distribution and relief; and between population and climate.

3 Explain the connections that you have found. Give reasons for your conclusions.

PRACTICAL KNOWLEDGE. **Tropical cyclones**

Odile causes the worst damage to electricity infrastructures in the history of Mexico.

The cyclone, which reached wind speeds of more than 200 km per hour, brought down 520 electric pylons and left 95% of the population of Baja California Sur without electricity.

The cyclone caused the death of at least four people and one person is missing. It also caused damage to urban infrastructures, roads, hotels and houses estimated at more than 900 million dollars.

El País, 22 September 2014 (adapted).

A tropical cyclone is a highly destructive natural phenomenon. It is a huge rotating system of thunderstorms that originates over tropical or subtropical waters. In the northern hemisphere, it circulates in an anticlockwise direction. In the southern hemisphere, it circulates in a clockwise direction.

Two conditions are needed for a tropical cyclone to develop:

- The surface temperature of the oceans must be higher than 26 °C. At this temperature, evaporation and condensation are intense, and they provide the necessary energy to generate strong winds and rain. This explains why most cyclones form at the end of summer, when the water is warm.
- Cyclones form in areas of low atmospheric pressure which attract high winds.

Cyclones move with the rotation of the Earth. As they move, they can change in size, speed and direction, according to atmospheric and oceanic conditions. When a cyclone comes into contact with colder waters or with land, it gradually weakens until it disappears altogether.

The centre of a cyclone is known as its 'eye'. It usually has a diameter of between 30 and 60 km, but may be as much as 200 km.

Satellite picture of Hurricane Katrina reaching the east coast of the United States in August 2005.

THE EFFECTS OF TROPICAL CYCLONES

- **Strong winds**. When the sustained speed of the wind is less than 60 km per hour, we refer to a tropical depression. If it is between 60 and 119 km per hour, it is a tropical storm. If it is higher than this, it is a hurricane or typhoon.
- **Storm surge**. This occurs in coastal areas when the sea level rises dangerously and there are huge waves.
- **Torrential rain**. This can cause flooding and landslides.
- **Tornados**. These are strong winds forming a violent, rotating column.

All these effects can result in serious personal and material damage, especially in coastal areas.

1 Answer the questions.

- Where are almost all cyclones formed?
- What is the climate like in this area? Why is it favourable to the formation of cyclones?

2 Use the Internet to find out about Typhoon Haiyan. Write a report which includes the following information:

- When it happened.
- The ocean where it originated and the countries that it affected.
- The characteristics of the cyclone.
- The effects it had in human and economic terms.

3 Imagine you lived in an area at risk from cyclones. How would it affect your life? What precautions would you take?

2 World population

FIND OUT ABOUT:

- The growth of world population
- Population ageing
- Migrations
- The population of Spain

KNOW HOW TO:

- Interpret a choropleth map
- Create and interpret a population pyramid

Most populated countries

In milions of inhabitants

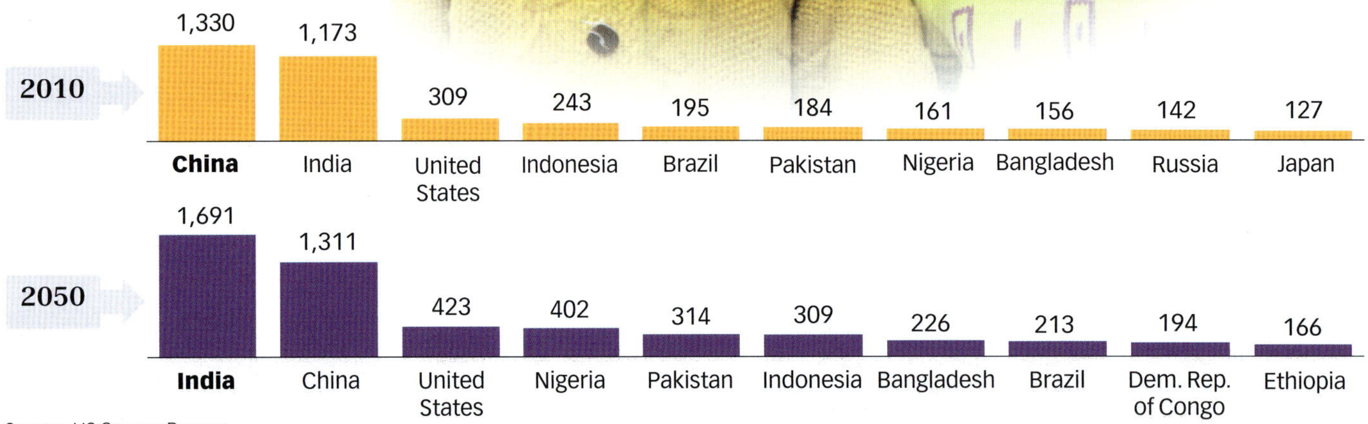

Source: US Census Bureau.

Where does most of the world's population live?

Today the world has more than 6,800 million inhabitants. 60.3% of these people live on the continent of Asia.

In 2010, the world's most populated countries were China, India, the United States, Indonesia, Brazil, Pakistan, Nigeria, Bangladesh, Russia and Japan.

It is estimated that over half of the world's population will still live in Asia in 2050.

? WORK WITH THE IMAGE

- What is the Earth's population?
- What percentage of the population lives in Asia?
- Which are the most populated countries in the world? On which continent are they?
- Which countries will be the most populated in 2050? Are they the same as in 2010?

Population distribution by continents

In millions of inhabitants, 2010

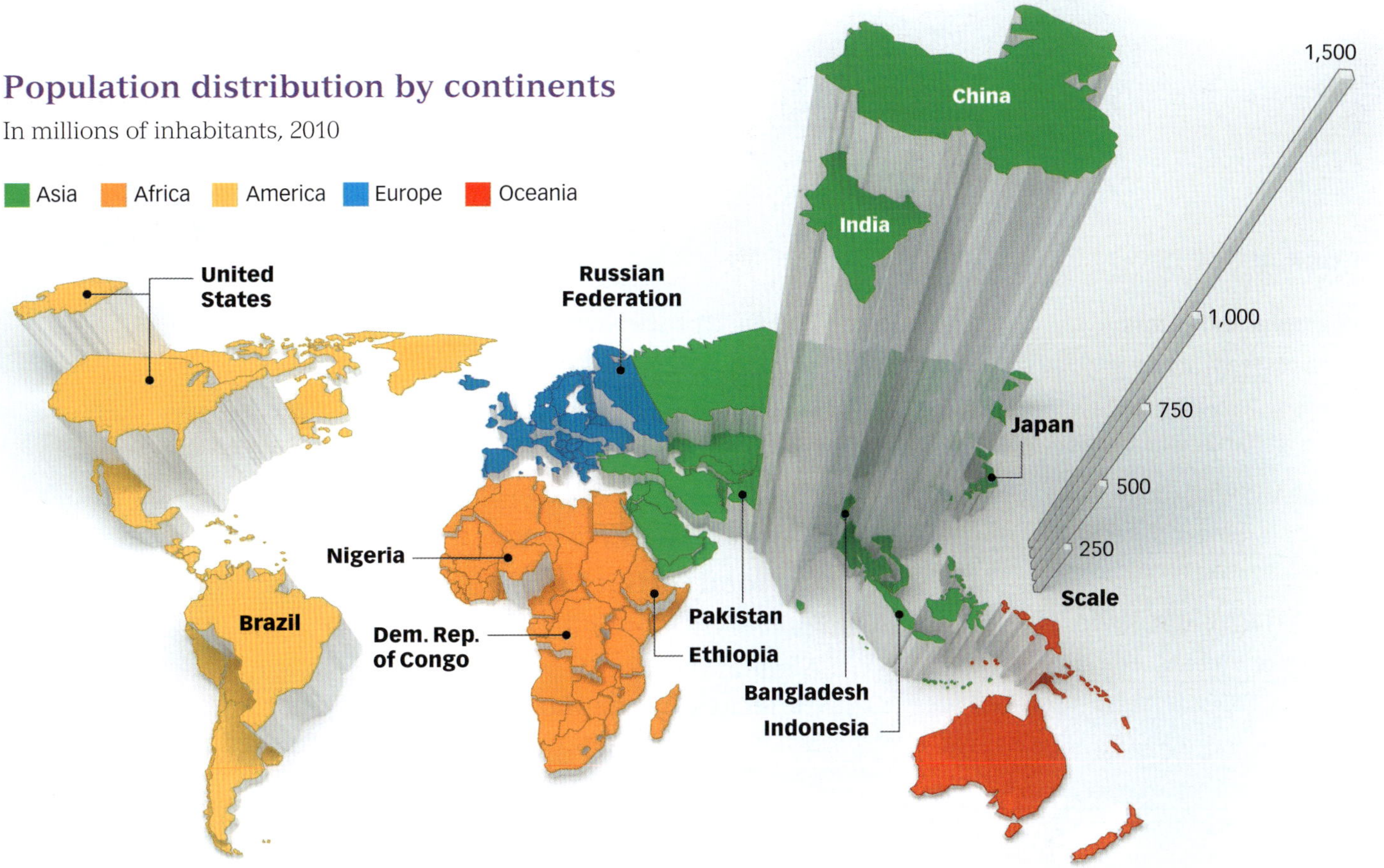

Source: US Census Bureau.

World population by continents

In millions of inhabitants

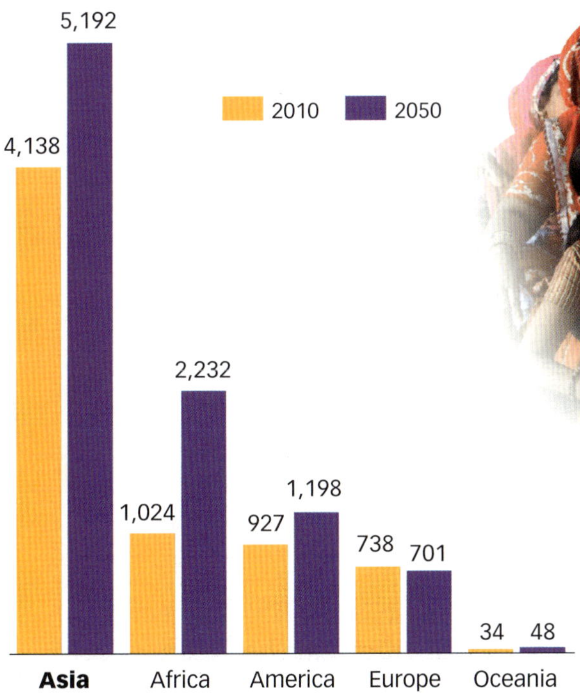

Source: US Census Bureau.

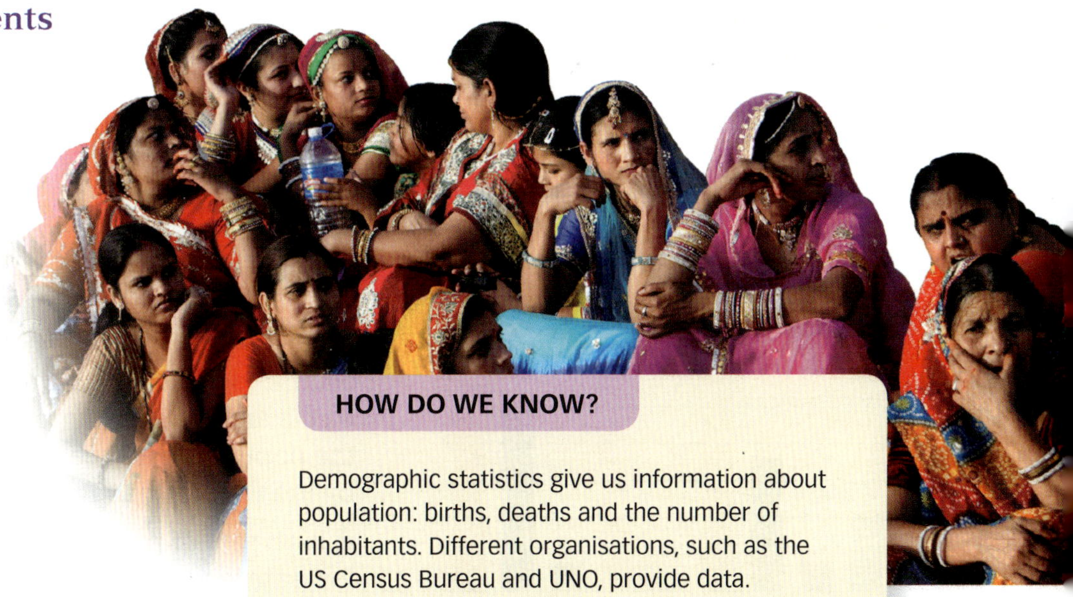

HOW DO WE KNOW?

Demographic statistics give us information about population: births, deaths and the number of inhabitants. Different organisations, such as the US Census Bureau and UNO, provide data.

- What percentage of the world's population lives in Europe? And in America?
- Which continent will have the biggest increase in population by 2050? Will any continent lose population?

1 How do we study population?

We use many different **indicators** to study population.

The birth rate and the fertility rate

The **birth rate** is the number of births in a population in a specific period. It is measured by the **crude birth rate (CBR)**.

The world birth rate is now 20 per 1,000 (‰), but the figure varies according to a country's level of development.

The **total fertility rate (TFR)** is an estimate of the average number of children that a woman has during her lifetime.

A minimum of 2.1 children per woman allows **generational replacement** to take place. But in developed countries today, the rate is only 1.6 children per woman.

The death rate

The **death rate** is the number of deaths in a population in a specific period. It is measured by the **crude death rate (CDR)**.

The death rate has fallen since the early 20th century. It is now about 8‰, but there are great differences between countries.

As a consequence of the fall in the death rate, there has been an increase in **life expectancy**. This is the average number of years that a person expects to live. Life expectancy in the world is about 70 years, but there are differences between women and men, and between different countries.

The rate of natural increase

Natural population growth is the difference between births and deaths. If more people are born than die, the population increases, and vice versa. It is measured by the **rate of natural increase (RNI).**

A high rate of natural increase is over 2%, and a low rate is under 1%. Today, the world's natural growth rate is around 1.2%, but there are big differences between countries.

However, real population growth is also influenced by **net migration**. This is the difference between the number of emigrants and immigrants in a specific period of time.

THINK ABOUT IT

Demographic indicators

CRUDE BIRTH RATE (CBR)

$$\text{CBR} = \frac{\text{number of births}}{\text{number of inhabitants}} \times 1{,}000$$

CRUDE DEATH RATE (CDR)

$$\text{CDR} = \frac{\text{number of deaths}}{\text{number of inhabitants}} \times 1{,}000$$

CHILD MORTALITY RATE (CMR)

$$\text{CMR} = \frac{\text{number of deaths under one year}}{\text{number of live births}} \times 1{,}000$$

RATE OF NATURAL INCREASE (RNI)

$$\text{RNI} = \frac{\text{number of births} - \text{number of deaths}}{\text{number of inhabitants}} \times 100$$

POPULATION DENSITY

$$\text{Population density} = \frac{\text{number of inhabitants}}{\text{surface area}}$$

- Compare this table to the text. Which indicators are measured by a percentage (%), a figure per thousand (‰), and an absolute figure (like 1 or 2)?

DEMOGRAPHIC INDICATORS OF WORLD POPULATION

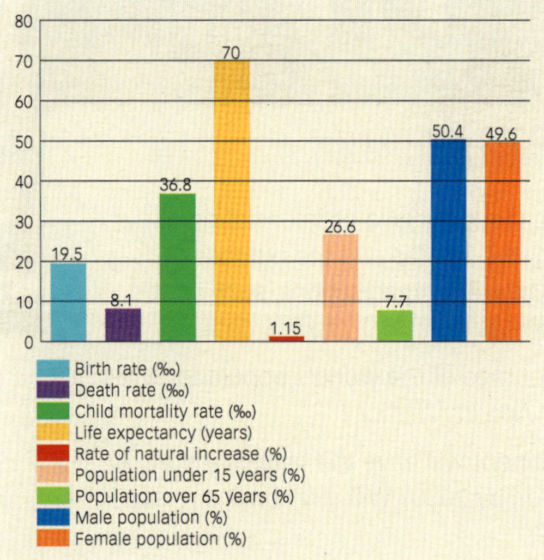

SOURCE: UNO.

Population density

The total number of inhabitants in a place is used to calculate **population density.** This is the ratio between the number of people and the size of the area that they occupy.

Population density is high when a lot of people live in a small area. If the same number of people lives in a very large area, then population density is low.

Population structure

A complete demographic study also examines the age and gender of a population.

- **Age**: the population is divided into young people (0–14 years), working age (15–64 years), and older people (over 65).
- **Gender**: the population is classified into males and females.

DID YOU KNOW?

What is a choropleth map?

A choropleth map, also known as an area-value map, shows how something is distributed across a region or throughout the world. It uses different colours or shades. If the value or density is lower, the colour or shading is lighter.

We ask the following questions to analyse a choropleth map:

- What space does it represent?
- What values are given?
- How are these values distributed?

ACTIVITIES

1. With a partner, define the following terms:
   - Birth rate
   - Death rate
   - Fertility rate
   - Life expectancy
2. Why is there a difference between the rate of natural increase and real population growth?

WORK WITH THE MAP

3. Study 'Did you know?', and analyse the choropleth map below.
   - What conclusions can you draw about world population?

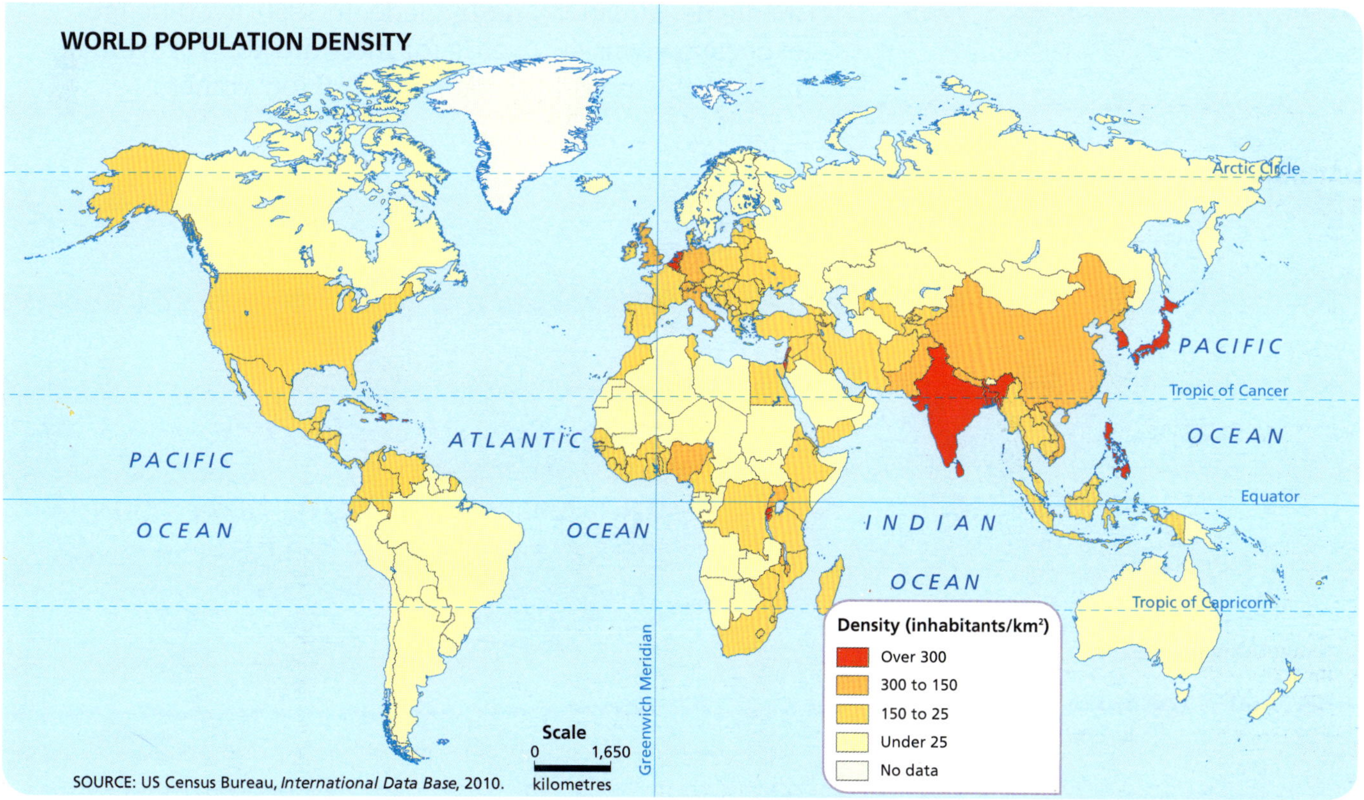

SOURCE: US Census Bureau, *International Data Base*, 2010.

GROWTH OF WORLD POPULATION

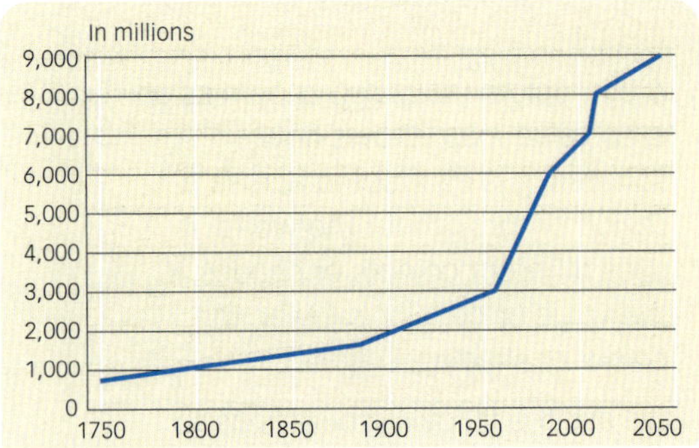

WORK WITH THE GRAPH AND THE MAP

1 Study the graph and the map. Answer the questions.

a What was world population in 1900? And in 2000?

b Will world population grow in the future?

c Where are the countries with the highest natural growth?

d Which countries have negative growth?

2 World population

The growth of world population

- World population grew very slowly before the 19th century. Birth rates were high, but death rates were also high.
- From the early 19th century, the death rate in developed countries fell due to improved medicine and better nutrition following improvements in agriculture. World population increased.
- In the 20th century, there was rapid population growth (or the **population explosion**).
- Today, the world's population is 7,000 million. By 2050, it is expected to reach 9,000 million.

Recent demographic trends

- The **birth rate** has fallen throughout the world.
- The **fertility rate** has also fallen. There was a rate of 5 children per woman in 1950, compared to 2.5 in 2013. In developed countries, the rate fell from 2.8 to 1.6.
- There has been a decline in the **death rate** in most regions. But in countries with ageing populations, this will grow again in the future.
- **Life expectancy** has risen because of advances in medicine, nutrition and hygiene. In 1950, average life expectancy was 46.7 years for men and 48.7 for women. In 2012, it was 68.1 for men and 72.6 for women.

NATURAL GROWTH OF WORLD POPULATION

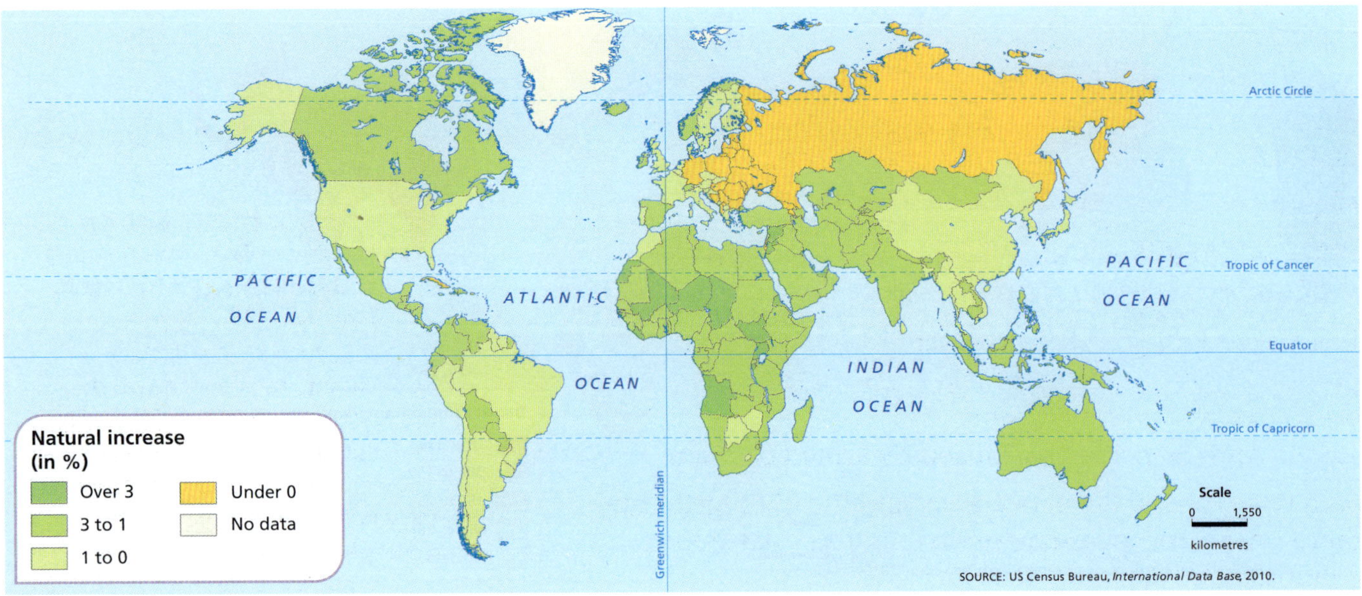

Regions with low population growth

Developed countries have low population growth, at around 0.2%. Some countries have negative growth, which means that they are losing population. For example, Germany had 82.5 million inhabitants in 2003 and 80.6 million in 2013.

These countries have a **low death rate**, but they also have a very **low birth rate** (11 ‰). The birth rate has declined sharply for the following reasons:

- There are more working women, who sometimes find it difficult to combine work and bringing up children.
- Contraception is widely available.
- Families delay having children.
- It is expensive to raise a family.
- Social values are changing.

As a result of the low birth rate and high life expectancy, many developed countries have an **ageing population**.

In recent decades, the birth rate has also begun to fall in many developing countries.

Regions with dynamic populations

Less developed countries have more dynamic populations. The natural rate of increase is around 2%. In some countries, such as Niger or Liberia, it is around 3%. Africa is the continent with the highest population growth.

The **birth rate** in these regions is very high, at 34 ‰:

- **Infant mortality rates** are high. Women have more children if they think that some of them may die.
- Many families work in **agriculture**. Their children work too and contribute to the family income.

The birth rate is also influenced by religious and political factors:

- Some **religions** are opposed to the use of contraceptives.
- In countries where governments implement **family planning policies**, the birth rate is usually lower.

Life expectancy is low in less developed countries. In some parts of Africa, it is less than 50 years. In these places, people do not have access to healthcare or clean water. They also suffer from the problems caused by armed conflicts.

As a result of low life expectancy and the high birth rate, there is a **young population** in many less developed countries.

THE POPULATION OF GERMANY AND NIGER

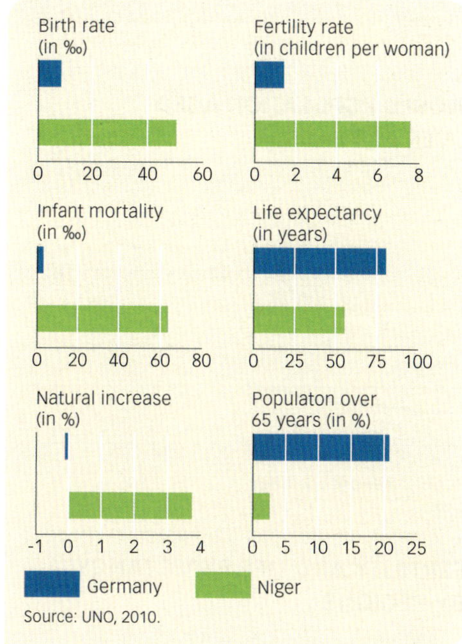

WORK WITH THE GRAPH

2 Compare the populations of Germany and Niger. How do you explain the differences?

3 Look for demographic data on another country. Is its population like that of Germany or of Niger?

THINK ABOUT IT

Why is overpopulation a problem in many countries of the world? With a partner, think about all the possible issues that arise in less developed countries, such as the capacity of agriculture to provide food for the whole population and the quality of life for large families on a limited income. Then share your ideas with the whole class, and discuss how these problems can be resolved.

3 Population ageing

WORLD POPULATION AGED OVER 60 YEARS

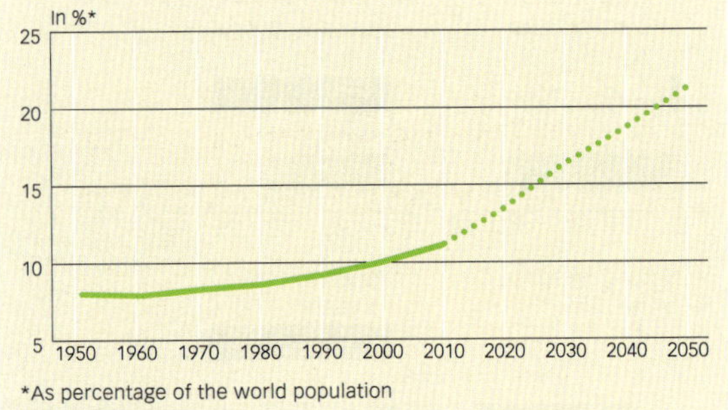

*As percentage of the world population
Source: UNO. 2010.

FERTILITY AND LIFE EXPECTANCY BY REGIONS

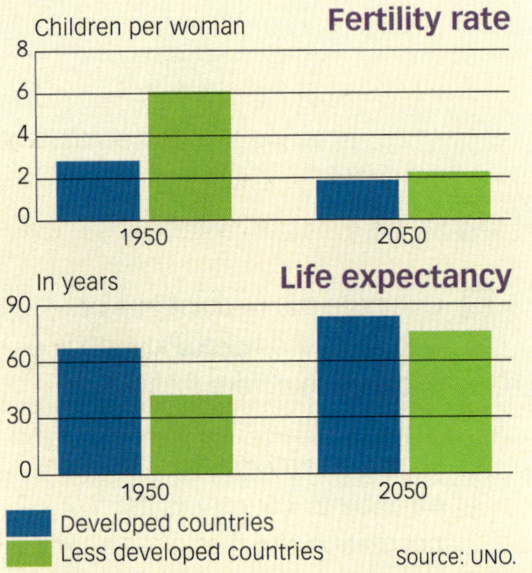

Source: UNO.

WORK WITH THE GRAPHS

1. What proportion of the world's population was aged over 60 in 1960? What is the proportion today?

2. Is the fertility rate only declining in developed countries?

ACTIVITIES

3. What is population ageing? What are its causes? What are its consequences?

Why does population ageing take place?

The population of a region gets older for two main reasons:

- **The decline of the fertility rate**. The number of children per woman has fallen all over the world.
  - In **developed countries**, many women go out to work and wait longer to have their first child.
  - In **emerging** and **less developed countries** child mortality has fallen. Consequently, many families no longer need so many children for economic reasons.
- **The increase in life expectancy**. People now live longer than in the early 20th century. This is due mainly to advances in medicine and improvements in diet, hygiene and working conditions.

Which regions have ageing populations?

- Today, life expectancy in **developed countries** is high: 78 years. Together with a low birth rate, this results in an **ageing population**. In developed countries, 17% of the population are aged 65 or over.
- The population is also growing older in some **emerging countries**. In Latin America, for example, life expectancy has increased by more than 20 years in the last five decades. At the same time, the birth rate has fallen.
- However, in **less developed countries** life expectancy is low and the birth rate is high. In Sub-Saharan Africa average life expectancy is 56 years, and the birth rate is 39 ‰. As a result, there is a young population: only 3% of the population is aged 65 or over.

The consequences of population ageing

Population ageing has economic and social consequences:

- **Generational replacement** does not take place, so population falls.
- Governments need to **spend more** on health care, pensions and retirement homes. But the taxes required for state spending are paid by a smaller working population.
- Older people often have lower incomes and **save** more. An economy will not expand if consumption drops.
- Ageing populations require more health and leisure services. In young countries, there is greater demand for consumer goods and educational services.

What is a population pyramid?

A **population pyramid** is a bar graph which shows the age and sex structure of a population.

AGE GROUPS		0-4	5-9	10-14	15-19	20-24	25-29	30-34	35-39	40-44	45-49	50-54	55-59	60-64	65-69	70-74	75-79	80+
Male	Population (%)	2.1	2.2	2.5	2.6	3.1	3	3	3	4.1	4.5	3.9	3.6	2.9	2.5	2.8	1.8	1.8
Female	Population (%)	2	2.1	2.3	2.4	2.9	2.9	2.9	2.9	3.9	4.3	3.8	3.4	3	2.7	3.2	2.3	3.6

These **data** are represented in the following bar graph:

The vertical axis shows the population in 5 year age groups. Each group is called a **cohort**.

Population pyramid of Germany

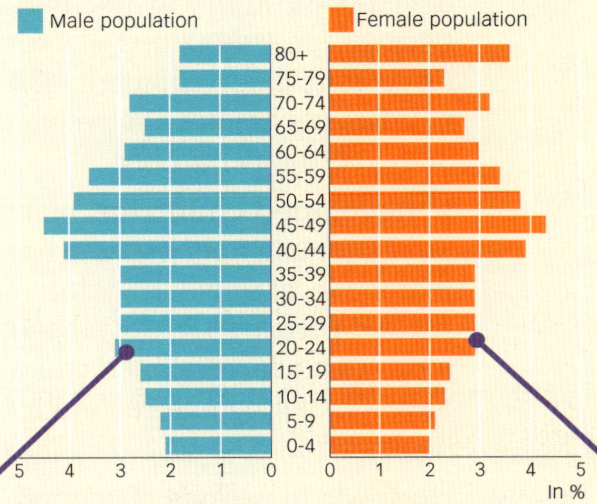

Source: US Census Bureau, *International Data Base*, 2011.

The left side of the graph shows the number of **males** in each age group as a percentage of the total population.

The right side of the graph shows the number of **females** in each age group as a percentage of the total population.

The shape of the pyramid helps you to interpret the data:

EXPANDING PYRAMID

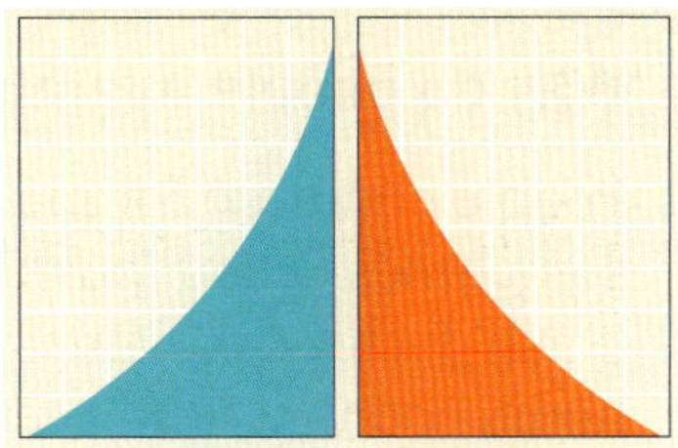

This pyramid is very wide at the base, indicating a high proportion of young people. It is characteristic of less developed countries.

STATIONARY PYRAMID

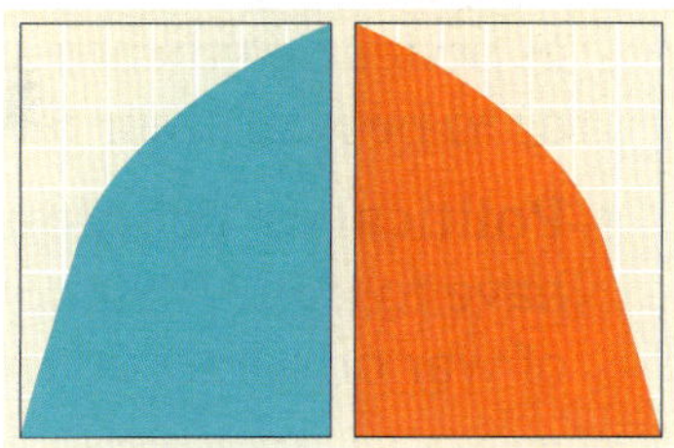

The base (young population) and the centre (adult population) are similar. The birth rate and the rate of natural increase are falling. This pyramid is characteristic of emerging countries.

CONTRACTING PYRAMID

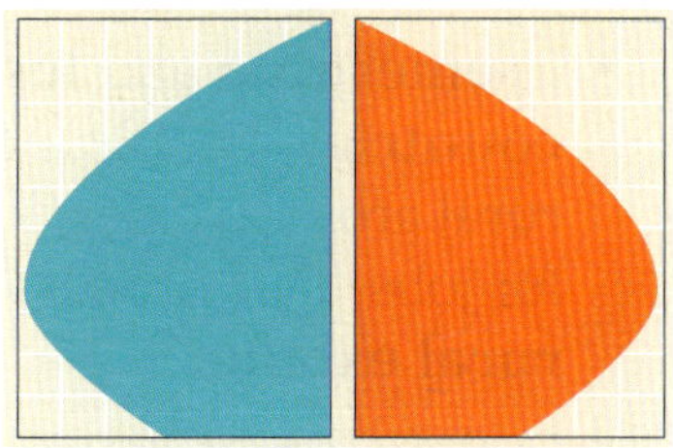

This barrel-shaped pyramid is wide in the centre, indicating a low birth rate, a low death rate and high life expectancy. It is characteristic of developed countries.

WORK WITH THE GRAPH

4 Study the population pyramid of Japan and answer the questions.

a Which type of population pyramid did Japan have in 1990?

b Is it predicted that Japan's population will change by 2050? How?

c What will the consequences be for Japanese society?

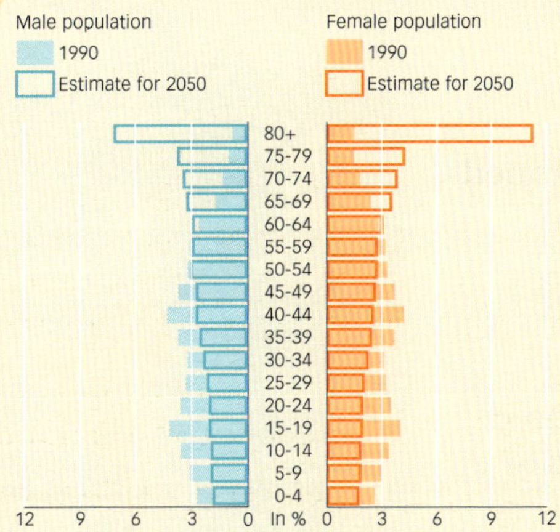

Source: US Census Bureau, *International Data Base*.

4 Migration

What is migration?

Migration is the movement of people from one place in the world to another for the purpose of taking up residence:

- **Emigration**: an emigrant is a person who leaves their place of residence.
- **Immigration**: an immigrant is a person who arrives from somewhere else.

Why do people migrate?

People emigrate for many reasons. These are sometimes called 'push' and 'pull' factors. 'Push' factors explain why people leave their home country. 'Pull' factors are what attracts them towards their new place of residence:

- **Economic**. Job opportunities, working conditions and higher wages are a frequent cause of emigration.
- **Social**. People emigrate in search of better healthcare and education, or to be near their family.
- **Political or religious**. Many people have to leave their countries because of war or persecution.
- **Natural disasters**. Some people leave their place following earthquakes, floods and droughts.

Internal migration

In 2013, more than 700 million people lived in another area of their own country.

- In **emerging** and **less developed countries**, people move from rural areas to cities, looking for jobs. This process is called **rural exodus**.
- In **developed countries**, retired people sometimes leave cities to live in quiet places with a mild climate.

International migration

In 2013, more than 230 million people lived in a foreign country.

In the past, many people emigrated from **less developed countries** to countries with more developed economies, which required a large unskilled workforce.

Today many people emigrate from a less developed country to an emerging country because it needs skilled workers to promote its economic development. 43.1% of migratory movements are now between less developed and emerging countries.

In **developed countries**, many immigrants are unqualified people who accept jobs that the local population does not want. Retired people from developed countries move to other developed countries that have a better climate, economy or health care.

However, in periods of economic recession, people also emigrate from developed countries. Well qualified younger people who are looking for work emigrate to both developed and emerging countries where there are opportunities for skilled workers.

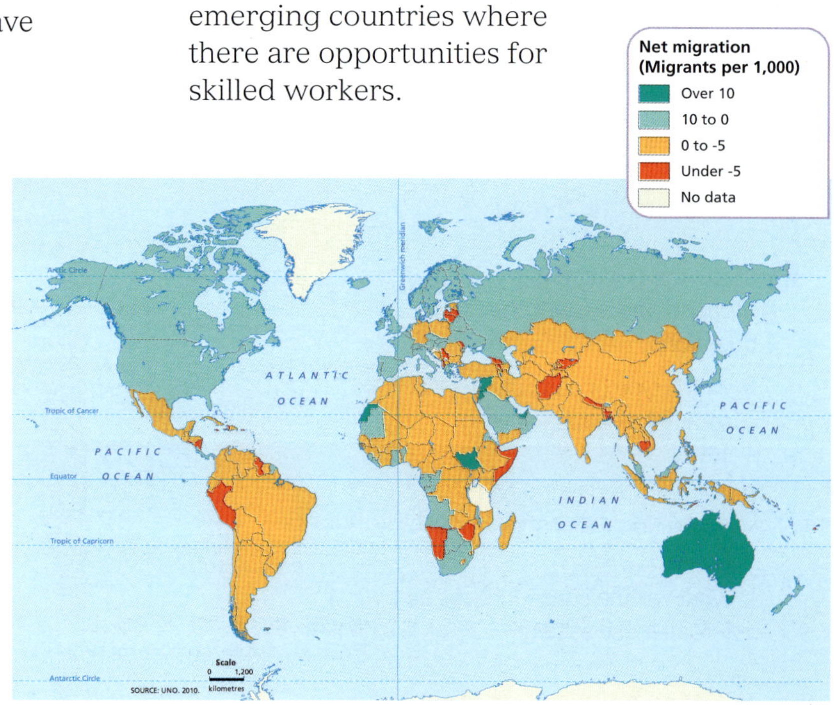

The characteristics of international migration

- Most migrants are **young adults**. Both **men** and **women** emigrate. Most emigrants from the Middle East are men. Many women emigrate from Oceania and Eastern Europe.
- **Unskilled** migrants often accept badly paid jobs that the local population is unwilling to do.
- Skilled migrants, with good **qualifications**, find higher paid work in both developed and emerging countries.
- Many migrations are between countries that are near each other or have a similar culture. People also leave poorer countries to live in more developed countries.
- As a result of the **economic crisis** that began in 2007, there has been small-scale migration from developed countries to emerging countries. There has been falling employment in developed regions and increasing demand for qualified workers in the emerging countries of Asia, Latin America and Africa.

The consequences of mass migration

- **Demographic**. Emigrants are often young adults. The countries they leave lose their young population. The countries they go to gain a younger population.
- **Economic**. Unemployment falls in a country when many people emigrate. The emigrants send money home to their countries of origin. Those countries receiving immigrants gain new workers who pay taxes and contribute to pensions and social services.
- **Sociological**. Many immigrants find it difficult to adapt to their new countries. This is made more difficult by racist attitudes and xenophobia. Another problem is the existence of mafias who control illegal immigration.
- **Cultural**. The arrival of people from another country promotes cultural interchange and widens people's experience and knowledge.
- **Environmental**. In countries where the population grows too quickly, there are problems of pollution, noise, provision of services and deterioration of neighbourhoods.

THINK ABOUT IT

Many governments have immigration laws. People who do not have the right of legal residence are considered illegal immigrants. Consequently, they can be deported. Millions of people cross borders illegally into regions like North America, the European Union and Australia. Mafias control illegal immigration in many parts of the world.

- Why do you think people try to migrate illegally? What do you think happens to them after their arrival in the new country?

WORK WITH THE MAP

1 Compare the map on the opposite page to a political map of the world. Find three countries that have positive net migration. Then find three countries that have negative net migration. How do you explain the differences?

ACTIVITIES

2 Think about the effects of migration. Then copy and complete the table in your notebook.

	Positive effects	Negative effects
In countries of origin	......	......
In destination countries	......	......

3 Would you like to emigrate? Why? Which country would you like to live in? Would you be an emigrant if you went to live in another country for two months?

Why is a doctor in Spain advertising in German?

5 The population of Spain

Population distribution

At the beginning of 2014, Spain had 46.5 million inhabitants. This was 9.2% of the population of the European Union. Spain was the fifth most populated country after Germany, France, the United Kingdom and Italy.

Spain has been losing population since 2012 because **negative net migration** has not been balanced by natural growth. This population decline is expected to continue. It is estimated that Spain will have a total population of 44.1 million inhabitants in 2023.

Population is unevenly distributed. Average density is 92 inhabitants per km^2, but there are big differences between one region and another. In general, the interior of the peninsula has the lowest population density, especially the Autonomous Communities of Castile and León, Castile-La Mancha, Extremadura and Aragón. The most densely populated areas are the coast, the Balearic and Canary Islands, Ceuta, Melilla and the Madrid region.

Low population growth

The **natural growth** of Spain's population is very low, at just over 0.1%. Both the birth rate (9.1‰) and death rate (8.3‰) are low. However, there are big regional variations. In Melilla, for example, natural growth is high, while it is negative in the Principality of Asturias.

As in other developed countries, the birth rate will continue to fall. Spain has a low proportion of women of childbearing age (15-49 years). Fewer immigrants (who are often young adults) are coming to Spain, which means that fewer children are born. At the same time, emigration from Spain to other countries is increasing.

The National Institute of Statistics predicts a slight increase in the fertility rate from 1.34 to 1.41 children per woman. However, this is still below the level at which generational replacement takes place.

POPULATION DENSITY IN SPAIN

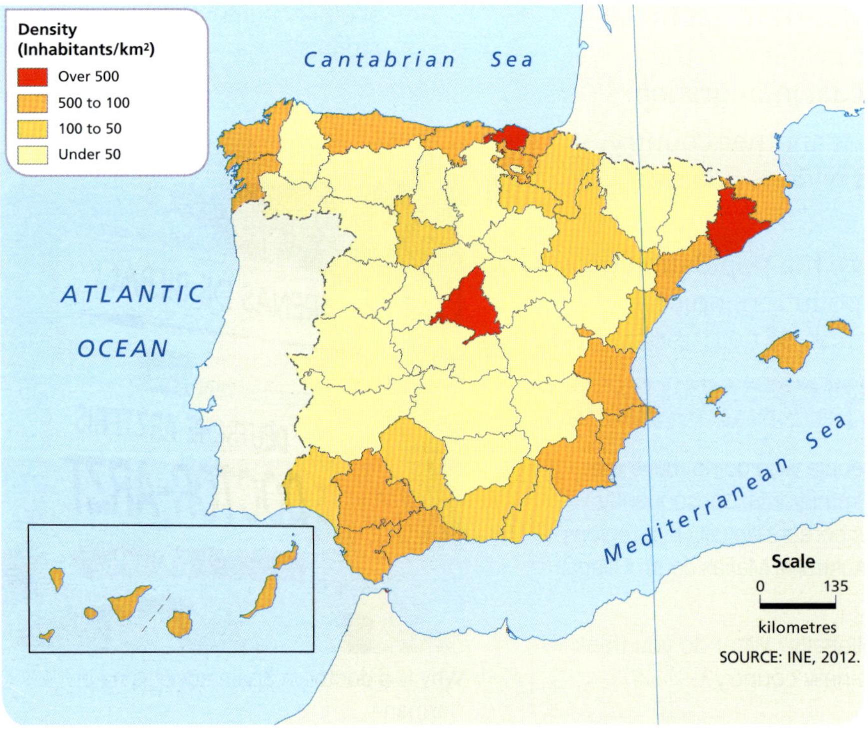

Rate of natural increase in Spain

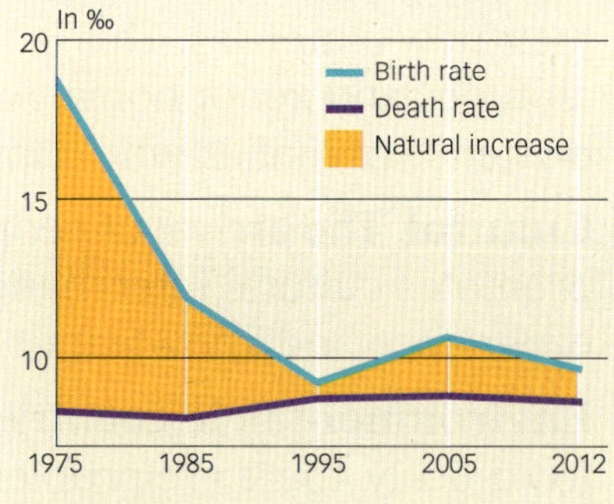

WORK WITH THE MAP

1. How is the population in Spain distributed? Which provinces are the most densely populated? Which are the least populated?

An ageing society

The population of Spain is gradually ageing as a result of the falling birth rate and **high life expectancy** (83 years), one of the highest in the world. This will have major economic, social and cultural consequences.

Population ageing became significant by the late 1980s, when the population aged over 64 reached 13% of the total. Today, the figure is 18%, while the population aged under 15 is 15%. This trend will continue in the future, since adults and elderly people form the largest age groups.

Migration flows

Traditionally, Spain has been a **country of emigrants**. The last large-scale emigrations were in the 1950s and 1960s. After the recession of 1973, many people returned, and in the late 20th century, the trend changed. Spain became a **country of immigration**.

Since 2010, however, immigration has declined as a consequence of the economic crisis. There are fewer job opportunities for immigrants, and many immigrants have returned to their countries of origin. At the same time, there has been an increase in emigration to other countries, especially by qualified young people looking for work.

As a result, there was **negative net migration** in Spain in 2013 (-256,849 people), and the country's total population fell.

Population change

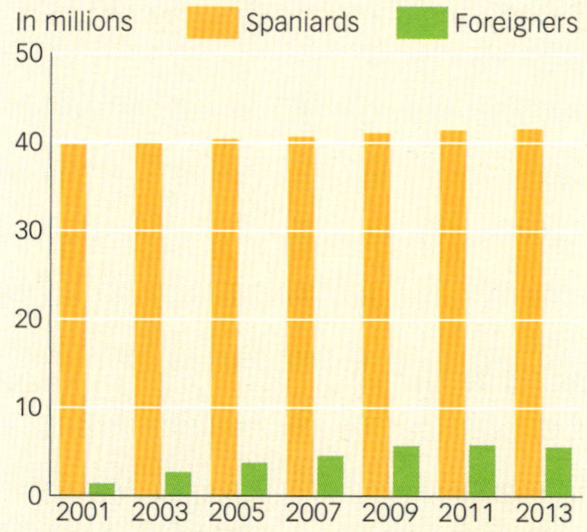

Source: INE, *España en cifras, 2013.*

Origins of immigrants

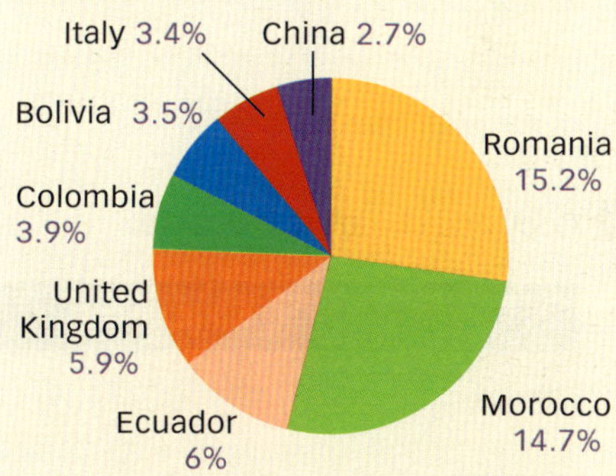

Source: INE, *España en cifras, 2013.*

POPULATION PYRAMID OF SPAIN

Source: INE, *España en cifras 2013.*

WORK WITH THE GRAPHS

2 Study the population pyramid.

a Is it an expanding, stationary or contracting pyramid?

b What are the main characteristics of Spain's population?

3 When did immigration from other countries to Spain increase significantly? Has it continued? Where do most immigrants to Spain come from?

1 Copy and complete the diagram.

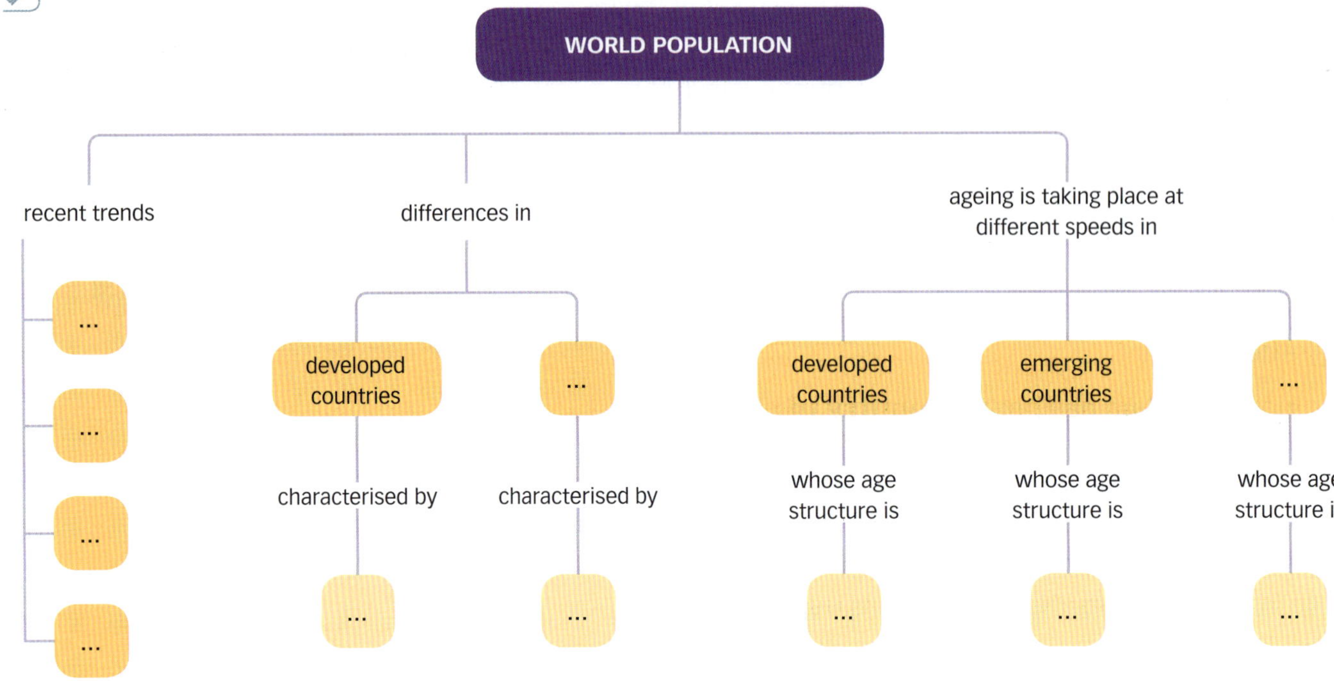

2 Copy and complete the table.

MIGRATIONS	
Types	......
Factors	......
Characteristics	......
Consequences	......
Main flows	......

3 Define the following terms.

- Life expectancy
- Net migration rate
- Rural exodus

4 Explain the differences between the following terms.

- Birth rate and fertility rate
- Death rate and life expectancy
- Rate of natural increase and real population growth
- Young, working age and older population
- Internal migrations and international migrations
- Population and population density

5 Explain the consequences of population ageing.

- Demographic: ....
- Social: ....
- Economic: ....

6 What are the consequences of the following in India?

- Thousands of doctors emigrate to other countries.
- There is a rural exodus.

7 Study the graph and answer the questions.

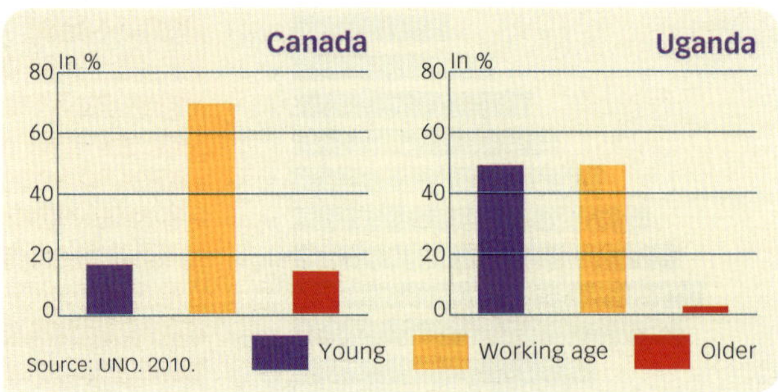

a Are they young or ageing populations?

b Describe their natural growth in the future. How do you know?

c Look for more information about the two countries. Does other evidence support your answers?

8 Analyse the population pyramid. Explain how Spain's population will change by 2050.

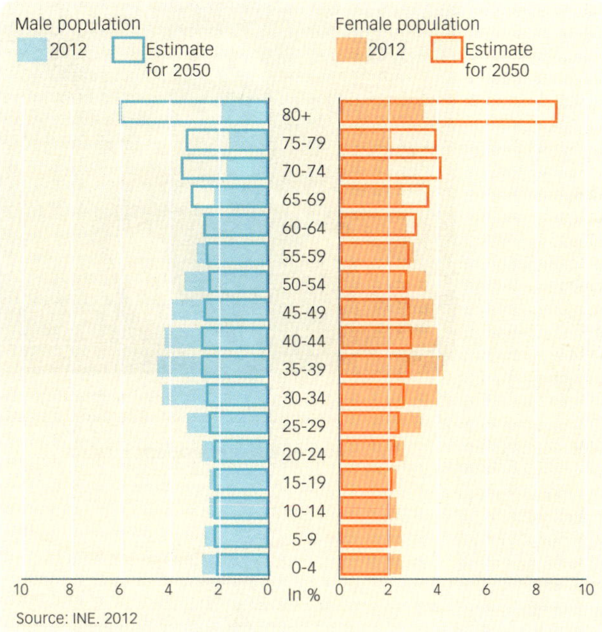

9 Compare the data for the two countries. Which is a developed country and which is a less developed country? Give your reasons.

DATA	COUNTRY A	COUNTRY B
Birth rate	39.8‰	9.7‰
Fertility rate	5.47‰	1.67‰
Death rate	14.5‰	10.24‰
Child mortality rate	90.2‰	2.8‰
Life expectancy	54.7 years	80.8 years
Population under 15 years	46%	16%
Population over 65 years	3%	18%
Rate of natural increase	2.5%	–0.5%

FIGHTING OBESITY

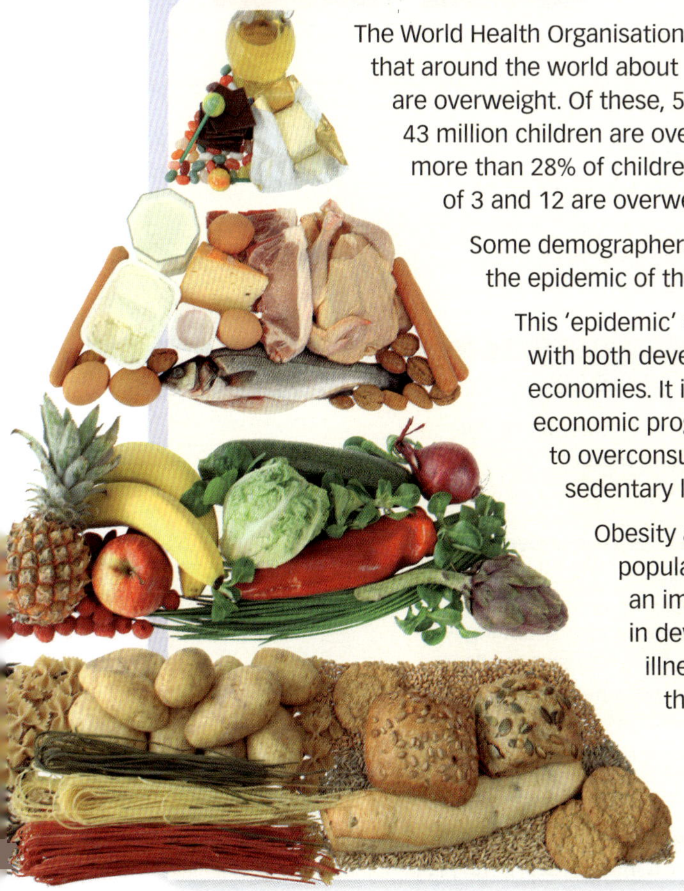

The World Health Organisation (WHO) estimates that around the world about 1,500 million people are overweight. Of these, 500 million are obese. 43 million children are overweight. In Spain, more than 28% of children between the ages of 3 and 12 are overweight.

Some demographers are calling obesity the epidemic of the 21st century.

This 'epidemic' affects countries with both developed and emerging economies. It is the result of economic progress which has led to overconsumption and sedentary lifestyles.

Obesity affects the population because it is an important risk factor in developing certain illnesses. According to the WHO, 2.6 million people die every year from illnesses related to obesity.

10 Answer the questions:

- What is the difference between 'overweight' and 'obese'?
- Is obesity a demographic problem? Explain your answer.

11 What is your response to the following text?

> In the year 2000, there were more people in the world classified as obese – around a thousand million – than people suffering hunger (800 million).
>
> E. Ontiveros and M.F. Guillén, *Una nueva época. Los grandes retos del siglo XXI*, 2012.

12 What measures can governments take to deal with this problem? And individuals?

13 Look at the photograph.

- Why is the food divided into different groups?
- Is this the kind of food you eat? Why?

THINK LIKE A GEOGRAPHER. **Analyse a text about ageing**

Before you begin, look at the photo and the title.
Where is the text from? What do you think it is about?

Spain, an ageing country

For every two people of working age, there is one who does not work. This is the most worrying statistic in the latest census carried out by the National Institute of Statistics (NIS) and published today using data from 2011. The report reveals that the average age of the Spanish population has increased by 1.5 years in the last decade, from 40 years in 2001 to 41.5 in 2011. This means that the dependency ratio has also risen. This is the relation between non-working people (those under 16 and over 64) and those considered active (from 16 to 64). In 2010, this rate was 0.485 and in 2011 it rose to 0.500. In other words, today there are two people of working age for every child under 16 or retired person.

This would be worrying if the economic climate were good, but it is especially alarming at a time of recession. "Five years ago, when the State received more money, it could compensate for this increase in the dependency ratio and maintain pensions. Now the situation has changed and people of working age have to pay more for those who are not working", says Jeroen Spikjer, researcher at the Centre for Demographic Studies at the Autonomous University of Barcelona. Moreover, spending on health and dependency increases as the population ages and the Government makes cuts in precisely these areas. It is difficult to be optimistic about the future.

El País, 14 December 2012

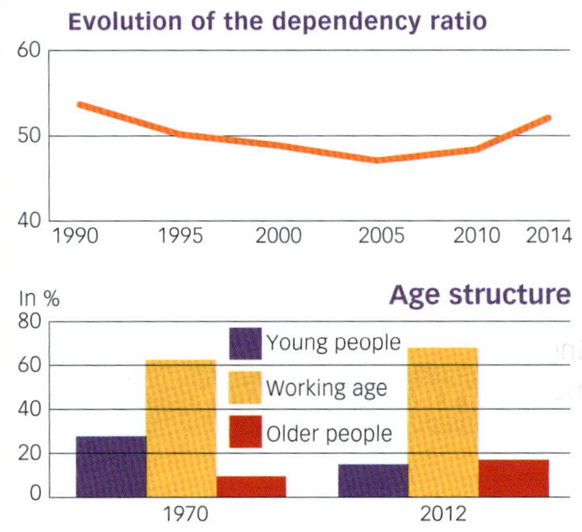

Source: INE and Eurostat.

1 Analyse the article.

- a What problem does the article describe?
- b How much did the average age of Spain's population increase between 2001 and 2011?
- c What is worrying about the demographic changes reported in the article?
- d What is the dependency ratio? Has it changed in recent years?
- e Does state spending increase when a population gets older? Why?
- f Can you think of a better title?

2 Think about the problem.

- a What are the likely consequences of population ageing in Spain?
- b Is population ageing a greater problem during a recession or a period of economic growth? Explain your answer.
- c Is population ageing only a problem in Spain? Give other examples.

3 What can we do about it?

- a Think of ways of dealing with the problem of Spain's ageing population. Copy and complete a table in your notebook.

HOW CAN WE...?	
Increase the number of young people	.....
Increase state revenue.	.....
Reduce state spending on pensions and services for older people.	.....

- b Cross out the measures that you disagree with. Tick (√) the five best measures.
- c Compare your ideas with a partner. Then share your ideas with the rest of the class.

PRACTICAL KNOWLEDGE. **The populations of China and India**

Birth control policies and population growth

China is the most highly populated country in the world with more than 1,300 million inhabitants. It is followed by India with more than 1,200 million. However, this situation will change according to estimates made by the United Nations Organisation (UNO). India's population will equal China's in 2028, when both countries will have 1,450 million inhabitants. After that, India's population will continue to grow rapidly.

In 1980, the Chinese government introduced a policy of 'one child per couple' when the population reached one thousand million inhabitants. This policy brought overpopulation under control. China's fertility rate was 6.4 children per woman in 1965, but today it is only 1.6. However, the negative consequences have included forced sterilisations and abortions, and baby girls who were killed at birth or abandoned. In 2013, the policy was relaxed to allow many families to have a second child.

Today, India has a fertility rate of 2.7 children per woman, which is higher than China's. Some regional Indian governments are worried about the rapid growth of their populations. They have suggested measures to slow down the rate of growth. In the state of Kerala, in the south of India, a law has been proposed which would limit the number of children per woman to two. Breaking the law would be punished by high fines or a prison sentence.

A poster on behalf of China's one-child policy: "One couple, one child".

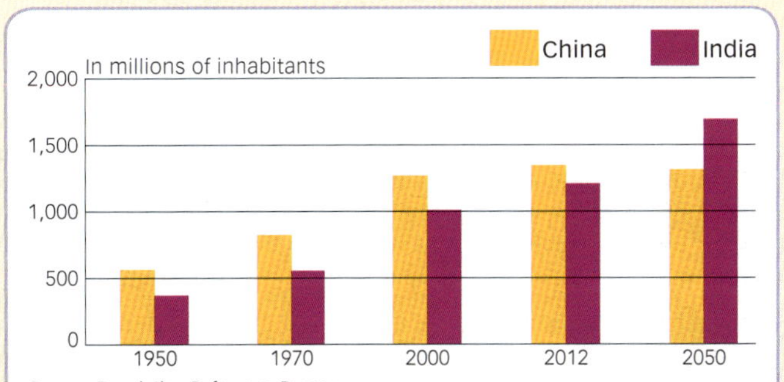

Source: Population Reference Bureau.

1 Analyse the demographic situation of India.

a If present demographic trends continue, why does the UNO estimate that India will have a bigger population than China after 2028?

b Why do you think local Indian governments are worried about this demographic prediction?

2 Analyse the population pyramids.
Do they show that India's population will continue to grow after 2028 and that China's population will fall?

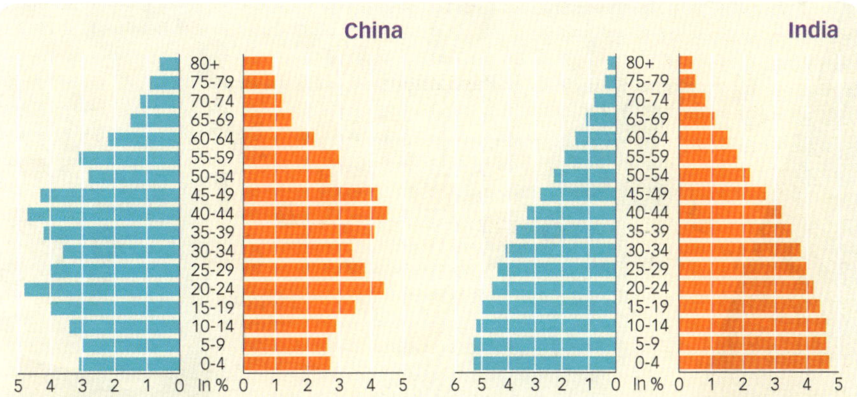

Source: US Census Bureau, *International Data Base*, 2011.

3 Think about the issue.

- The Health Department of the Indian state of Rajastan promotes voluntary sterilisation in exchange for a car. What do you think about this as a way of controlling the number of births in the area?

3 Cities

FIND OUT ABOUT:

- The structure and functions of cities
- Urban development
- The global network of cities
- Urban problems
- The urban network of Spain

KNOW HOW TO:

- Analyse and compare graphs and maps
- Analyse the growth of a megacity

Oslo: a sustainable city

Oslo is one of the richest cities in Europe and the most populated in Norway. More than 600,000 people live in the city and 1.5 million in the metropolitan area.

Oslo aims to be one of the most eco-friendly cities in the world through careful economic, social and cultural development.

A greener city

About two thirds of Oslo is made up of protected woodland, hills and water. 95% of its citizens have access to a green area within 300 metres of their houses.

Oslo wants to improve its biodiversity by avoiding the introduction of invasive species and protecting its fauna and flora. It is creating new green spaces and water courses.

* **A healthier city**
Oslo wants its citizens to generate less waste and make the city a cleaner place. Areas of urban decay will be cleaned up and recycling will be extended.

* **Less pollution**
By 2030 Oslo plans to reduce its greenhouse gas emissions by 50%. Oil-fired heating will disappear by 2020. Priority will be given to electric cars.

? WORK WITH THE IMAGE

- Compare your town to Oslo. What policies should be implemented by the local authorities to make it a sustainable living area?
- Do you practise recycling in your own life?

HOW DO WE KNOW?

Every year, the European Commission gives the European Green Capital Award to the city with the best ideas for improving the environment and quality of life of its citizens. These prizes are an example of good practice which other cities can follow.

- Investigate. Have any Spanish cities received the Green Capital Award? Why?

1 The structure and functions of cities

What is a city?

Cities share the following characteristics:

- **Population size**. They have large numbers of inhabitants and are densely populated. Their size varies: in Spain, a city is a locality with more than 10,000 inhabitants.
- **Urban space**. Cities are built-up areas and often have high buildings. The space is used for housing, businesses and transport infrastructures.
- **Administrative status**. Urban centres have played a key role in human history. Their official status as 'cities' often originates in their historical importance.

There may be one or more cities in an **urban** (or **metropolitan**) **area**. This is an area that has a population density of more than 150 inhabitants per km^2.

The functions of cities

Cities often have a wide range of functions:

- **Residential**: all cities have this function.
- **Political**: administrative institutions are based in provincial, regional and national capitals.

  Some cities, like Brussels and New York, are the seats of international organisations.
- **Commercial**: all cities play a role in their regional economies, while some cities also have international importance.
- **Industrial**: large industrial cities, like Shanghai in China, are found around the world.
- **Cultural**: cities like Oxford play an educational role, while Rome and Mecca have religious significance.
- **Tourist**: they have cultural attractions, beaches, etc.

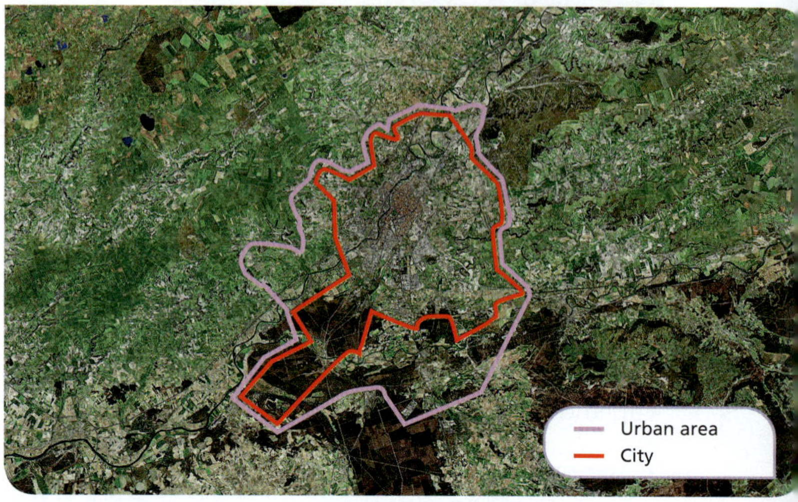

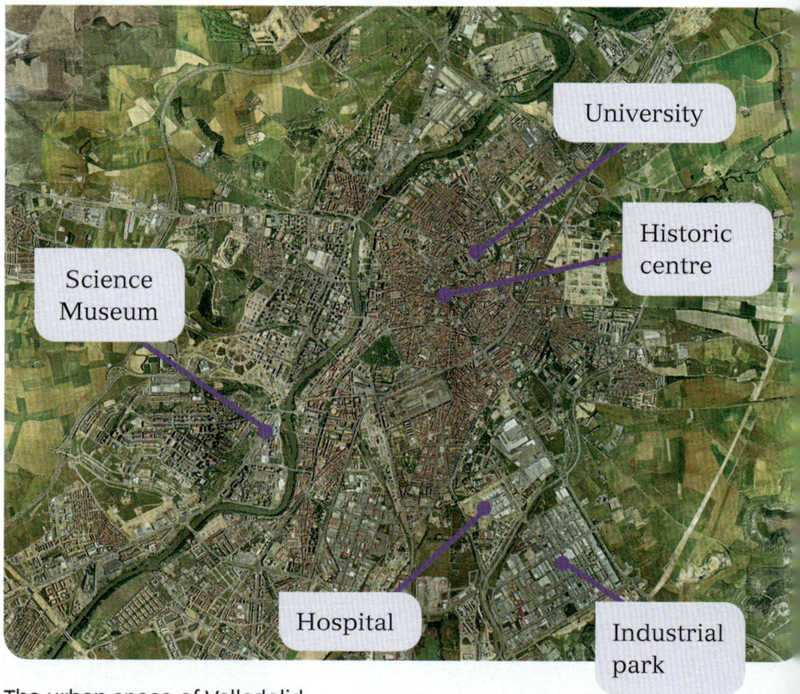

The urban space of Valladolid.

WORK WITH THE PHOTOS

1 Study the aerial photos of Valladolid and answer the questions.

a Which covers more space: the city or the urban area?

b What types of land use are there in the urban area?

c How extensive are the green spaces of Valladolid?

The structure of cities

Cities generally have several main areas.

- **The central zone** is a key area in a city:
  - It often includes the **historic centre**, where the city originated. Official buildings, like the city hall, are often located there.
  - The **Central Business District (CBD)** is where most commercial and financial activity takes place. Banks, businesses and offices are located in the CBD.
  - Many **services**, such as hospitals, museums and university campuses, are found in the central zone, but they may also be located on the outskirts of a large city.
- **City extensions** (*ensanches*) were built as residential areas in the 19th century as cities expanded.
- **The outskirts** are where the city is often still expanding.
  - **Residential areas** provide housing for people who work in the city or travel there to use its services. In developed countries, the **suburbs** usually have schools, shopping centres and good transport facilities. However, in less developed countries **shanty towns** are poor residential areas.
  - **Industrial parks** and service areas are often located near the main roads on the outskirts of cities where land is cheaper and there is easy access.
- **The rural urban fringe** is beyond the outskirts. It is an intermediate area with a range of activities and population groups.

The Gothic Quarter in Barcelona.

The city extension in Barcelona.

The outskirts of Barcelona.

DID YOU KNOW?

What is a diffuse city?

A diffuse city is not continuous and compact but contains rural areas within its metropolitan boundaries. New districts have been built on the outskirts of cities:

- Many companies move their main branches away from the city centres.
- Roads and public transport have improved.
- There has been greater demand for a healthier lifestyle, and for more green spaces.

- In groups, study the aerial view of your Autonomous Community on Google Maps. Find three "diffuse cities".

THINK ABOUT IT

A neighbourhood is the basic unit of urban life and is made up of a number of streets. Each neighbourhood shares a common history. It usually has a number of services so its inhabitants can meet most of their needs within their district.

- Think about where you live. How would you define your neighbourhood? What about the neighbourhood near the school? Share your ideas in groups.

2 The spread of urbanisation

Urban growth

Today, just over half of the world's population lives in cities. According to predictions, this proportion will rise to 70% by 2050.

In the past, urban population was mainly concentrated in developed countries. However, today it is increasing rapidly in emerging and less developed countries.

Urban growth has led to an increase in the number of **megacities** (urban areas with more than 10 million inhabitants).

There were only five megacities in 1975 but today there are more than 20. By 2025, there will be over 30 of them, mainly in emerging and developing countries.

Developed countries

Cities expanded greatly when industrialisation took place in Europe and North America in the 18th and 19th centuries. Today, in developed countries:

- A very high percentage (over 80%) of the total population lives in cities.
- Urban population is still growing, but at a lower rate than in emerging and less developed countries.

Three factors explain the increase in the urban population of developed countries:

- A wide range of **jobs** and **services** is found in cities. Agricultural work is not well paid.
- **Young people** arrive in the cities and bring up their families there.
- As a result of **globalisation**, cities have become important centres for the distribution of goods and services to a world market.

GROWTH OF URBAN POPULATION

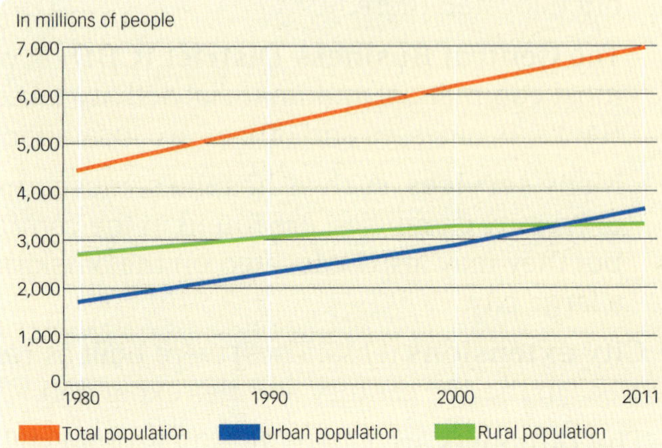

SOURCE: World Bank, 2011.

The CBD in Kuala Lumpur (Malaysia).

WORK WITH THE MAP

1 With a partner, study the map on the opposite page and answer the questions.

a Where are the five largest urban concentrations in the world? Which countries are they in?

b Are the most populated cities in the world mainly on the coast or inland? How do you explain your answer?

c Where is the greatest concentration of highly populated cities? Why?

Emerging and less developed countries

Urban growth began later in emerging and less developed countries than in developed countries. However, today the rate of growth in these regions is much more rapid than in developed countries. Nearly half of their total population lives in cities.

This rapid rate of growth is due to the following factors:

- People **migrate** from rural areas to the cities in search of jobs, higher wages and better services.
- The rate of **natural growth** is higher in cities than in rural areas. This is because younger people go to live in cities and have more children.
- As a result of **globalisation**, international companies often set up businesses in cities in emerging countries. This creates job opportunities.

DID YOU KNOW?

Slums

Every year, millions of agricultural workers leave behind the harsh conditions of their villages and move to the cities in search of a better life.

Many immigrants are soon disillusioned. They cannot find work and they have to live in precarious conditions in fast growing slum areas. These areas lack basic services such as running water, sewers, electricity and schools.

Slum in Kibera, Nairobi (Kenya).

ACTIVITIES

2 What differences are there between urbanisation in developed and less developed countries?

3 Look on the Internet for more information about Kibera or another slum in a less developed country. Make notes and prepare a report.

THE WORLD'S LARGEST CITIES

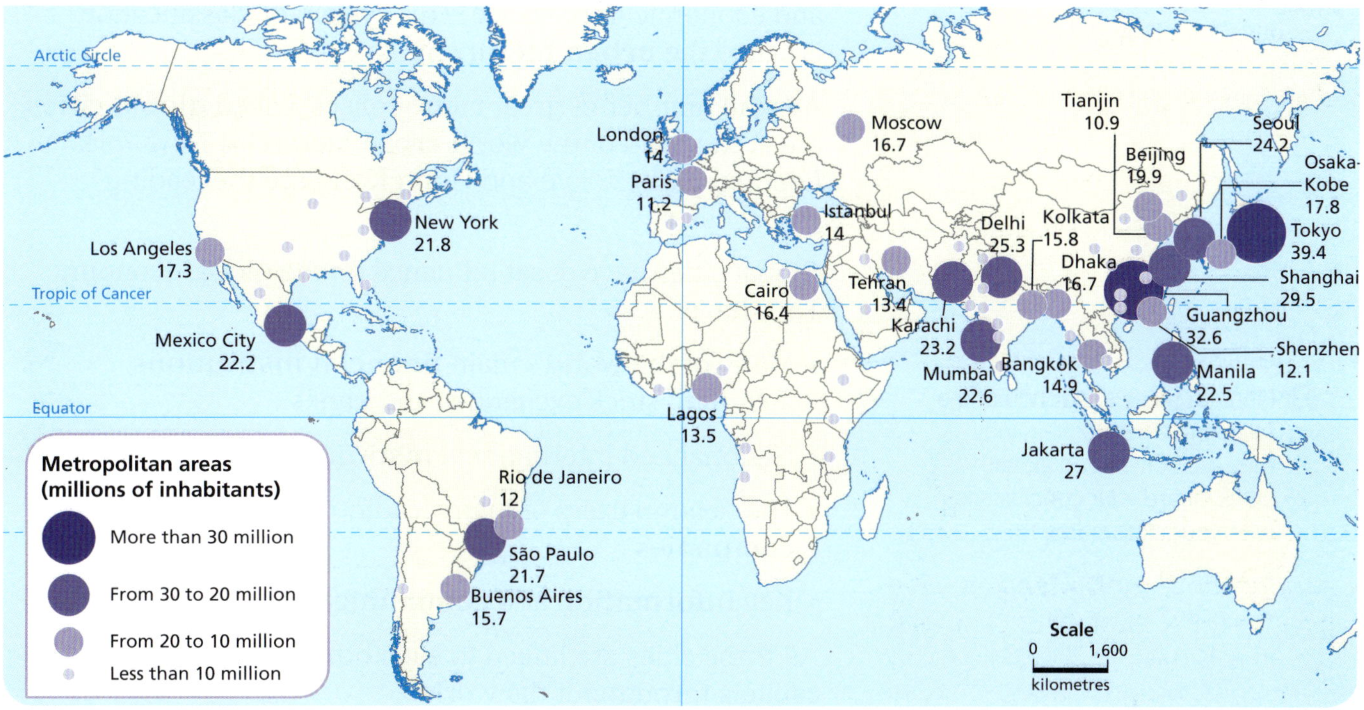

3 The worldwide urban network

TRANSPORT NODES

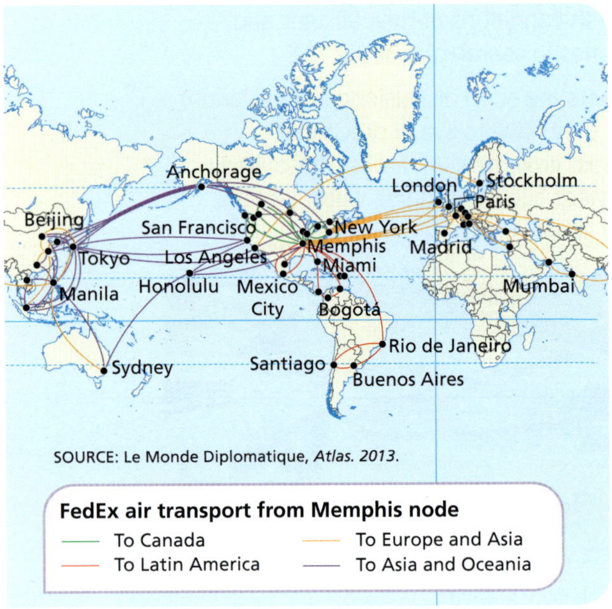

WORK WITH THE MAP

1. Look on the Internet to find out what FedEx does.
2. Which cities are linked to each other? Are they global cities?

ACTIVITIES

3. Find out more about two of the following cities. Explain why they are considered global cities. Share your conclusions with your classmates.
   - New York
   - London
   - Hong Kong
   - Tokyo
   - Paris
   - Shanghai

The influence of cities

Cities around the world are linked to each other in many ways and form a worldwide urban network:

- Large **companies** have their headquarters in big cities. But their factories, warehouses and smaller branches are often located in smaller urban areas, and in other countries. Consequently, their influence spreads through the international economy.
- **Political institutions**, at local, regional, national and international level are all based in cities.
- **Transport networks** connect cities to each other and to their rural surroundings.
- The **media** that produce television, the written press and the Internet are mainly based in cities. In this way, cities promote the **spread of information**.
- Cities are **centres of innovation** in many fields, including the economy, culture and lifestyle trends. Universities and company research and development facilities are based in cities.

The urban hierarchy: global cities

Cities can be classified according to several criteria, such as their influence, functions, population size and economic importance. This system of classification is called the **urban hierarchy**.

A small number of great metropolises, called **global cities**, are at the head of the world urban hierarchy. New York, London, Paris, Tokyo and Hong Kong are the leading global cities.

Global cities have great influence on the global economy. They have:

- Many of the world's main **financial institutions**, including stock exchanges and banks.
- National and international **institutions**.
- The headquarters of many leading **multinational companies**.
- Key **information** and **communications** industries.

All these cities are linked to each other, and to national capitals throughout the world.

LEADING CITIES	
Business	Tokyo, New York, London, Hong Kong, Shanghai
Human capital	New York, Boston, Tokyo, Sydney
Information exchange	Paris, New York, Zurich, Stockholm
Cultural	London, New York, Shanghai
Political	Brussels, Washington, Geneva, New York

LEADING EMERGING CITIES	
Business	Mexico DF, Lagos, Bogotá, Casablanca, Nairobi
Human capital	Karachi, Lima, Chongqing, Bangkok, Istanbul
Innovation	Cape Town, Karachi, Ho Chi Minh City, Lima, Lagos

GLOBAL CITIES AND EMERGING CITIES

The urban hierarchy: national and regional cities

There is also an urban hierarchy within countries:

- **National metropolises**. National capitals and other important cities play an important political and economic role in their own countries. They are the seat of national institutions, and leading companies establish their headquarters there. Many countries have more than one national metropolis. A national metropolis may also be a global city.
- **Regional cities.** These cities play an important role in their own region. They are usually well connected to the national metropolises by transport networks.
- **Small cities and towns**. These urban areas only play a role at local level.

WORK WITH THE MAP

4 Are the main global cities found in developed or emerging countries? Why?

5 Which global cities of the European Union play a major role internationally? In which fields?

6 What factors are playing an important role in the growth of major cities in emerging countries?

ACTIVITIES

7 What do we mean when we say that there is a hierarchy among cities?

8 In what ways do cities promote the spread of information?

9 Do you think it is a handicap for a country not having a global city?

4 Urban problems

Linfen (China). As a result of mining and industrial activity, this city has the worst atmospheric pollution in the world.

Pollution

- Cities contribute to **global warming**. Vehicles and heating systems use fossil fuels like coal and oil. These produce greenhouse gases which trap heat at the Earth's surface.
- An **urban microclimate** (or **heat island**) is when a city has a slightly higher temperature than the area around it. In city centres, there are many vehicles and tall buildings (which retain the heat), and fewer green areas. Heat accumulates during the day, and temperatures remain high at night.

Traffic

There is a high concentration of traffic in cities, which has the following results:

- High **spending** is required for road networks and traffic management.
- Large **spaces** are set aside for car parks and traffic infrastructures.
- People spend many hours each week **travelling** round the city. **Traffic accidents** are frequent.
- High fuel consumption leads to an increase in **pollution**. There is also considerable noise pollution.

Waste management

Cities create a great amount of waste because they have large populations and high consumption.

- In **developed countries**, urban waste management is achieved through controlled landfill areas, incinerators, compost facilities and recycling plants.
- In **less developed countries**, illegal waste dumps are created in urban areas. This waste often goes straight into rivers and seas.

TRAFFIC DENSITY IN BARCELONA

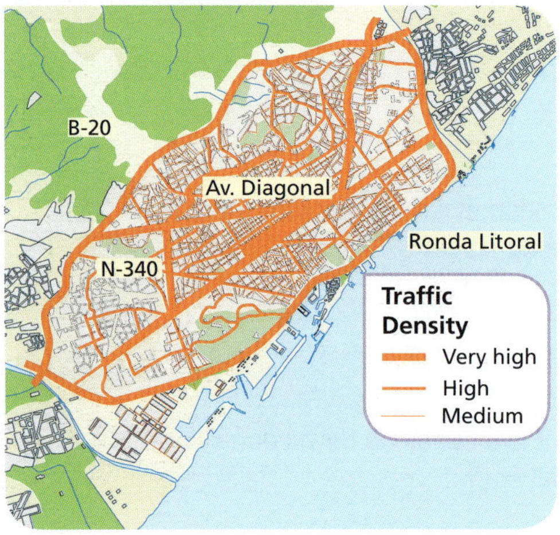

WORK WITH THE MAP

1 Where is the greatest traffic density? Do you think there is a connection between traffic density and the urban microclimate?

Residential segregation

In general, urban residents live in areas that reflect their income. House prices vary greatly from one part of a city to another. They are affected by:

- The quality and type of **housing**, and the proximity of green areas.
- Distance from **work**, and the availability of **services** like schools and health centres.
- **Environmental** issues, such as traffic, the noise level and air quality.

The separation of groups of people into different areas according to their income is called **residential segregation**:

- People on low incomes often live in deteriorating districts or at a considerable distance from the city.
- A neighbourhood suffers from **social exclusion** if it has a high level of poverty and crime. This deprives people of basic opportunities like finding a good job and living in healthy conditions.

In less developed countries, **shanty towns** have grown on the outskirts of big cities. These poor districts lack basic facilities such as running water and street lighting.

Other urban challenges

- **Resource management**. Cities generally consume far more resources (such as water, energy and food) than rural areas. It is the responsibility of city councils to guarantee urban supply.
- **Infrastructures**. City councils often acquire debts because infrastructures like bridges or tunnels are very expensive to build.
- **Housing**. People find it difficult to find good accommodation as a result of **property speculation**, which causes house prices to rise.
- **Social integration**. It is important that people from different backgrounds and cultures live together in mutual respect.

ACTIVITIES

2 What is an urban microclimate?

3 What are the consequences of intensive use of transport?

4 With a partner, list the main problems of your city (or the nearest city). Then think of measures that the council can take to deal with them.

Favelas (slums) in the city of Rio de Janeiro in Brazil.

DID YOU KNOW?

Living in a cemetery

The City of the Dead in Cairo (Egypt) is like any other city with its bars, shops and schools. It has a million inhabitants. But this district is unusual – it is a cemetery!

Some tombs were built with rooms and even gardens so that relatives could mourn there. But there was so little available housing in Cairo that poorer people started to live in these rooms: in return, they looked after the graves.

5 Spanish cities

Urbanisation in Spain

Urbanisation began later in Spain than in some European countries, and it was not always a rapid process. It took place in three stages:

- **Early growth**. From the mid-19th century to the mid-20th century, the growth of cities was linked to **industrialisation**. Expanding industrial cities like Bilbao and Barcelona attracted migrants from rural areas. Other important cities were Madrid, Valencia and Zaragoza.
- **Rapid urbanisation**. From 1955 to 1980, the expansion of the Spanish economy caused a wave of internal migration. About one third of the Spanish population migrated from rural areas to the growing cities.
  - Urban population continued to grow in industrial regions like Madrid, Catalonia and the Basque Country. At the same time, it also increased along the north-west coast and in the Ebro valley.
  - Urban growth reflected the dynamism of Spain's **tourist industry.** For example, the population of Málaga almost doubled in this period, rising to half a million inhabitants. Urban population grew rapidly all along the Mediterranean coast and on the Balearic and Canary Islands.
- **Recent changes**. The pattern of urbanisation has changed since 1980:
  - Migration from rural to urban areas has continued but at a much slower rate.
  - Many companies that are based in large cities have begun to move some of their activities to neighbouring areas with good communications. This has led to the growth of new **metropolitan areas** and the development of the model of the **diffuse city**.

The urban system in Spain

The **urban system** is a series of administrative, economic and transport links between urban centres. Towns and cities also form **subsystems** because they have close links with each other.

URBAN SYSTEM IN SPAIN

Subsystem	Centre
Central: cities of Castile-la-Mancha and Castilla.	Madrid
Basque-Ebro Valley: many cities in the north-west.	Bilbao-Zaragoza axis
Catalonia: Catalonia and the north of Castellón Province.	Barcelona
Levant: Communities of Valencia and Murcia.	Valencia
South-west: western Andalucia, south Extremadura, Ceuta and Melilla.	Seville-Córdoba-Málaga
Balearic Islands.	Palma
Canary Islands.	Las Palmas de Gran Canaria, Santa Cruz de Tenerife
Galician coast: cities on the Atlantic coast.	A Coruña-Vigo axis
Asturias: the triangle Gijón-Avilés-Oviedo.	Oviedo

ACTIVITIES

1. Describe the process of urbanisation in Spain. What were the three stages in this process?
2. What position does your city (or nearest city) occupy in the Spanish hierarchy? Explain your answer.

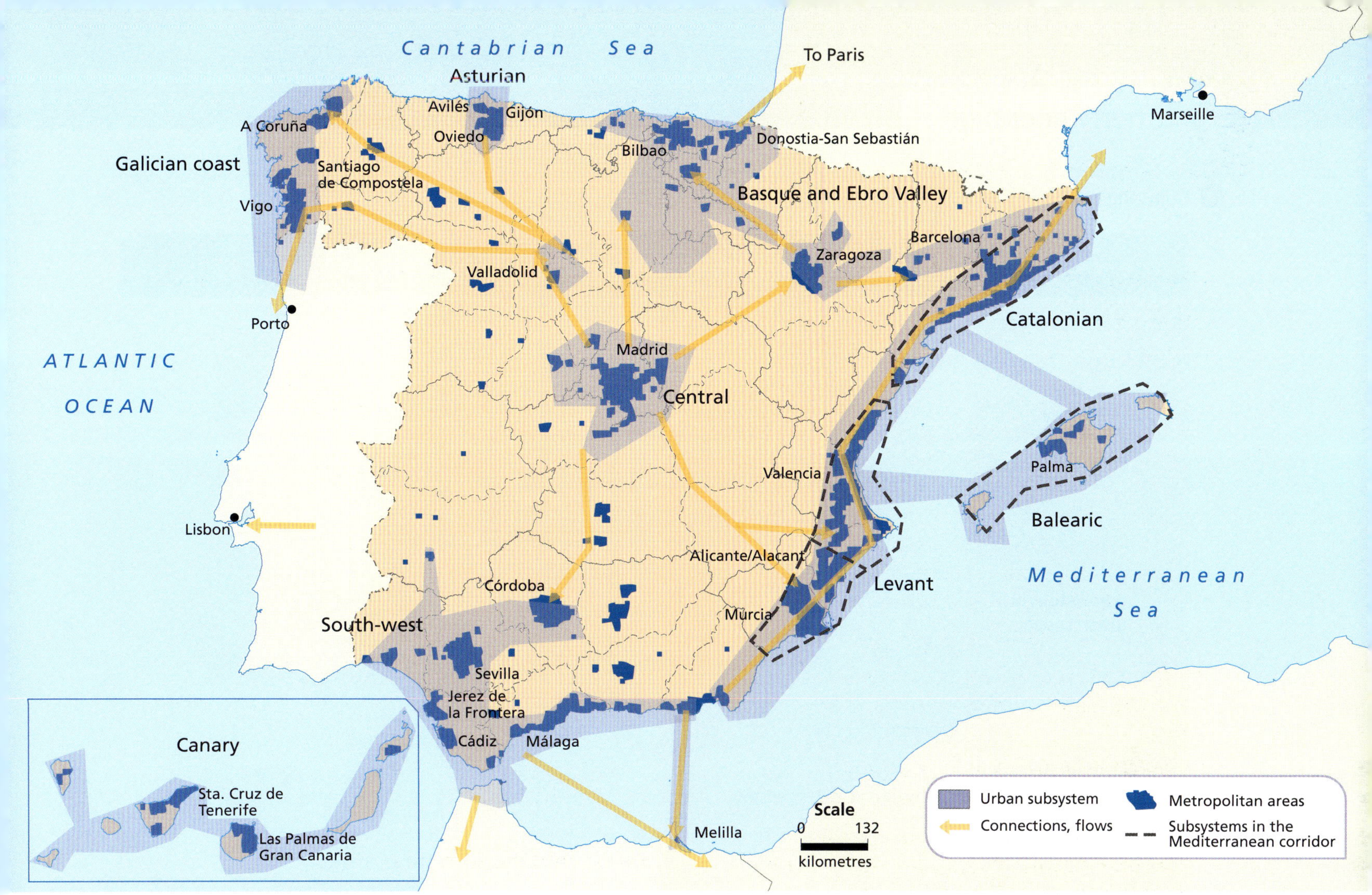

THE SPANISH URBAN SYSTEM

The urban hierarchy in Spain

The hierarchy of Spanish cities reflects a long process of urbanisation. With the exception of Madrid, there are not many large cities in the interior of the peninsula:

- **National metropolises** (Madrid and Barcelona) have populations of over three million inhabitants. They have great economic, political and cultural influence over the rest of the country. Technologically advanced industries and many highly specialised companies are based there, and both cities have strong links with the international economy.
- **Regional metropolises**. Valencia, Bilbao, Zaragoza, Seville and Málaga have populations of between 500,000 and 1,500,000 inhabitants. They perform important urban functions and have great regional influence.
- **Regional cities** like Valladolid, Alicante, Murcia, Palma and Cádiz have between 200,000 and 500,000 inhabitants. They are well-connected to the larger cities and perform administrative, industrial and commercial functions in their Autonomous Communities.
- **Medium-sized cities**, like Burgos, have between 50,000 and 200,000 inhabitants. They are sometimes provincial capitals.
- **Small towns.** These provide basic services at local level.

WORK WITH THE MAP

3 Identify the urban subsystems in Spain. Which are the main cities in each? What place do they hold in the urban hierarchy?

4 Are most urban centres in the interior of the Peninsula or on the coasts and islands? Explain your answer.

1 Copy and complete the table.

THE URBAN SPACE		
The city	Main characteristics	......
	Functions	......
	Problems	......
Organisation of urban space	Zones	......
	Diffuse city	......
Process of urbanisation	Developed countries	......
	Emerging and less developed countries	......
The urban hierarchy	World	......
	National	......

2 Is a 'city' the same as an 'urban area'? Explain your answer.

3 Define the following terms.

- CBD
- Urban hierarchy
- Global city
- City extension
- Diffuse city
- National metropolis
- Metropolitan area

4 What were the main stages of urbanisation in Spain?

5 Analyse the map of the city of Valencia (on the right). Which functions and economic activities can probably be found in each zone?

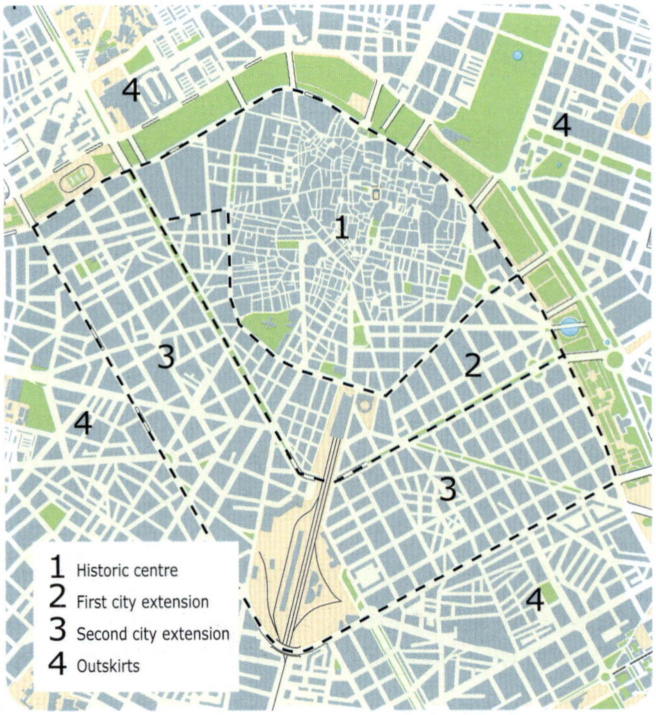

6 Analyse the graph showing the figures for urban population and answer the questions.

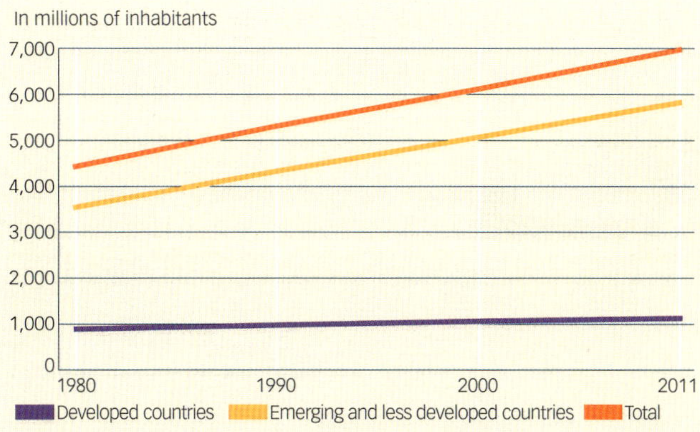

SOURCE: World Bank, 2011.

a Describe the evolution of urbanisation in the world.

b In which countries is urban population growing most rapidly today? Why?

c Are these the most urbanised countries in the world today?

7 Prepare a table on the urban hierarchy in Spain. Include at least one city in each category and give its population.

CATEGORY	NAME OF CITY	POPULATION
National metropolis	....	....
....	....	....

SOCIAL EXCLUSION IN CITIES

Many big cities in developed countries have depressed areas where social exclusion is common. In France, this type of area is called a *banlieue*.

These areas developed in the 1960s. Immigrants arrived in large numbers to take advantage of the jobs created by the economic boom. They lived in large blocks of flats on the outskirts of the city. Services were poor and there was overcrowding. People who obtained jobs and money began to leave the area and moved to other parts of the city.

The result was:

- Those left behind became poorer.
- Buildings began to deteriorate.
- Services remained poor.
- Unemployment was high and social problems grew.

8 What is a *banlieue*? Why are there sometimes social problems in the *banlieues*?

9 In 2005, young people in the *banlieues* began a protest. Find out what happened and write a short report.

10 Form groups. You all live in the *banlieue* shown in the photo. Choose a role (shopkeeper, mayor, woman with children...), and think about how the problems in your area affect you personally.

- Hold a council meeting to discuss what measures need to be taken.

PRACTICAL KNOWLEDGE. **The growth of a megacity: Lagos (Nigeria)**

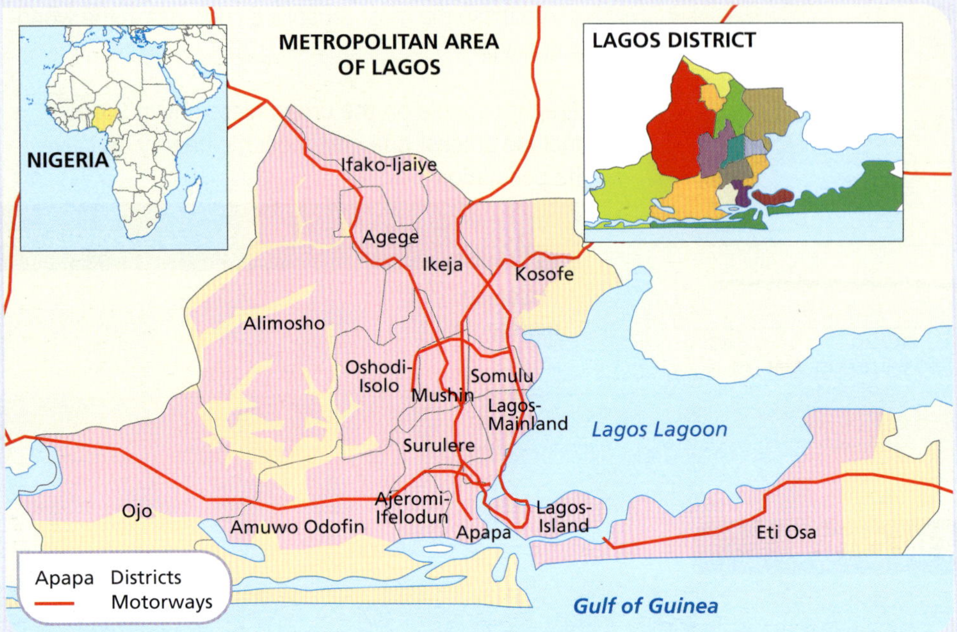

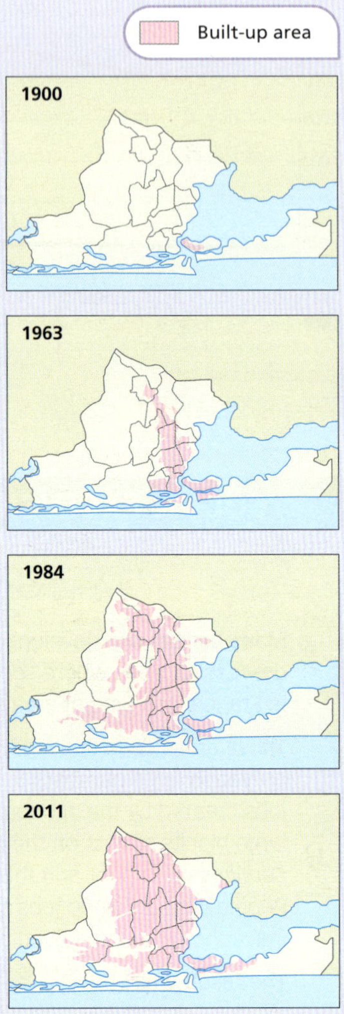

The urban expansion of Lagos (1900-2011).

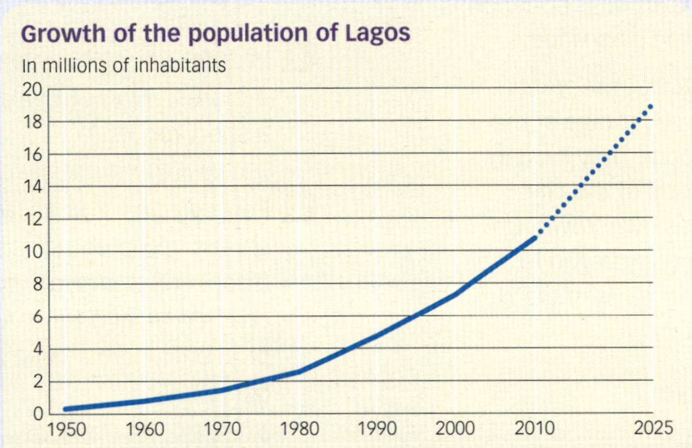

Lagos is one of Africa's fastest growing cities. This port city is the economic heart of Nigeria. It attracts workers from the rural interior where wages are much lower.

Every day, 2,000 people arrive in the city looking for somewhere to live. Many have very little money, and they end up living in a slum or on the streets.

70% of the population lives in slum areas. Many of these live on the water in improvised boats or 'floating houses' at the mercy of tropical storms and floods.

1 With a partner, analyse the documents.

a Can Lagos be considered a megacity?
b When did it begin to grow rapidly?
c Will this continue in the future?
d What population is the city likely to have in 2025?

2 Look at Lagos on Google Maps or Bing Maps.

a Describe the city.
b Are all the buildings in Lagos of the same size and quality?
c Does this tell us anything about Nigerian society?

TECHNIQUES. Use Internet resources

A satellite view of the world

We can see a lot of the images taken by NASA (National Aeronautics and Space Administration) on the Internet. These satellite photos help us to learn a lot about our planet. By analysing them, scientists can predict natural disasters and measure the impact of environmental problems.

Worldview is an interface provided by NASA which enables us to compare different images and information. You can see it at: : http://earthdata.nasa.gov/labs/worldview/.

Worldview works with layers. You can add a layer and compare the results. This enables us to make studies of an area.

Tap on the symbol and begin your journey. We recommend that you begin with a tour of the world's cities lit up at night: **Earth at Night**.

You can link and download your images, change the projection, etc.

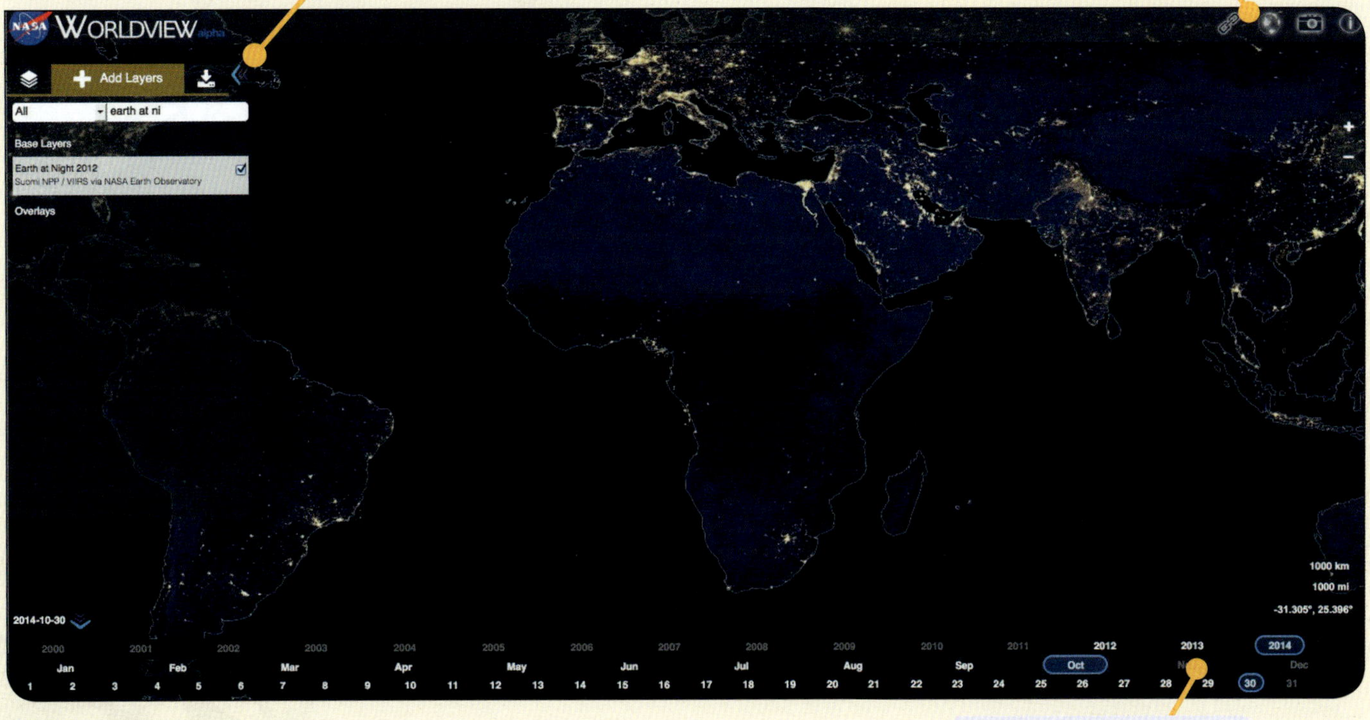

You can see changes over time by selecting different dates.

1 Use Worldview to look at Earth at Night.

a Which areas of Earth are most brightly illuminated? Do they correspond to the most developed urban areas?

b What do the poorly illuminated areas have in common? Identify at least three things.

c Is there any relation between the most densely populated areas and the most brightly illuminated? Explain why.

d Change the projection. What changes do you see in the way the Earth is represented?

2 Look for "NASA Earth Observatory" on the Internet.

a With a partner, choose two images that you find especially interesting.

b Make notes on everything that you can learn from the images. Look for background information on the places in the photos.

c Share your findings with your classmates.

4 The primary sector

FIND OUT ABOUT:

- The primary sector
- Natural and man-made features of agrarian landscapes
- Agriculture, livestock and fishing
- The primary sector in Spain

KNOW HOW TO:

- Interpret a topographic map

The cultivation of oranges

A Planting

Planting takes place in the spring. The trees should be planted at some distance from each other because their roots spread.

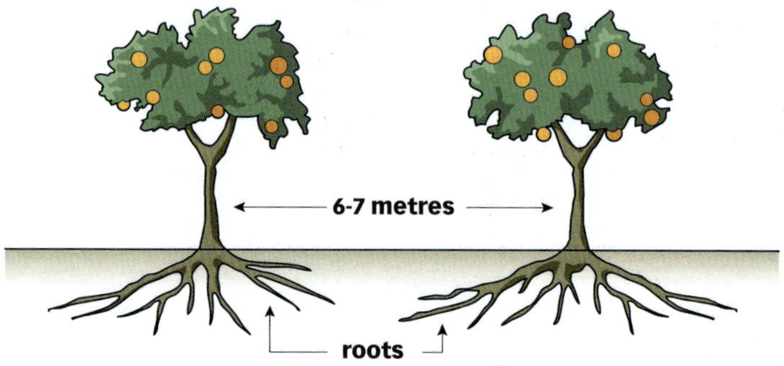

B Tending

Orange trees need a lot of water. They need many hours of sunlight and mild temperatures.

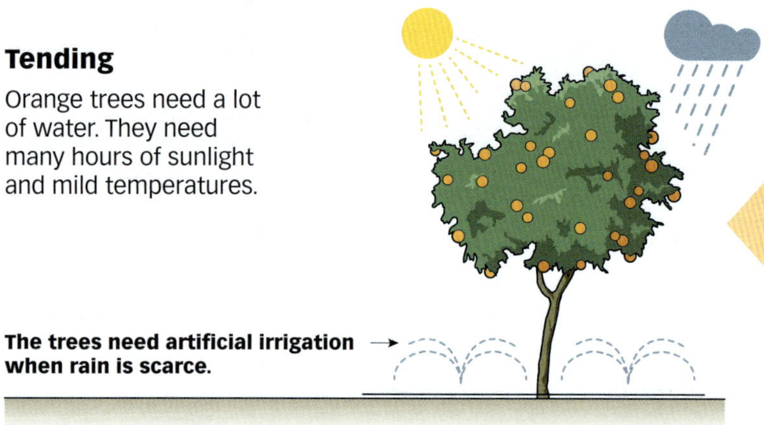

C Harvesting

The oranges are often harvested by hand. In Spain, the harvest is usually between November and May.

? WORK WITH THE IMAGE

- What are the characteristics of the orange harvest?
- Which are the main orange producing countries? Are they also the main consumers of oranges?
- Which of Spain's Autonomous Communities produce most oranges? Is climate an important factor?

Main world producers of oranges

In thousands of tons

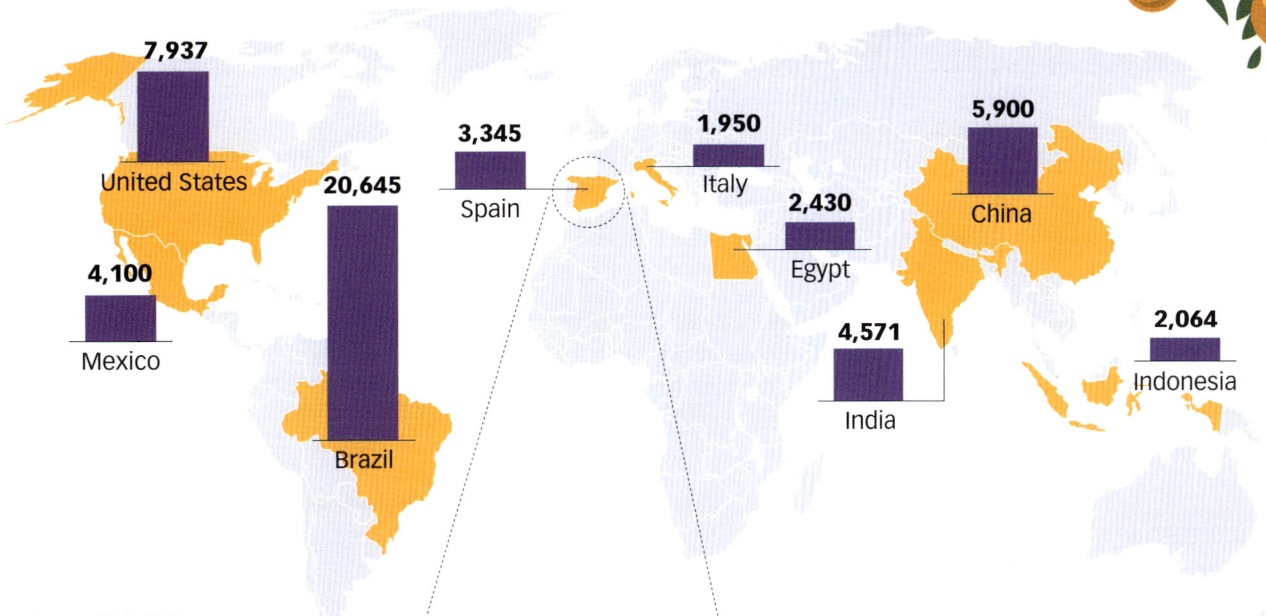

Source: FAO, 2011.

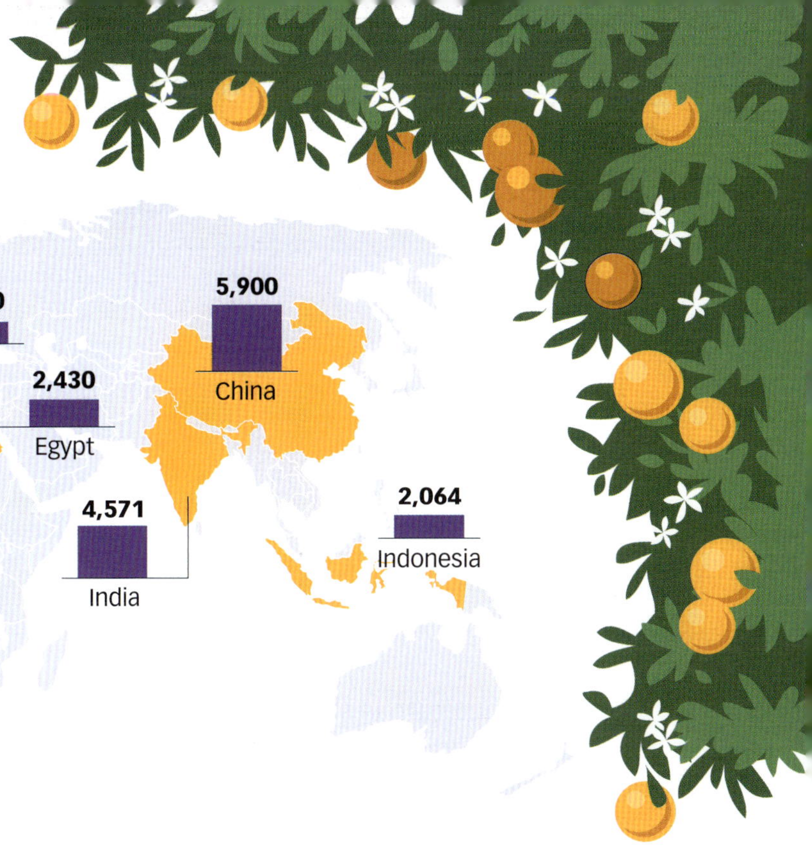

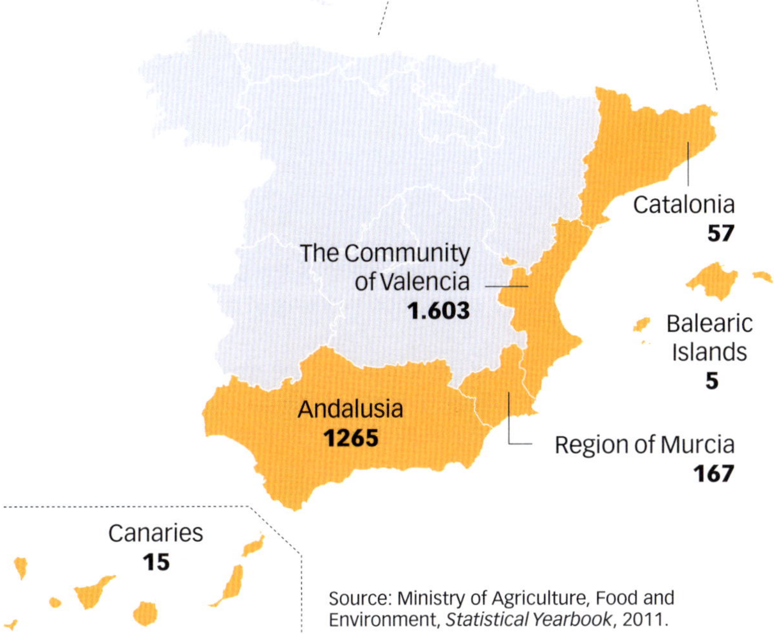

Source: Ministry of Agriculture, Food and Environment, *Statistical Yearbook*, 2011.

Main world consumers

Data in thousands of tons

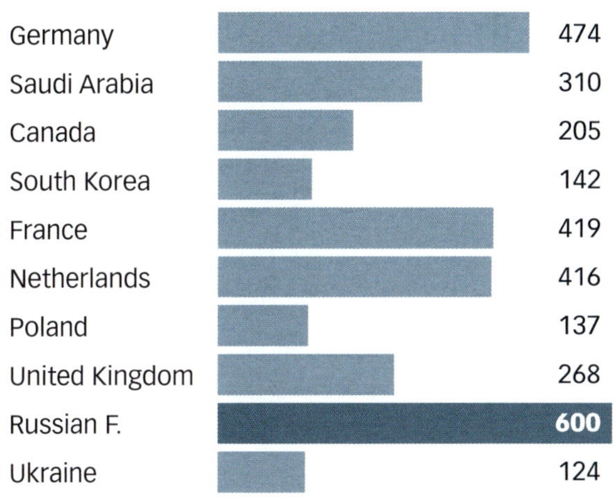

Source: FAO, 2011.

Production and export in Spain

Data in thousands of tons

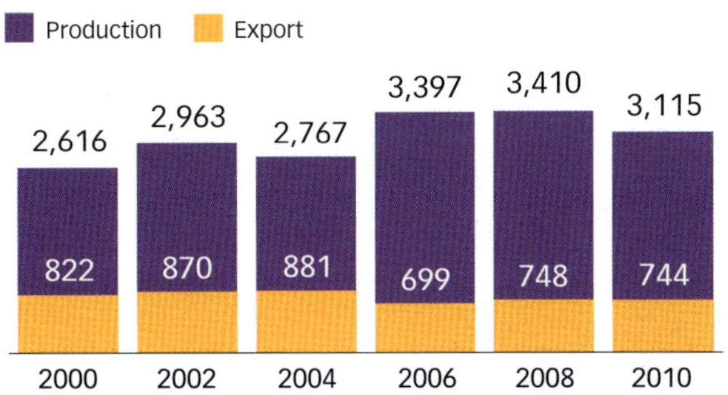

Source: Ministry of Agriculture, Food and Environment, *Statistical Yearbook*, 2011. Ministry of Economy and Competition. Statistics of overseas trade.

HOW DO WE KNOW?

The Food and Agriculture Organisation of the United Nations (FAO) provides us with statistics and varied information about primary sector activities: agricultural and livestock production, exports and imports, fishing activities and prices of agricultural products.

- Which information on this page is obtained from the FAO?
- Which other statistical sources have been used for the maps and graphs on this page?

1 The primary sector today

What is the primary sector?

The **primary sector** involves activities that obtain basic (or primary) resources directly from nature. It consists of:

- **Agriculture**
- **Livestock farming**
- **Forestry**
- **Fishing**

Today, agriculture is still vital to the economy of less developed countries. Before their economies expanded, most developed countries had an **agricultural revolution**, when new techniques and technology were introduced in agriculture.

The importance of the primary sector today

About 40% of the world's active population works in the primary sector. However, this percentage is gradually decreasing. At the same time, the percentage of the active population working in industry and the services sector has increased.

In 2012, the agricultural sector contributed about 3% of the world's production (**GDP**). However, there are great differences between regions:

- In **developed countries**, between 1% and 5% of the active population works in the primary sector. The sector contributes less than 4% of GDP (1.5% in USA and Japan). Productivity is high due to mechanisation and technological progress.
- In **emerging** and **less developed countries**, about half of the active population works in the primary sector, which produces over 10% of GDP. In the poorest countries, about 80% of the active population works in agriculture. This creates over 50% of their GDP. However, they have low productivity because they still use traditional farming methods.

POPULATION WORKING IN THE PRIMARY SECTOR

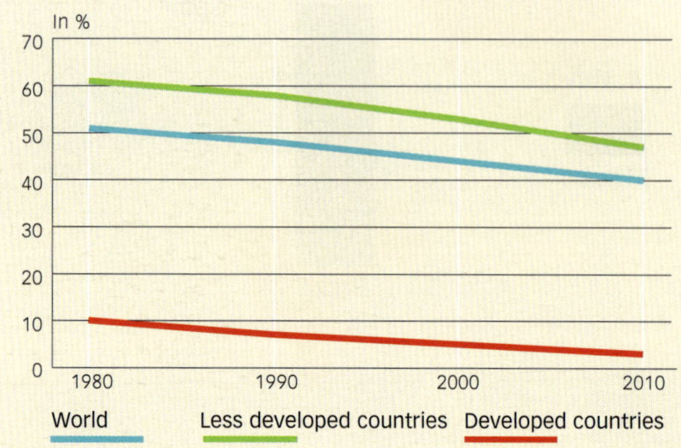

SOURCE: FAO, 2013.

WORK WITH THE GRAPH AND THE MAP

1. Describe the evolution of the proportion of the world's population working in the primary sector.
2. Compare the graph and the map on the next page. In which parts of the world do many people still work in the primary sector? What do these countries have in common?

ACTIVITIES

3. Copy and complete the following table in your notebook.

THE PRIMARY SECTOR		
	Developed countries	**Less developed countries**
% of active population	.....	.....
% of GDP	.....	.....
Technology	.....	.....
Productivity	.....	.....

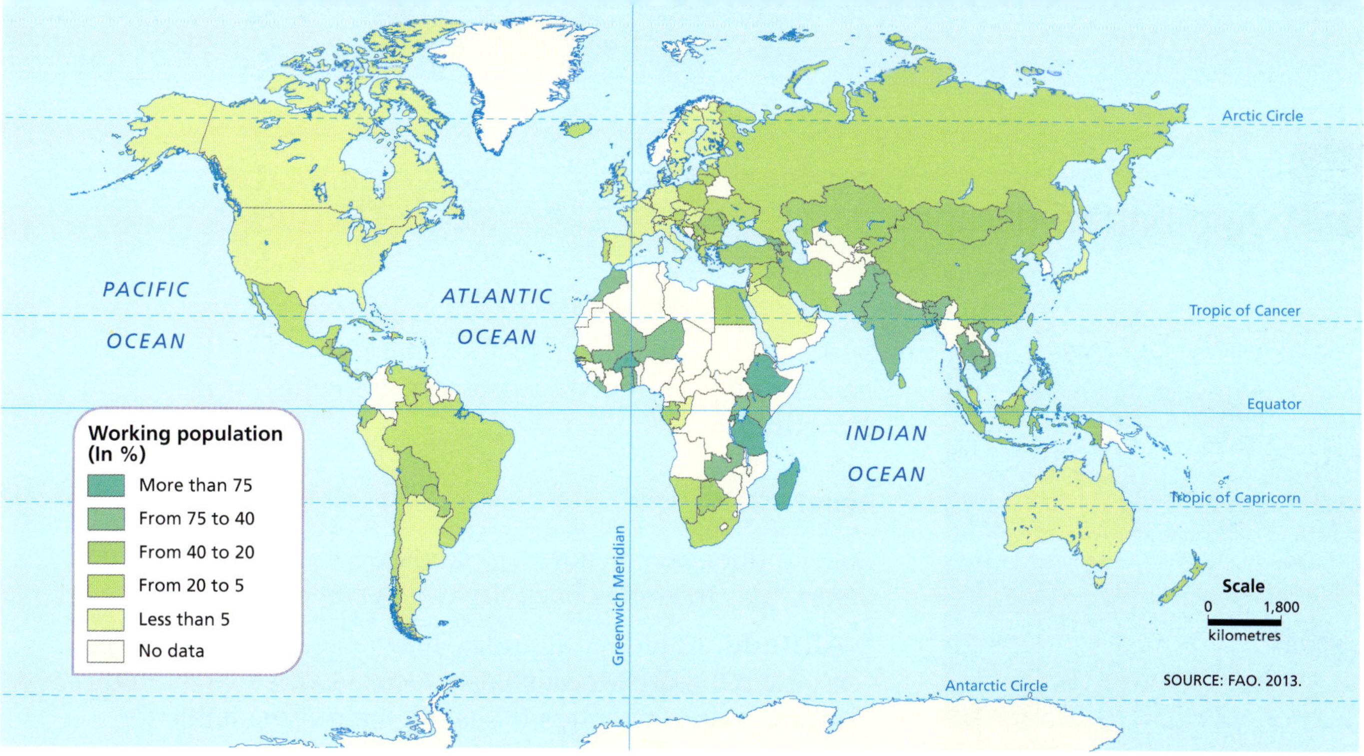

WORKING POPULATION IN THE PRIMARY SECTOR BY COUNTRY

DID YOU KNOW?

Economic systems

An **economic system** is the way that goods and services are produced, and benefits are distributed:

- **Traditional or subsistence economy**. Families produce almost everything they need. They sell their surplus at a local market, where they also acquire other products. This system is found in less developed societies.
- **Capitalist or market economy**. The state intervenes little in economic activity. The economy is based on free competition, private property and profits.
- **Communist or centrally planned economy**. The state controls all aspects of the economy, including production and profits. This system is not widespread today.

In practice, many countries have **mixed economies**. These are basically market economies, but also have a certain amount of central planning.

- Think about your own country. What kind of economic system does it have?

Gross Domestic Product (GDP). This is the value of all the goods and service produced in a country over one year. It is a good indication of its wealth.

THINK ABOUT IT

What is China?

China's economy has grown rapidly for several decades. It has a huge manufacturing sector, and today it is the world's leading trading nation. At the same time, it has a low per capita income, and many regions are rural and not developed.

Officially, China is ruled by the Communist Party and has a "socialist market economy": it combines an important state sector with a market economy.

- With a partner, look for some figures on China's population and economy on the Internet. Is China a developed or an emerging country? Does it have a capitalist, communist or mixed economy?
- Debate your ideas with the class.

2 Agrarian landscapes

A tea plantation in India. It is possible to grow crops on hillsides by building terraces.

Both natural and human factors shape agrarian landscapes. The most important natural factors are **relief**, **soil** and **climate**.

Relief

Relief influences the agrarian landscape because it determines the type of activity which can be carried out.

- **Altitude**. At higher altitudes, temperatures fall, and this restricts the growth of some species. Land at an altitude of less than 200 metres is usually the most suitable for agriculture.
- **Gradient**. It is possible to mechanise many farming techniques on flat land. But it is difficult to cultivate land at a steep gradient of more than 10°. Most farming is therefore carried out on plains and in valleys.

Soil

Plants absorb the nutrients from the soil that they need to grow. Each plant requires a particular type of soil:

- **Porosity**. This determines the amount of **oxygen** that the soil contains and its ability to retain **water**.
- **Chemical and biological composition**. This determines its richness in **mineral nutrients** and **organic matter**.

TYPES OF SOIL

Clay. Soil with a high clay content is rich in nutrients. However, it may absorb too much water and needs good drainage.

Sandy soil does not retain nutrients or water. It needs fertilisers and frequent watering.

Humus. Organic matter makes the soil rich in nutrients.

Cultivation in the desert (Israel). The plants receive water through an expensive system of drip irrigation.

WORK WITH THE PHOTO

1 The geographic coordinates of this irrigated land are 32° 35′ N, 35° 14′ E. Look for it on Google Maps. Can you see any cultivated land in the desert?

Cultivation under plastic in Almería. This system prevents irrigation water from evaporating and maintains humidity.

Climate

Each plant requires certain conditions of temperature, sunlight and moisture. Throughout human history, people have selected the plants that they cultivate, choosing those that are suitably adapted to the local climate.

There are four main climatic factors:

- **Extreme cold**. In general, plants cannot grow at extremely low temperatures. In fact, many plants cannot grow at temperatures below 10 °C.
- **Excessive heat**. Few plants grow at very high temperatures, above 45 °C.
- **Humidity**. Heavy, frequent rainfall favours the growth of weeds and contributes to soil erosion.
- **Dry conditions**. Plants need a certain amount of water. Agriculture is more difficult in places where there is scarce, irregular rainfall.

It is almost impossible to practise agriculture at the Poles, in deserts or in tropical rainforests without the use of expensive modern technology.

ACTIVITIES

2 What types of farming are you familiar with (for example, in your region)? Identify three factors that influence this type of farming.

WORK WITH THE PHOTOS

3 Compare the photos of agriculture in India, Israel and Spain. Which activity do you think is most profitable? Why do people practise the other types of agriculture if they are less profitable?

3 Human influence on the agrarian landscape

Agricultural holdings

Agricultural holdings are units of agricultural production with a single proprietor, which may be either a company or a person. Their main activity is agriculture, livestock farming or forestry. They may be large estates, medium-sized farms or smallholdings.

A holding consists of one or more **plots**, or pieces of land. The shape, size and borders of the plots influences the landscape. The plots can be small (less than 1 hectare), medium (between 1 and 10 hectares) or large (more than 10 hectares). They may also be regular or irregular in shape.

- When the plots are not surrounded by fences, the landscape consists of plots that are not separated from each other. This **open-field** landscape is common on the southern sub-plateau of Spain. In this type of landscape, the farmworkers' houses are usually grouped together.
- When the plots are separated by fences, hedges or walls and they are linked by a network of roads, they form a landscape of **closed fields**. The farmworkers live in isolated houses near the fields or in villages. This type of landscape is found in Cantabria in Spain, and along the Atlantic coast in Europe.

DID YOU KNOW?

Rural settlements

Human settlements are distributed in the countryside in two main ways:

- **Dispersed settlement.** Houses are spread out across the countryside. Cultivated land and fields surround each house.
- **Nucleated settlement**. Buildings are grouped together to form a village. The village is surrounded by fields and farmland.

Dispersed settlement in Gipuzkoa.

An open-field landscape in Valencia.

Closed fields in Normandy, France.

Types of farming

Farming may be intensive or extensive:

- **Intensive farming** uses the land to the maximum of its capacity. Modern techniques are used in developed countries, which achieve a high level of productivity. In less developed countries, farmers practise traditional agriculture, using large numbers of farmworkers. Productivity is much lower.
- **Extensive farming** is the cultivation of crops like wheat and barley on large areas of land. It is possible to grow considerable quantities of crops by using farm machinery in large, flat fields.

Two different types of watering system are used:

- **Irrigation crops**. Water is taken from artificial sources such as irrigation channels, flooding and sprinkling.
- **Rainfed crops**. This system depends on rainfall, but it is often supplemented by modern irrigation techniques in times of drought.

One or more crops may be grown on an agricultural holding:

- **Monoculture**. A single type of crop is grown. This type of large scale mechanised agriculture produces large quantities of a product at a competitive cost.
- **Polyculture**. Several types of crop are grown, usually on a smallholding. They are often for the consumption of families working the land.

Other features of agrarian landscapes include the following:

- **Livestock farming**. The size, form and boundaries of the land, and the way it is used, shape the agrarian landscape.
- **Forestry**. The species of tree (conifers, hardwood, etc.) defines the landscape.

Intensive irrigation agriculture (Morocco).

Extensive rainfed agriculture (Canada).

ACTIVITIES

1 Are open-field landscapes associated with dispersed or nucleated settlement? And closed field landscapes?

2 Investigate forestry and write a short report. Explain a) the types of products that it provides and b) the threat to the environment from deforestation.

WORK WITH THE PHOTOS

3 Compare the photos.

a What are the main characteristics of these agrarian landscapes?

b How big are the plots? What is their shape?

c Do they show monoculture or polyculture?

4 Agriculture in developed regions

DID YOU KNOW?

The agricultural revolution

The agricultural revolution refers to a rapid process of change in the ownership and use of the land. It began in the United Kingdom at the end of the 18th century.

Agricultural production increased, and so food improved. As a result, the mortality rate decreased and the population grew. However, the new techniques meant that there was less demand for the work of farm workers who often left the countryside to live in cities.

Commercial agriculture

The agricultural revolution that began in Europe and North America at the end of the 18th century initiated a transition from subsistence farming to large scale **commercial agriculture**.

The main goal of commercial agriculture is to obtain maximum profit. This is achieved in different ways:

- **Specialisation**. Usually, only one product is cultivated on a farm. This simplifies the use of machinery and labour, and increases productivity.
- **Mechanisation**. Machines partly replace human labour. This speeds up the production and reduces labour costs.
- **Modern farming methods**:
  - Fertilisers improve soil fertility, so more crops can be grown each year.
  - High-yield seeds increase production.
  - Greenhouses, artificial soils and new irrigation systems are used in places with dry climates and poor soil.
  - Biotechnology produces new varieties of **genetically modified** (GM) crops.
- **Storage, transport and commercialisation**. Today, products can be carried long distances at a low cost.

Commercial agriculture requires large **investments**. However, the increase in productivity offsets the costs and leads to greater profits.

Advanced farming techniques

Hydroponic systems

This technique allows cultivation to take place in an artificial growing medium like rice husks, sand or sawdust. The water used in irrigation contains all the nutrients that the plant needs.

Sanding

A fine layer of manure is on top of the soil and a thick layer of sand is spread over that. The sand filters the water and heats up rapidly, transmitting heat to the lower layers. In this way, the plants absorb nutrients more easily.

Drip irrigation

This technique delivers water to the plants through a series of closed tubes which drip-feed water to each plant. The plant does not waste energy in obtaining water and its natural growth cycle is shortened. Water is saved because each plant receives only the amount it needs.

Benefits and costs of commercial agriculture

High productivity leads to increased efficiency. As a result, commercial agriculture has a beneficial effect on prices, marketing and distribution. However, there are also costs:

BENEFITS AND PROBLEMS OF COMMERCIAL AGRICULTURE

BENEFITS

- **Lower prices.** High productivity leads to lower prices. However, lower prices are also the result of subsidies in wealthy countries, which make it difficult for less developed countries to compete.
- **Specialised transportation**. Large quantities of agricultural produce are carried over long distances.
- **Preservation techniques**. These also enable produce to be distributed a long way from where it is grown.
- **Efficient distribution**. Wholesalers travel to the places where the produce is grown.

PROBLEMS

- **Pollution**. Pesticides and chemical fertilisers pollute the soil and the water of the aquifers. Their recovery is very slow. This contamination affects our food and health.
- **Water use**. Modern irrigation systems use a lot of water and reduce the reserves held in aquifers.
- **Rural exodus**. The increasing use of machinery has led to a decline in the number of jobs in rural areas. Many younger people have left to live in cities.

ACTIVITIES

1 Study the table above, and look for more information on the Internet. Write a report on commercial agriculture.

a Explain what commercial agriculture is.
b Continue with a summary of its benefits.
c Discuss its disadvantages.
d End with a short personal conclusion.

2 Type "organic farming" into a search engine on the Internet.

- What is it?
- What are the advantages and disadvantages of this type of agriculture?

DISTRIBUTION OF COMMERCIAL AGRICULTURE

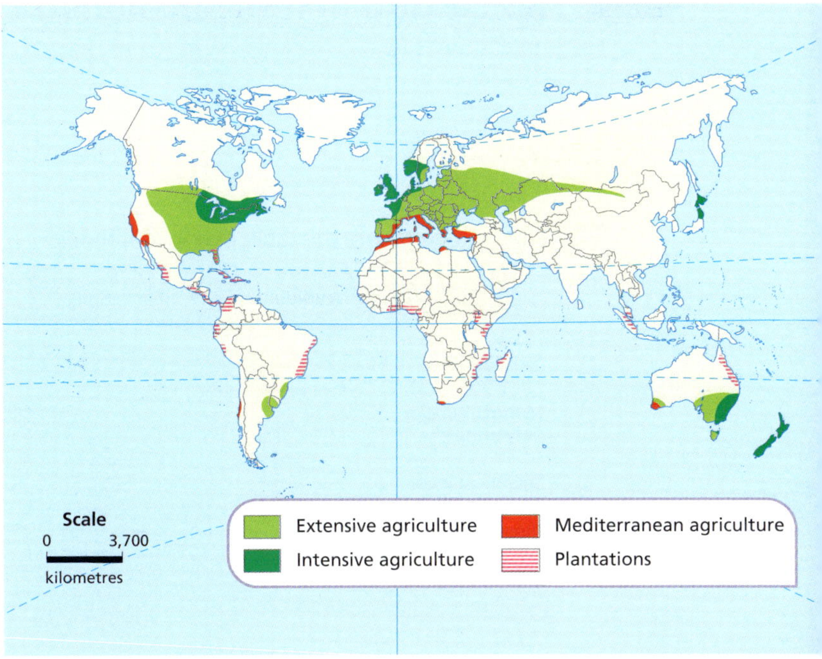

WORK WITH THE MAP

3 In which regions is commercial agriculture widespread? Find out more about agriculture in three of these regions.

WORK WITH THE GRAPH

4 What conclusions can you draw about agriculture in less developed countries from their use of fertilisers?

USE OF FERTILISERS

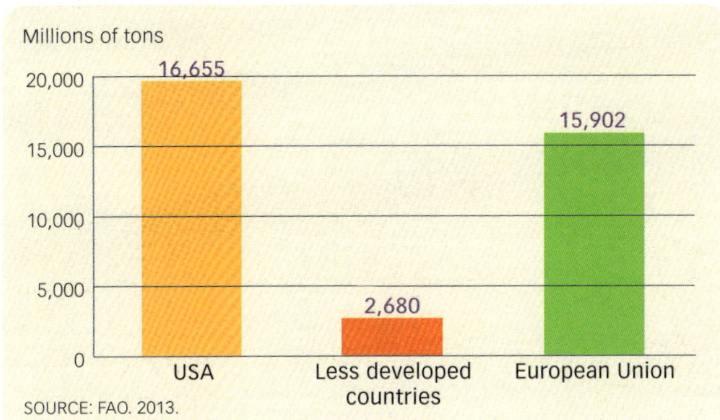

SOURCE: FAO. 2013.

5 Agriculture in less developed regions

ACTIVITIES

1. Copy and complete the table in your notebook.

Type of agriculture	Traditional	Plantations
Size of farms	.....	.....
Monoculture or polyculture?	.....	.....
Technical level	.....	.....
Disadvantages	.....	.....
Productivity	.....	.....

Sedentary rainfed cultivation.

Irrigated monsoon cultivation.

What is traditional agriculture?

Traditional agriculture is a type of **subsistence agriculture**, which is practised on small plots of land. Families farm the land to obtain food for themselves.

Traditional agriculture has the following characteristics:

- **Polyculture**. Farmers cultivate several crops at the same time to make sure there is enough food all year round.
- **Manual labour**. Families often provide most of the workforce, using old-fashioned farming techniques and implements. As a result, productivity is low.
- **Physical factors**. Crops are only abundant when the weather is good.

Types of traditional agriculture

- **Shifting cultivation**. This is practised in areas of Africa, South America and south-west Asia. The main crops are cereals and tubers. Trees are cut down and burned in the forest, and the ash is used to fertilise the soil. After a few years, the soil is no longer fertile and the farmers move on to new areas.
- **Sedentary rainfed cultivation**. This is practised in savannah regions using a system of rotation. Cereals are grown in one area, and products such as nuts and tubers are grown in another. A third part is left fallow and is used to graze livestock.
- **Irrigated monsoon cultivation**. This is practised in eastern China, south-east Asia and on the coast of India, where monsoon rains cause extensive flooding. Rice is cultivated in the flooded land. Small plots of land are tended by many workers.

Plantation agriculture

Plantations are large agricultural estates, which are usually located in **less developed tropical countries**, mainly in Africa, Central and South America, India and south-east Asia. The earliest plantations were established by European colonists between the 16th and the 19th centuries. Crops could be obtained that did not grow in Europe, such as cocoa, coffee and cotton.

Today, plantations produce **coffee** in Brazil and Ethiopia, **tea** in India and **rubber** in Indonesia and Malaysia.

The number and size of plantations in developing countries has grown as a result of economic **globalisation**. Plantations are now either directly controlled by multinational companies, or by local producers who sell their products to them.

Plantation agriculture is a form of **commercial agriculture**. It requires large extensions of land, considerable capital investment, a good transport network, abundant cheap labour and modern machinery and techniques. Other characteristics are high productivity, monoculture and production for export.

Benefits and disadvantages of plantation agriculture

- **Benefits for multinational companies.** Production costs are low because workers in less developed countries are paid low wages.
- **Benefits for less developed countries**. Jobs are created, which stimulates local economies. Transport and marketing infrastructures are improved. Modern farming techniques and more efficient production processes are introduced.
- **Disadvantages for less developed countries**.
  - Plantation crops are exported. Most of the profits go to foreign companies, while local workers have a low standard of living.
  - Many small farmers must leave their land and work on plantations because they cannot compete.
  - Companies aim for maximum output so exploitation becomes very intensive. This causes severe environmental impact.
  - Decisions are made based on the sales price of products. If prices drop, wages drop too, and operations sometimes end. This causes unemployment and poverty.

A COFFEE PLANTATION

6 Livestock farming and fishing

Extensive livestock farming.

Intensive livestock farming.

Livestock farming

Livestock farming consists of raising animals for food (such as meat, milk and eggs) or other economic purposes (like leather). Some animals are raised as a source of labour, or for breeding. **Cattle**, **sheep** and **pigs** are the most common animal species.

Some farms concentrate exclusively on stock-breeding. Others, however, combine crops and stock-breeding. New technologies have helped to increase production significantly. Scientific research in the last century has led to **selective breeding**, which improves productivity and develops the resistance of livestock to disease. New systems for cooling and freezing enable farmers to transport products for long distances in perfect condition. However, mechanisation and modern technology are not used to the same extent in all types of livestock farming.

Extensive livestock farming

Extensive livestock farming is practised in both developed and developing countries.

- It is typical of **large farms**, where animals graze in **natural pastures**.
- It requires **little investment** in labour or in facilities or feed for animals, but **productivity is low**.
- Typical livestock includes **cattle** and **sheep**. In developed countries, products are intended for sale; in less developed countries, they are for family consumption.

This type of livestock farming is found in areas with abundant pastures, such as the central and western United States, the Pampa of Argentina, eastern Brazil, Australia and the Republic of South Africa.

Intensive livestock farming

Intensive livestock farming is typical of developed countries. It has the following characteristics:

- Farms **specialise** in a single animal species.
- **Large investments** in facilities, technology, animal feed, etc. are required, but **productivity is high**.
- Sometimes the animals are raised in **stables** where diet, health and fattening are controlled. In other cases, animals are **semi-stabled**: they eat forage plants in pastures or they eat animal feed when there is not enough grass.
- Typical livestock include **cattle**, **pigs** and **poultry**. Production is intended for sale at markets. Intensive livestock farming is practised mostly in the eastern United States and Canada, Western Europe, south-east Australia and New Zealand.

This type of livestock farming has led to increased food production. However, it also damages the environment. Animal manure pollutes the soil and water. Large amounts of land are needed to produce food for so many animals. The methane gas released by large herbivores causes air pollution.

Fishing today

Fishing is the catching of fish and other animals from seas, rivers or lakes. The world fishing industry has expanded in recent decades, and today 85% of catches are for human consumption. The rest is used to make oil and fishmeal.

Much of the fish consumed in developed countries is imported from emerging and less developed countries. China has both the highest fish production and the highest consumption per capita of fish products.

Types of fishing

There are three main kinds of fishing:

- **Coastal fishing** takes place very near the coast in small boats using traditional methods. The catches are for home consumption and the local market. This type of fishing is an important source of food in developing countries.
- **Inshore fishing** is carried out near the coast in small or medium-size boats. Their equipment depends on the fish which is caught. The boats use refrigerated compartments or ice to preserve the fish, and deliver their catch each day.
- **Off shore fishing** is practised out at sea. It uses large ships equipped with modern technical instruments which can detect the shoals of fish. Many ships have installations where the fish are cleaned, cut up, packaged and frozen. They are **factory ships**, and they can be at sea for several weeks.

Fishing grounds

Fish are caught in **fishing grounds**, areas in which fish naturally concentrate. The best fishing grounds are along continental shelves. Sunlight and the nutrients deposited there by rivers favour the growth of plankton, which is food for many fish.

There are restrictions on where fishing can take place, which is controlled by international legislation allowing coastal countries the right to fish up to 370 kilometres from their shores. For fishing to take place outside this area, **international treaties** must be signed.

At present, fishing catches are excessive in relation to the amount of fish available. This leads to **overfishing** and the danger of different species disappearing. Governments have adopted different measures to encourage **sustainable fishing**. For example, they ban certain fishing techniques (such as trawling), limit the number of catches or establish periods when no fishing is allowed.

Many countries practise **aquaculture**, which is the breeding of fish in controlled environments. Most of the salmon and mussels consumed today come from **fish farms**. Aquaculture increases the supply of fish and seafood, and prevents species from becoming extinct.

7 The primary sector in Spain

The primary sector

The importance of the primary sector in Spain has declined since the mid-20th century. Today, 4.2% of the active population works in this sector, which constitutes 2.7% of GDP. However, the value of production grows each year, and the sector is still important.

THE PRIMARY SECTOR IN SPAIN
STRENGTHS
• Spain has one of the strongest primary sectors in the European Union. Its agricultural sector represents more than 11% of the EU total, and it is the leader in fishing production.
• Agricultural and fish products make up the raw materials of many of Spain's industries.
• The food industry contributes around 15% of Spanish exports.
WEAKNESSES
• An ageing population: more than 61% of the owners of agricultural holdings are aged over 55.
• A large number of small and medium holdings, which may not stay profitable in an increasingly competitive market.
• The low productivity of rainfed agriculture as a result of low investment.
• The low productivity of extensive livestock farming.

Agriculture

Agriculture has a leading role in the Spanish primary sector. Agricultural products represent more than 60% of total agrarian production, well above the European average.

- After France, Spain is the EU country with the greatest percentage of land used for agriculture (34.4%). About 20% of this land is irrigated, and the rest is rainfed.
- Spanish agriculture is characterised by the diversity of products. It includes fruit and vegetables, vines, olives and cereals.
- Most agricultural trade (both exports and imports) is with the rest of the European Union.

Livestock farming

Stabled intensive livestock production predominates in Spain. Holdings are characterised by strong investment, mechanisation and high output. They are located near the cities where their products are sold.

Livestock farming contributes 35% of the value of total agrarian production. Spain has the second highest number of sheep, pigs and goats in the EU.

- Pigs make up a third of animal production. Almost half are in Catalonia and Aragón.
- Cattle represent 28% of total production and are found in large numbers in Castile and León, Galicia and Extremadura.

Fishing

Spain has the largest fleet in the EU and the volume of its catch is the second largest. It is also the world's second largest consumer of fish.

- Spain's fishing industry was modernised after entry into the EU. However, many fishermen lost their jobs during this transition. Today, traditional and commercial fishing co-exist.
- The continental shelf of Spain is narrow and does not provide enough fish to meet the country's needs. Fishing treaties have therefore been agreed with other countries.

Regional diversity

In recent years, the agricultural sector has benefited from investment by the CAP (Common Agricultural Policy) of the EU.

- However, not all regions have benefited to the same extent. The Mediterranean and Andalusian coasts have found new markets for their fruits and vegetables. But other products, such as the cereals and vines of the interior, have declined.
- The regions worst affected by the problems of the agricultural sector have been partly depopulated.

Data on the primary sector in Spain

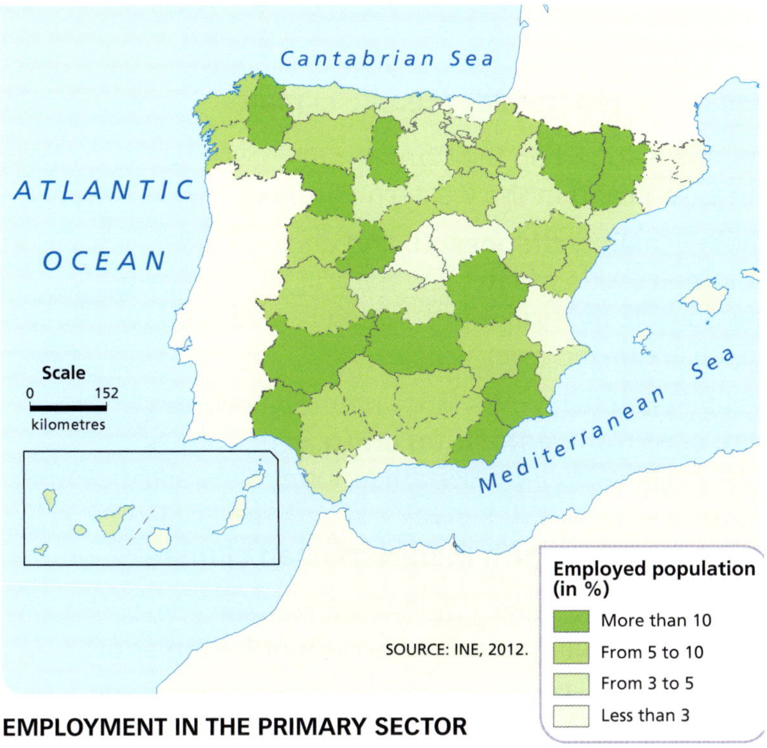

EMPLOYMENT IN THE PRIMARY SECTOR

Evolution of employment and contribution to GDP

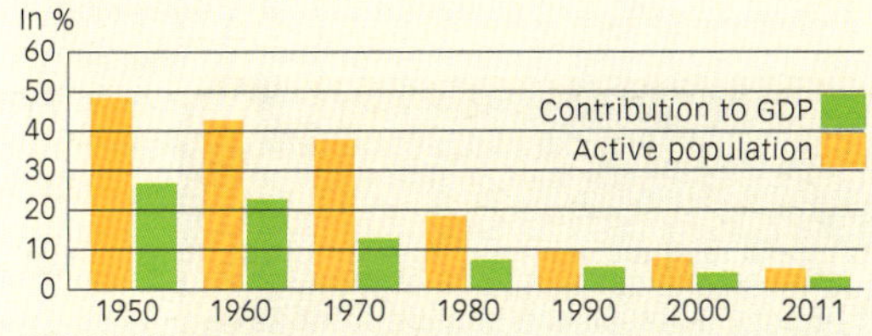

SOURCE: INE.

Classification of agrarian production

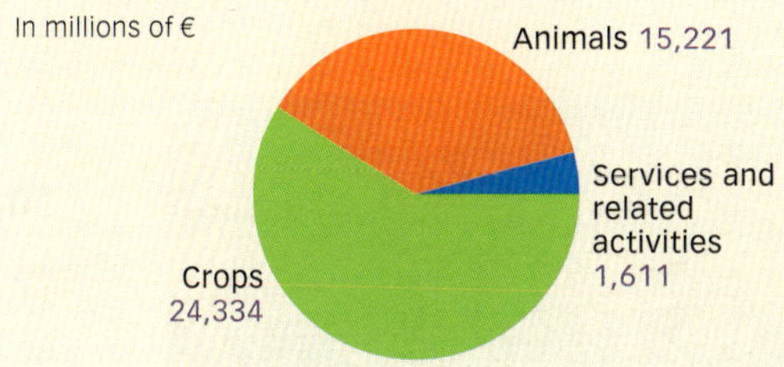

SOURCE: INE, 2012.

Cultivated land

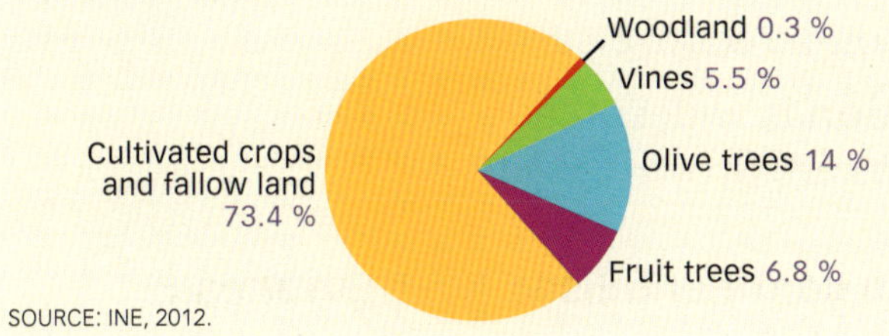

SOURCE: INE, 2012.

Growth of organic agriculture

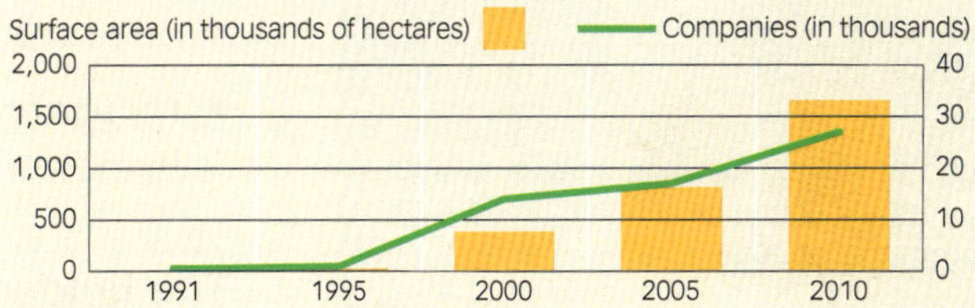

SOURCE: INE.

Livestock

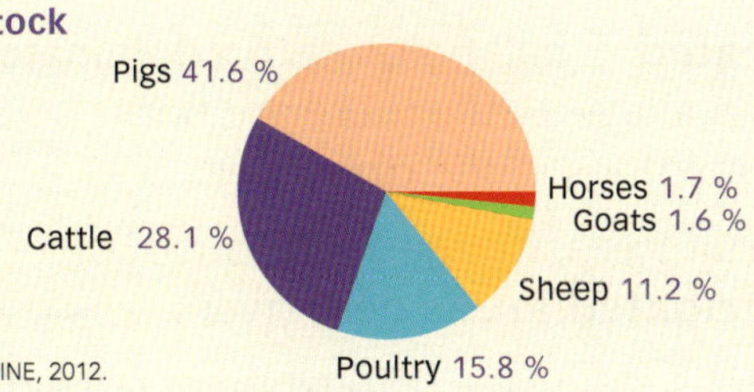

SOURCE: INE, 2012.

Spanish fishing catches by zone

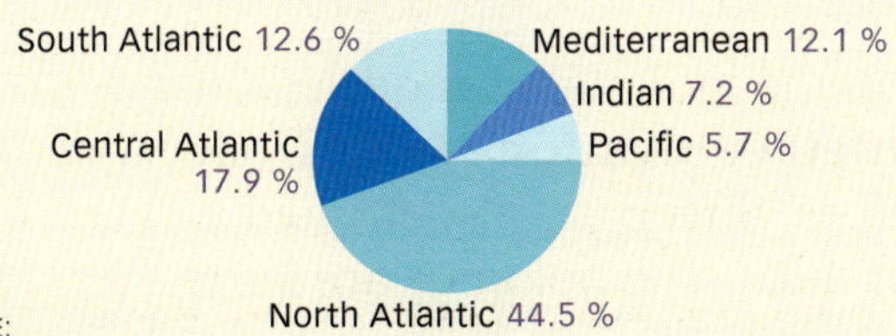

SOURCE: www.magrama.gob.es/es/

WORK WITH THE GRAPHS

1. Form groups of three. Choose two or three of the graphs or maps and explain their significance to the rest of the group.

ACTIVITIES

2. Look on the Internet for information about the Common Agricultural Policy (CAP). Why has it been changed many times since it was first created?

8 Agrarian landscapes in Spain

Humid oceanic landscape (Galicia) (A), warm Mediterranean landscape (Valencia) (B), continentalised Mediterranean landscape (Badajoz) (C).

ACTIVITIES

1. Where is each type of agricultural landscape found?
2. Describe the main characteristics of each type of landscape.

The landscape in the humid, oceanic climate

This landscape is found on the Cantabrian coast, in much of Galicia and in the western Pyrenees.

- Settlement is mainly **dispersed**, and enclosed smallholdings predominate. The natural environment is not favourable to agriculture, which is restricted to low coastal areas and river valleys.
- The oceanic climate favours large natural meadows. These are used for **livestock farming**, especially of dairy cattle. There is also a certain amount of forestry.

The landscape in the warm Mediterranean climate

This landscape is found along the Mediterranean coast, in the Guadalquivir valley and in the south of Extremadura.

- Settlements are **nucleated**. In the south, there are both large and small open holdings. Small closed holdings predominate on the Catalan, Levantine and Balearic coasts.
- **Rainfed cultivation** is practised in the Guadalquivir valley, as well as in coastal hills, where wheat, vines, olive trees and almond trees are cultivated.
- **Irrigated cultivation** is widespread elsewhere. Fruit and vegetables are grown extensively.

The landscape in the continentalised Mediterranean climate

This landscape occupies much of inland Spain.

- Settlements are **nucleated**. Small holdings predominate in the Duero and Ebro valleys, while there are both large and small open holdings in the drier areas.
- Cereals, vegetables, fruit trees, forage and industrial plants are grown in **irrigated areas** near rivers.
- In high flat areas, there is extensive cultivation of **cereals**, combined with extensive **sheep** farming. In Extremadura, clearings in the Mediterranean forests form *dehesas* or wide, open pastureland.
- In the areas between the mountain ranges and the plains, there is **rainfed cultivation** of vines, olive trees and almond trees. Sheep are also farmed extensively for their meat. The holdings are large and open.

The landscape of the Canary Islands

On the Canary Islands, agriculture is influenced by mountain relief, volcanic soil and the warm climate. Settlement is **nucleated**.

- On the **coasts** and in areas exposed to the cool, humid Atlantic winds, **export products** such as potatoes, bananas and tomatoes are cultivated. There are also plantations of tropical fruit such as mango, avocado and pineapple, and cultivation under plastic of flowers, cucumbers and peppers.
- In higher areas **inland**, traditional agriculture is practised, along with livestock farming of goats.

The humid mountain landscape

This landscape occupies the highest altitudes on the Iberian Peninsula.

- There is extensive **livestock farming** (cattle and sheep) for meat, and there are many pastures for grazing. **Forestry** is also an important economic activity.
- The relief and climate do not favour cultivation. This is mainly practised in low valley areas, where mixed crops are grown for family consumption and fodder is cultivated for cattle.

Landscape of the Canary Islands (La Palma).

Humid mountain landscape (Asturias).

1 Copy and complete the table.

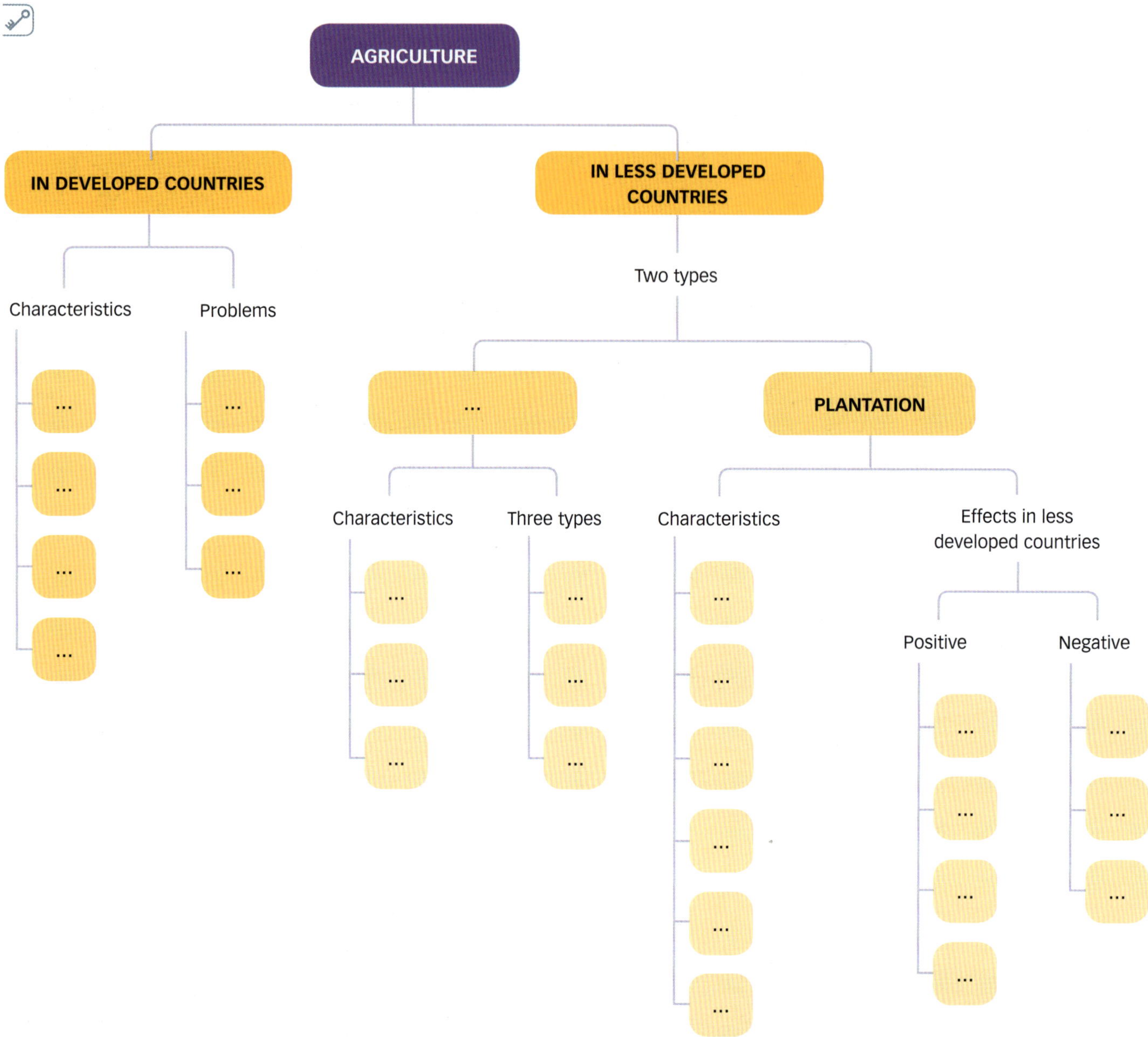

2 Explain the following terms.

- Primary sector
- Aquaculture
- Transhumance
- Fishing ground

3 Explain the difference between these terms. Listen and check your answers.

- Open-field and closed holdings
- Extensive and intensive livestock farming
- Inshore fishing and deep-sea fishing
- Commercial and subsistence agriculture

4 Copy and complete the following table in your notebook.

THE PRIMARY SECTOR IN SPAIN	
General characteristics	......
Agriculture	......
Livestock	......
Fishing	......

5 What does the photo show: agriculture in a developed country or a developing country? Explain your answer.

6 What kind of agriculture is shown in the illustrations? Where does the water come from?

7 In what ways have technological advances led to productive livestock breeding?

8 Find out why trawling is considered to be a destructive fishing technique.

SUSTAINABLE RURAL DEVELOPMENT

Sustainable rural development means satisfying the needs of people today without harming the ability of future generations to do the same.

Certain agricultural techniques, such as the use of fertilisers, pesticides or genetically modified crops, increase production but also harm the environment. They pollute the air, water and soil.

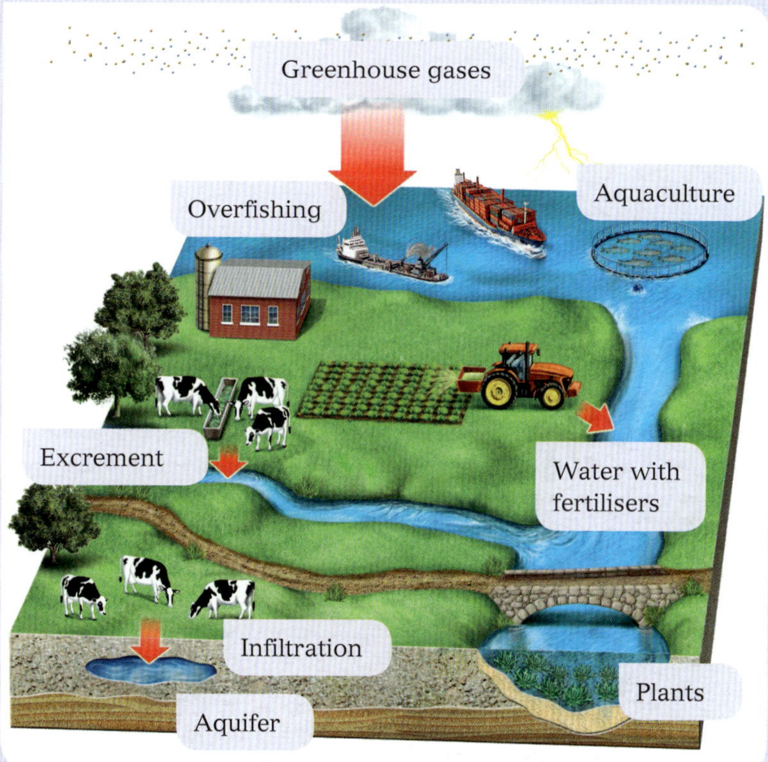

IMPACT OF THE PRIMARY SECTOR ON THE ENVIRONMENT

9 Which agricultural technologies help to increase food production?

10 How do these technologies affect the environment?

11 How can we combine an increase in food production to fight world hunger, with sustainable rural development? Discuss your ideas with your classmates.

TECHNIQUES. **Interpret a topographic map**

Relief is shown on a contour map through the use of **contour lines**. These lines join together points at the same altitude. The maps also give information about hydrography, vegetation, population, transportation routes and agriculture.

Contour maps are useful in identifying how land is used in a particular area.

In order to interpret a contour map, you should consider the following:

- **Title**. This helps to identify the area shown on the map. It often indicates the biggest settlement.
- **Scale**. This establishes the relation between the real dimension of the land and its representation on the map.
- **Map key**. This gives information about the meaning of the signs and colours used on the map. It will help you to identify rivers, buildings, roads, types of vegetation and agriculture.
- **Contour lines**. The separation between the lines will tell you if the land is flat or mountainous.

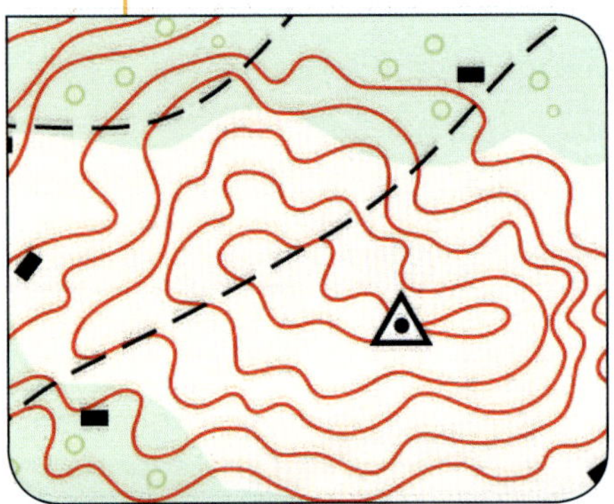

– If the lines are close together, the gradient of the land is steep.

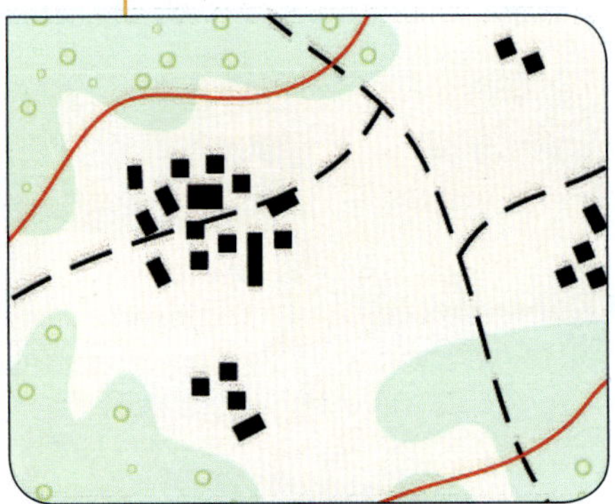

– If the lines are separated, the gradient of the land is not very steep.

- **Altitude**. The main contour lines show the height of the land above sea level.
- **Colours**. Areas at the same height are in the same colour. Green is usually used for low areas, orange for intermediate levels, brown for mountain areas, and purple or white for areas above 3,000 metres.

1 Analyse human settlement.
   a What types of settlement can you see?
   b Which is the largest? How do you know?
   c Are any isolated buildings shown? What are they?

2 Look at the scale.
   a What is the scale of the map? What does it indicate?
   b Why is it important to know the scale of a map?

3 Study the map key.
   a What symbols have been used?
   b Which colours have been used? What do they indicate?
   c Should any other information be given in the map key?

4 Analyse the contour lines.
   a Is this a flat or a mountainous area?
   b Is there an area where the contour lines are very close together? What does this mean?
   c Are these areas at a high altitude? How do you know?

5 Identify land use.
   a Which types of vegetation are found in this area?
   b Are they mainly found at low or high altitudes?
   c Do natural or artificial forms of land use occupy most space?

6 Relate agriculture to the natural environment.
   a What type of agriculture is practised in this zone?
   b Are there any cultivated fields? Where?
   c What factor or factors explain their location? Explain the link between agriculture and the natural environment.

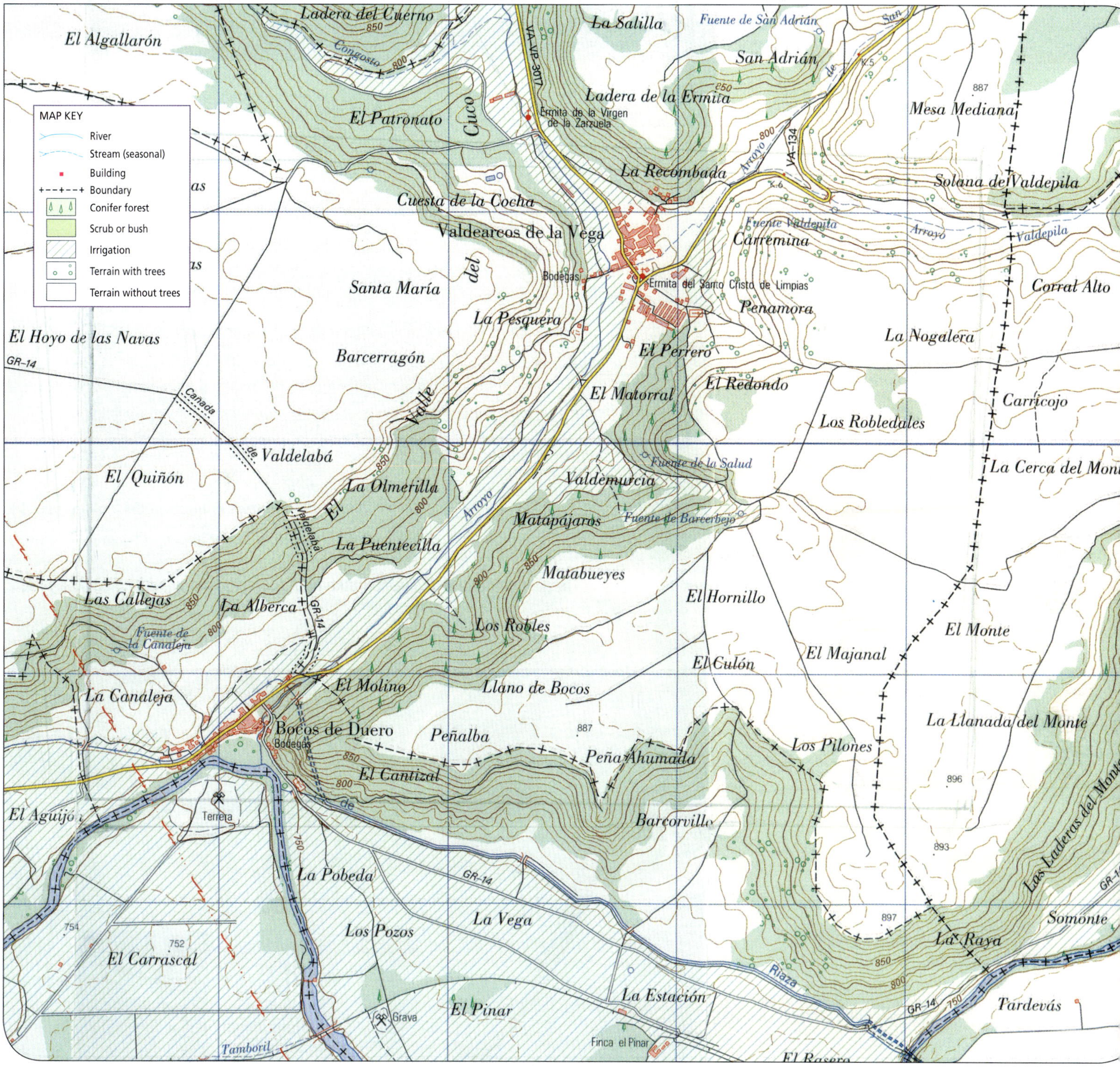

Topographic map. Scale 1:25,000.

7 Plan a trip using the topographic map.

a In groups of four, you are going on a trip from Bocos del Duero to the hermitage of the Virgin of Zarzuela in the north.

b Plan the route from your starting point. Check if there are marked roads that you can follow. If not, decide which direction you are going to take.

c Use the scale to calculate the distance (in a straight line).

d Look at the contour lines and decide whether the terrain is flat or mountainous. Will you need to do any climbing to reach your destination?

e Look at the map key. Work out what the landscape is like: its vegetation, cultivated areas and rivers.

5 Mining, energy and industry

FIND OUT ABOUT:

- Types of mining
- Renewable and non-renewable energy sources
- The characteristics of modern industry
- The industrial powers
- Mining, energy and industry in Spain

KNOW HOW TO:

- Assess the location of an industry
- Analyse sustainable energy projects

The strategic importance of rare raw materials

Industrial technology requires certain raw materials that are only found in a few places. As a result of their scarcity, they are considered 'strategic resources'.

Many of these resources are found in China. So China controls their supply, and could even prevent other countries from manufacturing high technology products.

Use of rare resources

These resources are used to make many objects that we use in our daily lives.

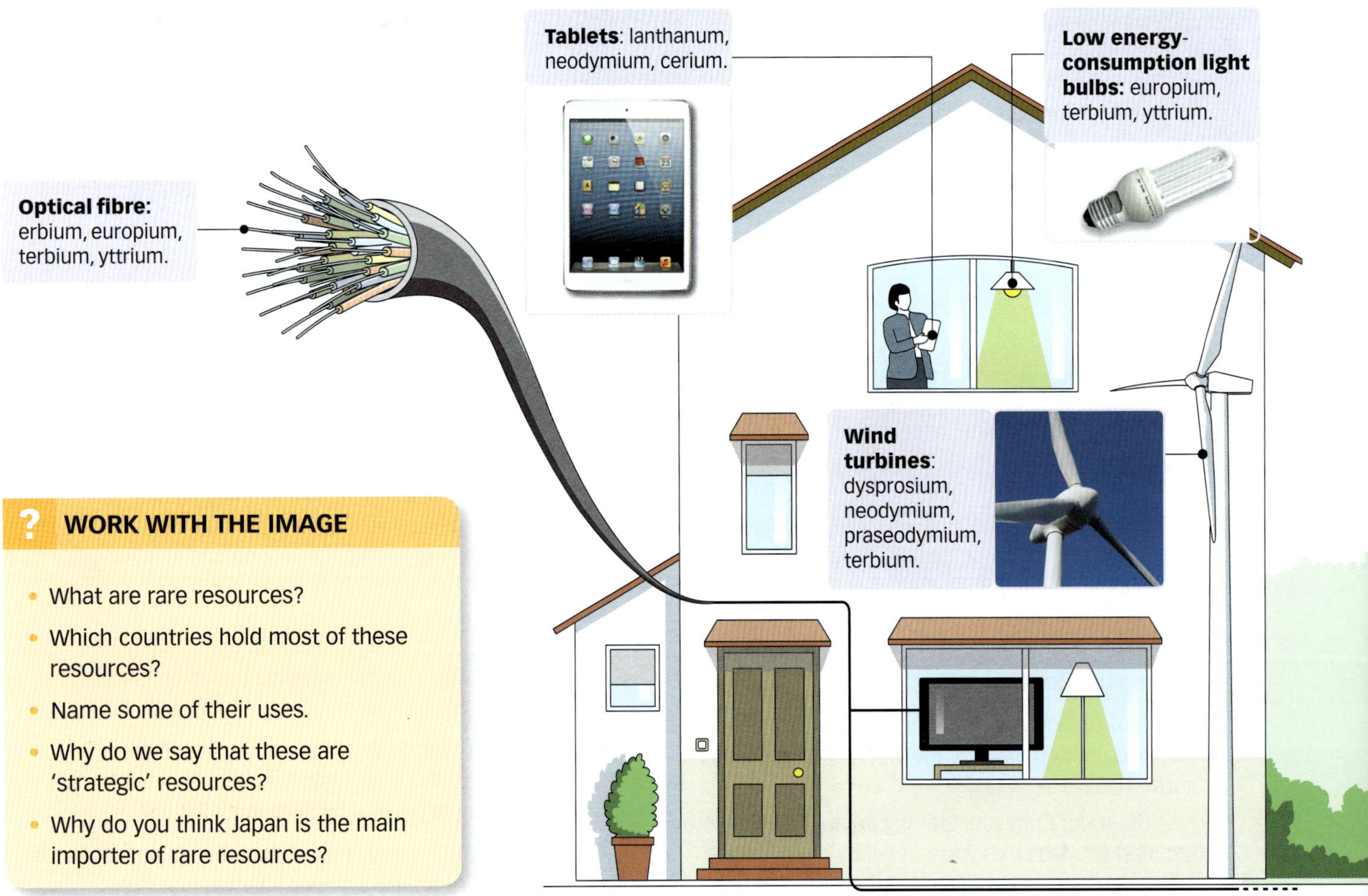

? WORK WITH THE IMAGE

- What are rare resources?
- Which countries hold most of these resources?
- Name some of their uses.
- Why do we say that these are 'strategic' resources?
- Why do you think Japan is the main importer of rare resources?

World producers of rare resources

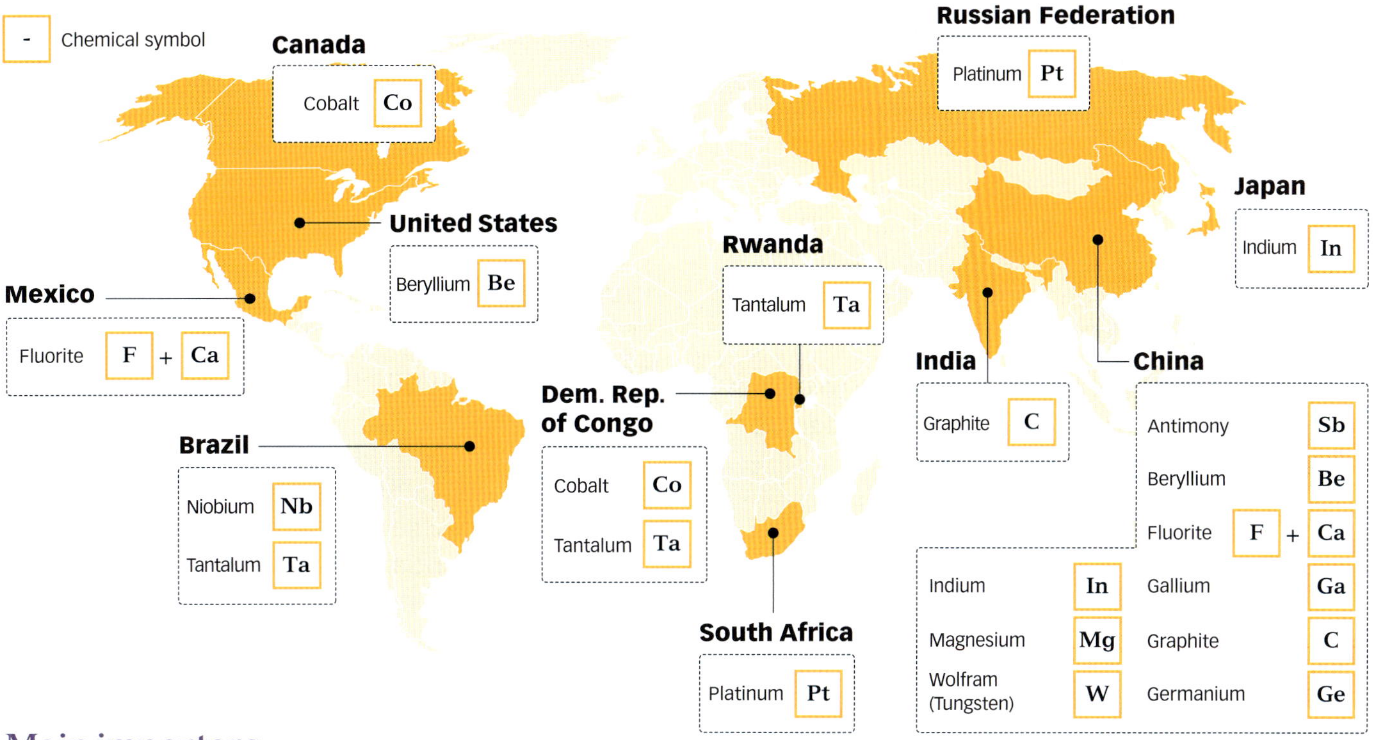

Main importers

In millions of dollars

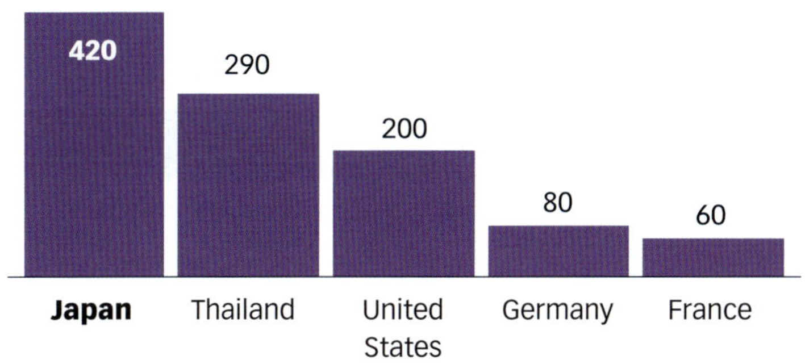

The mining of valuable minerals can cause armed conflict in developing countries. **Coltan** (which contains tantalum) is essential to the electronics industry. Rival groups are fighting for control of this resource in the Democratic Republic of Congo and Sierra Leone.

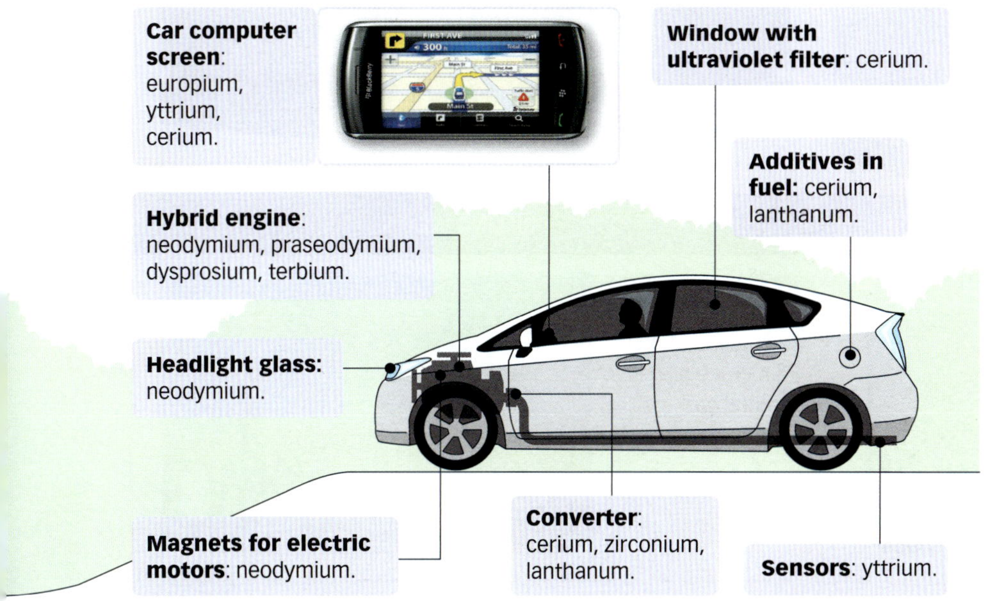

HOW DO WE KNOW?

Geologists investigate the distribution of rocks and minerals in the Earth's crust and they record the results on geological maps. These maps help different countries to decide where to look.

- Find out which institutions in Spain make geological maps.

1 Mining

Natural resources and their exploitation

Natural resources, such as minerals, fresh water and forests, are found naturally in the environment, and are essential for a country's economic development. Their exploitation depends on three factors:

- **Technology**. This makes it possible to obtain natural resources in various ways and from distinct locations.
- **Cost**. The benefits from a resource must be greater than the cost of its exploitation. New extraction techniques can lower costs.
- **Markets**. An unprofitable resource may become profitable if demand increases. However, the opposite is also true.

Consumption is very high in developed countries. Less developed countries are among the biggest producers of natural resources.

Mining activity

Minerals are a natural resource. A mineral is an inorganic natural substance, with a specific chemical composition, which is found in the Earth's crust. **Mineral deposits** are high concentrations of minerals in a particular area.

Mining refers to the different activities involved in extracting minerals from their deposits. It is a primary sector activity because products are obtained from nature. However, it is linked to the secondary sector because it provides essential raw materials for industry and construction.

There are four basic types of mining:

- **Surface** mining.
- **Underground** mining.
- **Underwater** mining.
- **Drilling** wells.

SURFACE MINING

This is the most common type of mining, and is used when the mineral is near the surface. **Surface mines** (A) are used to mine metals. **Quarries** (B) are used to mine industrial rocks such as granite.

UNDERGROUND MINING

This is used when the mineral is at great depth. It is expensive because wells and galleries must be excavated.

UNDERWATER MINING

Special boats called dredgers are used to mine the sea bed for minerals. The material is taken on board the boats.

DRILLING WELLS

Drilling wells are used mainly in the extraction of fuel like crude oil and natural gas.

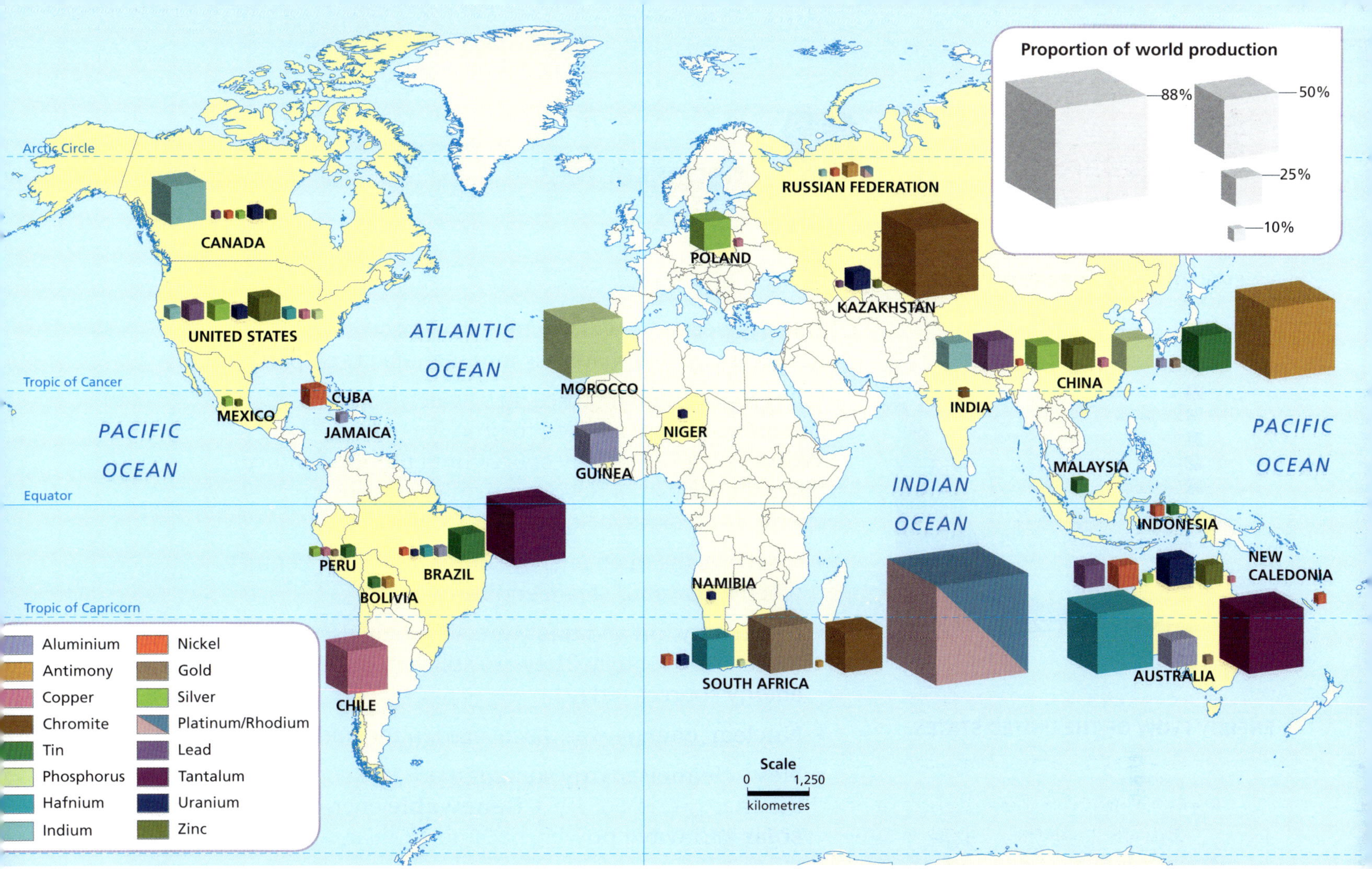

MAIN MINERAL DEPOSITS IN THE WORLD

Mining today

Mining was carried out in ancient times. However, it has developed greatly since the Industrial Revolution began in the late 18th century.

Today, countries need very large quantities of raw materials and energy to maintain their manufacturing industries. **Investment** in mineral resources has stayed high despite the world economic crisis which started in 2007.

Mining may cause serious problems:

- **Environment**. It destroys the landscape and pollutes water and soil. The use of less invasive techniques is not always possible because of the high costs involved.
- **Excessive mining**. It uses up deposits, and exhausts the land.
- **Conflict**. In less developed countries, mining can lead to combats to control the deposits, and expulsion of the population from mining areas.

WORK WITH THE MAP

1. Which countries have the greatest mineral resources?
2. Which countries have more than 20% of the reserves of a certain mineral?

ACTIVITIES

3. Form groups. Each group chooses two minerals, and finds out about them and how they are used. Prepare a short presentation for the class.

THINK ABOUT IT

Recycling

Discuss the following questions in groups:

What is recycling? Does it help the environment? What things do we recycle? Could we recycle more? What is e-waste? How can we reduce it? What problems are caused by mining? Which problems could be reduced by greater recycling?

2 Energy sources

EVOLUTION OF ENERGY CONSUMPTION

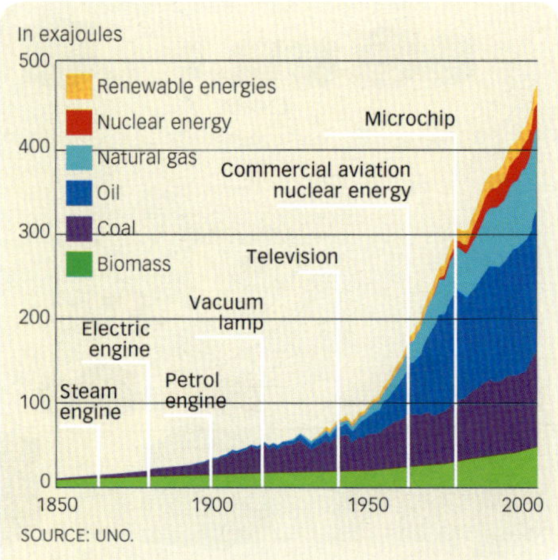

ENERGY FLOW OF THE UNITED STATES

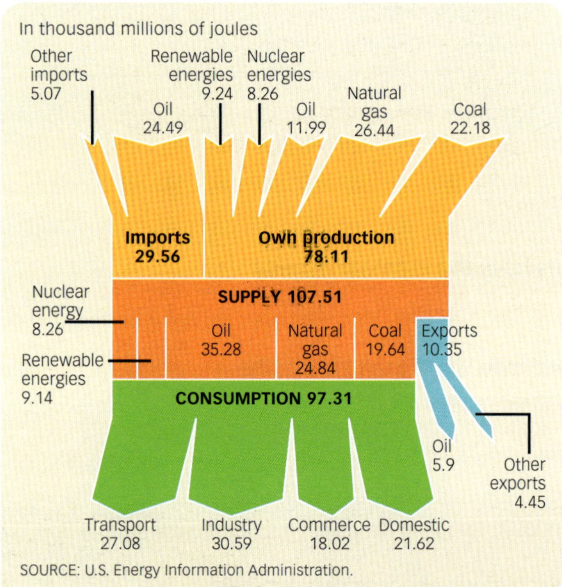

WORK WITH THE GRAPHS

1. After which date does energy consumption rise sharply?
2. Which energies increase the most? Why?
3. Where does most of the energy consumed in the United States come from?

History of energy

Energy performs work in industrial processes by transforming raw materials into manufactured products.

In Prehistory, human force, fire, water, wind and animals were all used to produce energy. **Wood** was a key energy source until the 18th century. Since the Industrial Revolution, many distinct energy sources have been used:

- **Coal** was the main energy source in the 19th century, when it powered steam engines.
- **Electricity**, **oil** and **gas** were all widely used by the late 19th century. They are still very important energy sources today.
- **Nuclear energy** was discovered in the mid-20th century.
- **New, cleaner** energy sources were looked for following the energy crisis of 1973. **Renewable** energies, like **wind**, **solar** and **wave power**, became popular.

In the late 20th and early 21st centuries, the world's energy demands have continued to grow due to population increase, the requirements of industry and the demands of industrial societies.

Classification of energy sources

Energy sources are **renewable** or **non-renewable**:

- **Renewable energies** are those which cannot be used up or which can recover naturally over time. Biomass (including wood), hydraulic, solar, wind, tidal and geothermal power are all renewable. They are also known as **alternative energies**.
- **Non-renewable** energies exist in limited amounts and cannot easily be replaced. Uranium, coal, oil and natural gas are all non-renewable. They are also called **conventional fuels**.

Energy sources can also be classified as **primary** or **secondary** according to where they are found:

- **Primary** energies are obtained from nature. They include fossil fuels like coal and oil, uranium, and renewable energies such as solar and wind power.
- **Secondary** energies are those produced from a primary source (like electricity) and fuels obtained from oil (like petrol).

WORK WITH THE GRAPHS

4 Name, from least to most, the types of energy produced and consumed in the world.

5 Why do you think oil takes first place in both graphs?

6 Why does electricity not appear in the first graph?

WORLD PRODUCTION OF PRIMARY ENERGY

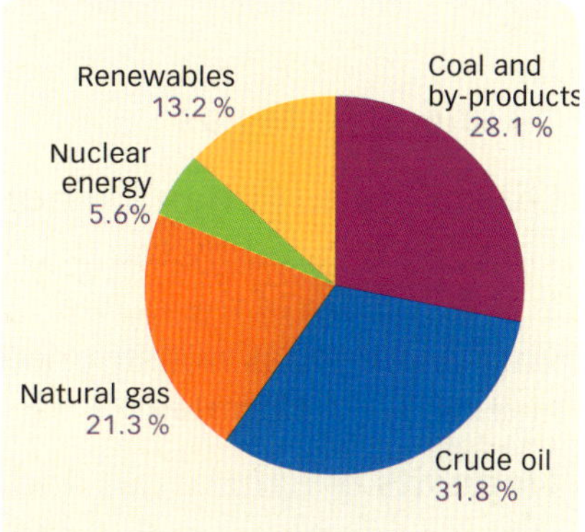

SOURCE: OECD, 2010.

WORLD ENERGY CONSUMPTION

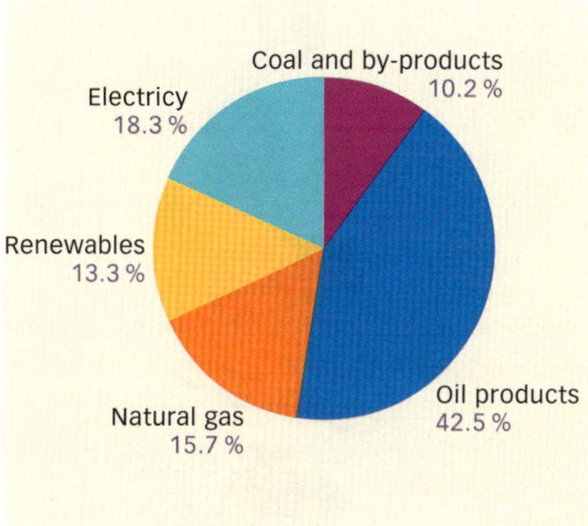

World production and consumption

There is a marked imbalance in the way that energy is produced and consumed in different parts of the world:

- The **OPEC countries** are among the main world suppliers of hydrocarbon fuels (oil and gas).
- Industrial powers, like **Japan** and the **European Union**, are essentially consumers, with high energy needs and a low level of production.
- The **United States** consumes and produces an enormous amount of energy. Thanks to new extraction techniques, it is increasing its oil and gas production.
- In recent years, the traditional consumers of energy have been joined by emerging countries such as **China** and **India**.

In the 21st century, energy consumption is expected to keep on rising. 80% of this energy will be from fossil fuels, which are non-renewable and polluting. This presents a challenge to sustainable development.

Careful planning will be required to:

- **Keep up with** increasing world demand.
- **Improve energy efficiency** by committing to renewable energies.
- **Invest in technology** to make renewable energy more efficient.
- **Develop** new forms of energy storage and transport.

DID YOU KNOW?

OPEC

The Organisation of Petroleum Exporting countries (OPEC) was founded in 1960. Today it has 12 members: Angola, Saudi Arabia, Algeria, Ecuador, United Arab Emirates, Libya, Nigeria, Iran, Iraq, Kuwait, Qatar and Venezuela.

Its purpose is to stabilise crude oil prices and obtain fair profits for its members.

ACTIVITIES

7 Why are some types of energy classified as 'primary' and others as 'secondary'? Give examples.

Oil rig.

BY-PRODUCTS OF OIL

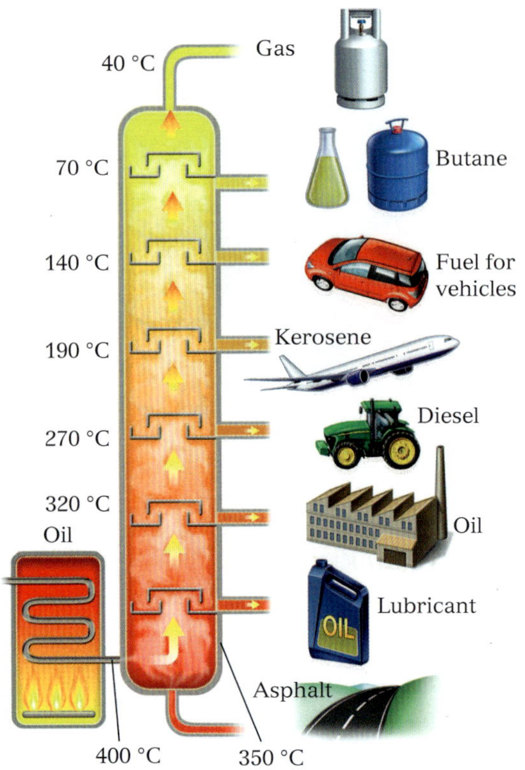

EXTENT OF OIL, GAS AND COAL RESERVES

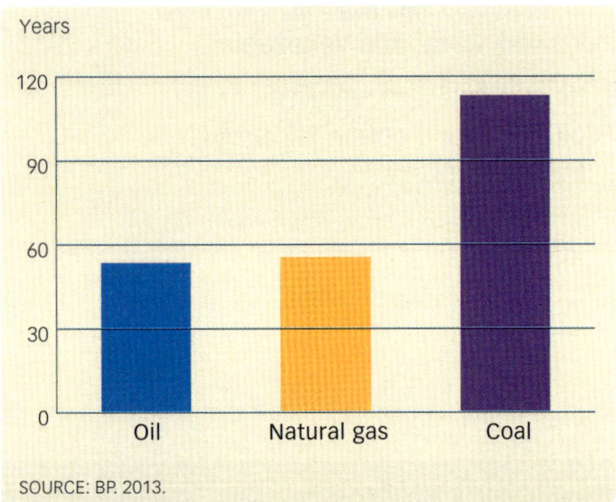

WORK WITH THE GRAPH

1. Name three by-products of crude oil.

2. When will the reserves of oil, gas and coal run out?

ACTIVITIES

3. Look for information on Internet about fracking.

3 Non-renewable energy

Oil and natural gas

Oil and **natural gas** are **non-renewable fossil fuels**. They are formed by the decomposition of living organisms at the bottom of the ocean over millions of years. This process has not stopped, but it is extremely slow and we are using these resources up very rapidly.

Deposits of natural gas and oil are often found in the same place, making their extraction easier. Crude oil provides numerous by-products, such as petrol, and is used to produce electricity and as raw material in industry. Natural gas has applications in all economic sectors, including the home.

The biggest **oil reserves** are in the Persian Gulf, Venezuela and Canada. **Gas reserves** are more diversified, but there are important reserves in the Middle East and the Russian Federation.

Demand for oil has tripled in the last fifty years. The biggest consumers are the industrialised countries and the most populated countries. The United States is the biggest consumer, followed by China and Japan. The consumption of **natural gas** has increased since the oil crisis of 1973, and it now represents a quarter of world energy consumption.

These resources cause the following problems:

- **Environmental damage**. This is mainly linked to oil, which causes atmospheric pollution and oil spills.
- **Dependence** on the energy-producing countries. This makes the industrial countries vulnerable to sudden price increases.

Coal

Coal is a **fossil fuel** formed by the decomposing remains of vegetation in shallow water. It was the main fuel in the 19th century and the first half of the 20th century. Since then, oil and gas have become more important, but coal is still used today.

- Coal is used to produce electricity. It is essential in iron and steel production, and the carbochemical industry. It can also be converted into liquid or gaseous fuel.
- Coal is the most abundant fossil fuel, and it is found all over the world. The main deposits are in **China**, the **Russian Federation**, and **North America**. It is estimated that reserves will last for another 100 years.
- It causes major **environmental problems**.

Uranium

Nuclear energy originates from a process called **fission**, which involves the disintegration of uranium atoms. Uranium is a chemical element, present in limited quantities in nature.

Uranium needs a process of **enrichment** before it can be used, requiring technology that is only available to a few countries. The energy liberated by uranium releases heat, which is then transformed into **electricity** at a **nuclear power plant**. Today, there are more than 430 functioning reactors in 30 countries, providing 15% of the world's electricity.

The **main producers** are developed countries such as the United States, France and Japan, which have been joined by China and India.

- The use of nuclear energy has caused **protest movements** which denounce the problem of storing radioactive waste and the danger of a nuclear accident.
- Advocates of nuclear power argue that it is a clean energy source, which does not release harmful gases into the atmosphere.

ACTIVITIES

4 What does 'fossil fuel' mean? What are fossil fuels?

5 Give one use of each form of non-renewable energy.

6 What are the problems related to the use of non-renewable energies?

7 Why do you think the Middle East is considered a 'strategic' area?

DID YOU KNOW?

How is electricity produced?

Electricity is the most common type of secondary energy. Electricity is generated from primary energies in hydroelectric plants (water), conventional power plants (coal), nuclear plants (uranium), wind farms (wind), and solar energy farms (the Sun). Electricity has many applications, and it has numerous advantages.

It is very versatile, because it can be turned into light, heat or movement. It is clean and silent as well and it is easy to transport.

However, it is not possible to accumulate electricity. Its production has to adapt constantly to avoid problems of supply or overload in the network.

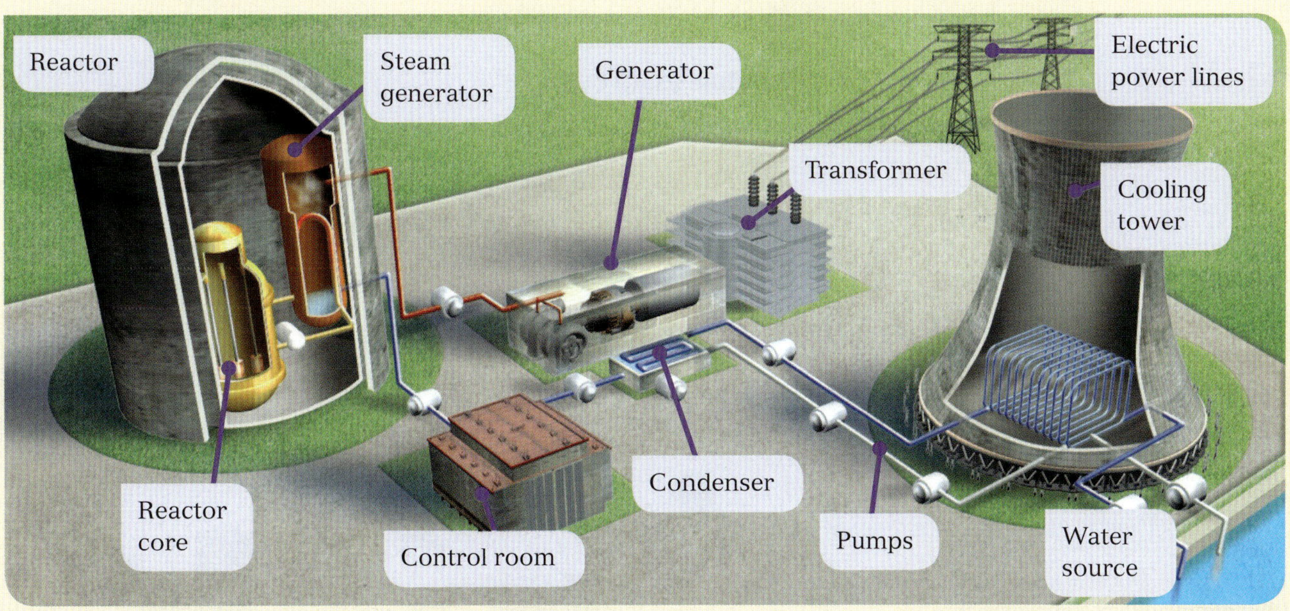

ELECTRICITY PRODUCTION IN A NUCLEAR POWER PLANT

4 Renewable energy

RENEWABLE ENERGY

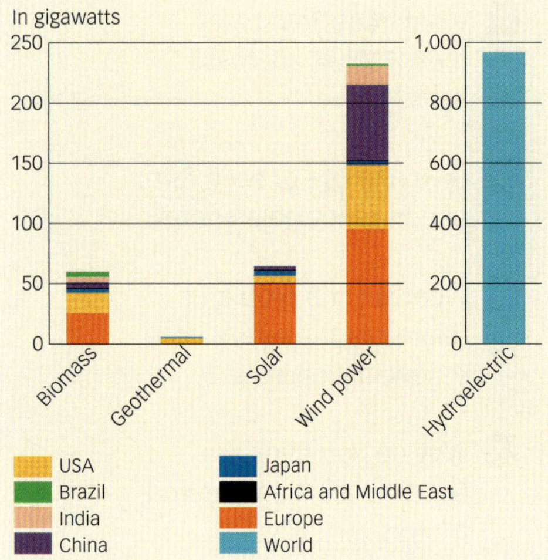

SOURCE: International Renewable Energy Agency, 2011.

WORK WITH THE GRAPH

1. What renewable energy sources are most used today?

Non-renewable energy sources cause problems, such as the depletion of resources and pollution. For this reason, technologies have been developed to obtain energy from more environment-friendly sources, such as the Sun, the heat from inside the Earth, the wind and the tides. These **renewable energies** will never be depleted.

In the 1970s, these energy sources were considered to be alternatives to conventional fuels and uranium. Consequently, they were called **alternative energy** sources. They are **hydroelectric**, **wind**, **solar**, **biomass**, **geothermal** and **tidal power**.

Today, the use of renewable energy sources is growing rapidly thanks to lower technological costs, more expensive fossil fuels, the payment of a tax for CO_2 emissions and the subsidies that governments provide.

There are two major **advantages**:

- These energies are available everywhere. Fossil fuels are only found in certain regions.
- They have little impact on the environment.

However, they also have some **drawbacks**:

- Supply is irregular.
- The energy produced is difficult to store.

By 2035, one third of electricity is expected to come from renewable sources and the use of renewable energy is likely to go on increasing.

THE BIODIESEL CYCLE

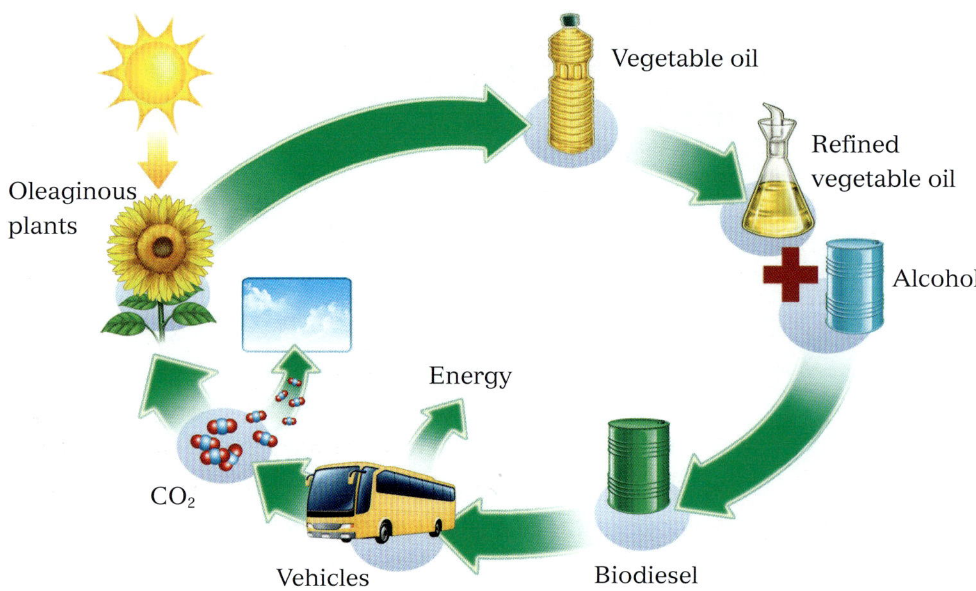

ACTIVITIES

2. What is renewable energy?

3. What are its advantages and drawbacks?

4. Copy and complete the table.

Energy	Source	Type of energy produced
Solar	Sun	Electricity and heat
...	...	...

5. Why are renewable energies the energies of the future?

MAIN RENEWABLE ENERGY SOURCES

Hydro-electric power

Hydroelectric power uses the force of falling water to generate electricity. At hydroelectric power stations, the water is held behind a dam, forming a reservoir. The force of the water released through the dam turns a giant turbine, which moves generators and makes electricity. This is the most commonly used renewable energy.

China, Brazil and Canada are the main producers. The largest dam in the world is the Three Gorges Dam on the Yangtze River in China.

Wind power

This is the second most commonly used source of renewable energy. **Wind** has become very important for producing energy. Wind turbines are becoming more and more efficient. They can make use of different wind speeds and their production capacity is increasing. China, the USA, Germany and Spain are the largest producers of electricity through wind power.

There are two main drawbacks to wind power: the impact of windfarms on the landscape and the birds which are killed when they fly into the turbines.

Solar energy

Today we can use the energy from the **Sun** by trapping its heat and light in solar panels and concentrated solar power systems.

- **Solar thermal energy** uses the energy from the Sun to heat water to below 80° C and use it for heating and other domestic purposes. It also generates steam which turns turbines and moves generators that produce electricity.
- **Photovoltaic solar energy** uses solar cells to transform sunlight directly into electricity.

Germany, Italy and Spain are the countries which produce the most electricity from solar energy.

Biomass energy

Biomass energy is produced by transforming **organic materials**, such as forestry, agricultural and urban waste, into energy.

The energy from biomass is used directly for heating or to produce electricity. In addition, biomass can be converted into synthetic fuels (ethanol, biodiesel, etc.) which are used in transport.

France is the largest consumer of biomass in the European Union and Brazil is the leading ethanol producer.

Geothermal energy

Geothermal energy uses the heat from the Earth. Its applications depend on temperatures:

- Below 90 °C: heating and hot water in homes.
- Between 90 °C and 150 °C: industrial uses and electricity production.
- Above 150 °C: water vapour is transformed into electricity.

Tidal energy

This type of power uses the energy that results from the rise and fall of the **tides**, which are caused by the gravity of the Moon pulling the oceans towards it.

It needs strong tides and expensive infrastructures, such as huge underwater turbines.

Energy can also be obtained from **waves**, **sea currents** or the **difference in temperature** between deep and surface water.

The largest tidal power station in the world is Sihwa Lake in South Korea.

5 The origins and development of industry

THE INDUSTRIAL PROCESS

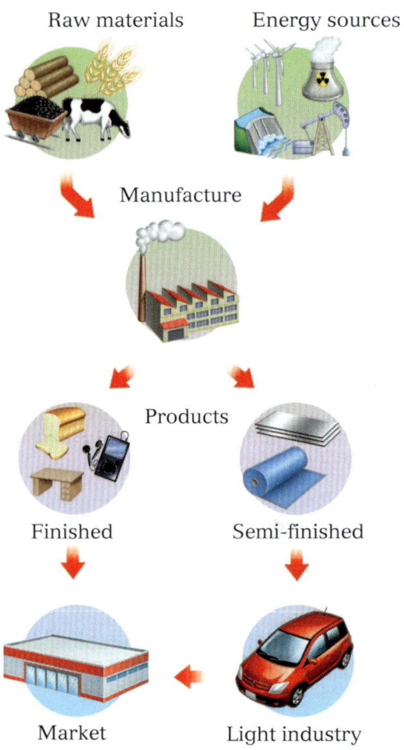

> WORK WITH THE ILLUSTRATION
>
> **1** Explain the industrial process. What is the difference between a finished and a semi-finished product?

A factory during the Industrial Revolution.

What is industry?

Industry is concerned with the transformation of raw materials into manufactured goods. It is the most important economic activity of the secondary sector.

Humans have made tools and instruments since Prehistory, but modern industry requires certain conditions:

- **Machines** and **energy** are used in the manufacturing process.
- Work is carried out in **factories** which employ large numbers of people.
- **Large quantities** of goods are produced.

Industry is linked to development, and industrialisation promotes economic growth. The richest and most developed countries have a well-established industrial network.

The origins of industry

For many centuries, people lived in rural areas and worked on the land. Families made the objects they needed at home (**domestic production**), or had them made in workshops (**craft production**), where a single worker made the entire product.

Industrial production began in **England** in the late 18th century. It was based on:

- **Technical innovation**. There were new inventions like the steam engine, spinning and weaving machines, and blast furnaces. The machines carried out part of the manufacturing process which until then had been done manually using coal-based energy.
- **Division of labour**. Production was divided into multiple tasks carried out by workers who were responsible for a particular phase of the process. This saved time and increased productivity.
- **Large-scale production**. The manufacture of many products pushed selling prices down and promoted international trade.

Industrialisation changed society and the economy. Large numbers of people moved from rural areas to work in factories, and cities grew quickly. The changes were so great that they became known as the **Industrial Revolution**.

In this first phase, industrialisation spread from England to a few other European countries, like **France**, **Germany** and **Belgium**.

The second stage of industrialisation

Mass production emerged in the USA at the end of the 19th century. It became known as the **Fordist model** because it was developed on the moving assembly lines of the Ford automobile factory.

Under this new efficient system, productivity increased enormously. Production costs were reduced, and products were made available to many more people. This was the beginning of an era of **mass consumption**.

The **USA**, **Western Europe** and **Japan** led this new revolution. A huge economic gap developed between the industrialised countries and other parts of the world – most of Asia, Africa and Latin America.

Types of industry

Industries are classified according to two criteria.

- **Degree of innovation and technological development**.
  - **Mature industries**. Today, textiles, iron and steel, and ship building are considered mature industries, although they were advanced industries in the early stages of industrialisation.
  - **Advanced technology**. Biotechnology, microelectronics, telecommunications and robotics are today's **cutting-edge industries**. They are areas which require large research and development expenditure.
- **The destination of products**. We differentiate between **basic** industries, **capital goods** industries (making products that are used to make other things) and **consumer goods** industries (for mass consumption).

A Ford T assembly line in 1928.

ACTIVITIES

2 What is industry? Why is it an indicator of economic development?

3 Look on the Internet for information about Henry Ford and the Ford Motor Company. Why was the moving assembly line such an important development?

TYPE OF INDUSTRY	WHAT DO THEY MAKE?	MAIN INDUSTRIES
Basic	They take raw materials from nature and convert them into semi-finished products (steel, aluminium, copper, fibres), which form the raw materials of other industries.	• Iron and steel • Metalworking • Petrochemical • Cement
Capital goods	They use the semi-finished products of the basic industries to make products for other companies.	• Machinery and engines • Equipment and material for transport • Building materials
Consumer goods	They make consumer products which can be bought directly from shops.	• Food • Textiles • Cars • Electronic goods • Light chemicals • Furniture

THINK ABOUT IT

The location of an industry

The following factors influence the location of an industry:

- Access to innovation and information technology
- Proximity to similar industries
- Transport and communications
- Capital
- Proximity to markets
- Availability of qualified work force
- Availability of raw materials
- Environmental factors

• Think about your own Autonomous Community. Is there an area that would be a good place to set up a company specialising in information technology?

A graphene screen.

WORK WITH THE PHOTO

1 What is graphene? Look for information about it on the Internet.

ACTIVITIES

2 Explain the main characteristics of the post-Fordist model.

6 The characteristics of modern industry

The new industrial model

Since the 1970s oil crisis, a new model of industrial production (known as the **post-Fordist model**) has emerged. It is based on:

- **New technologies** and industries such as information technology, telecommunications and microelectronics.
- A focus on each **consumer's preferences**. Many different models of the same product are manufactured to cover the varying requirements of consumers.
- A **flexible**, **automated** and highly specialised system of production.
- The participation of **emerging countries** in industrial development.

The key role of research

Industrial companies dedicate a high percentage of their capital to **research and development (R & D)**. Two innovations have revolutionised 21st century industry:

- **New materials**. This is vital in both traditional industries, like textiles and food, and in **cutting-edge industries**, like electronics and biotechnology.
- **Nanotechnology**. The prefix 'nano' means one billionth of something. Using this technology, tiny electronic devices and much faster computers have been built. New devices can travel through a person's blood stream, and diagnose illnesses.

The decentralisation of production

The **globalisation of the world economy** has led to strong competition between companies and countries. They have developed new strategies to reduce costs and stay competitive. The decentralisation of production takes place in different ways:

- A company locates the **different phases** of production in **different countries**. For example, research and design may be carried out in one country; financial and commercial control in another; and production in a third country.
- A company **subcontracts** a particular activity, such as marketing, to other specialised companies.
- A company makes **cooperation agreements** with other companies.

1 A company designs a new football in Germany and the USA.

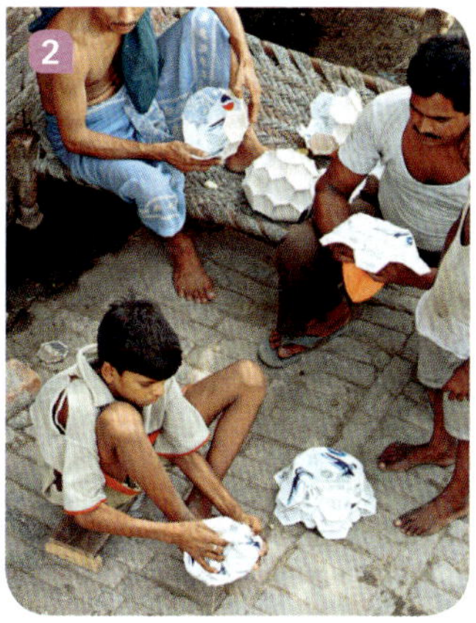

2 Production takes place in Asia.

3 Marketing and distribution is planned in Germany and the USA.

4 The football is sold throughout the world.

Automation and industrial tertiarisation

Modern industry uses **automated production systems**. This means that machines and robots carry out mechanical tasks which used to be performed manually in the past. Productivity has increased and production costs have fallen. However, there have been important consequences for the work force:

- **Employment**. Fewer workers are required.
- **Specialised expertise**. There is greater demand for technicians, managers and professionals who provide services connected to the industrial process. The number of unqualified workers falls.

Industries **insource services**, which are not directly related to manufacturing but are important in running the company. Traditionally, activities like marketing, logistics, design and research belong to the service (or tertiary) sector. However, today they are such an integral part of the industrial process that we refer to the **tertiarisation** of industry.

The growth of multinational industries

A **multinational** is a large company with **subsidiaries** in many countries. It is usually set up in one country before it moves into the international market.

The growth of multinationals is linked to the expansion of international trade and to advances in transportation, information technology and communications.

These companies are more powerful economically than some countries. They use the advantages offered by a country, such as tax breaks, low labour costs, and permissive environmental laws. They use advanced technology, invest in research, and can produce large quantities of a product at very low prices.

WORK WITH THE PHOTOS

3 Explain why part of this company's work takes place in developed countries and part of it in less developed countries.

ACTIVITIES

4 Why do you do think some jobs are being lost in developed countries?

DID YOU KNOW?

Offshoring, or industrial relocation

Offshoring is the relocation of some of a company's business activities in a different country. Thanks to improved information technology and communications this became widespread in the 1990s. At first, companies in developed countries relocated in emerging countries like China and India. But today these emerging countries are now relocating some of their production to new emerging countries like Indonesia.

7 The industrial powers

VALUE OF INDUSTRIAL PRODUCTION

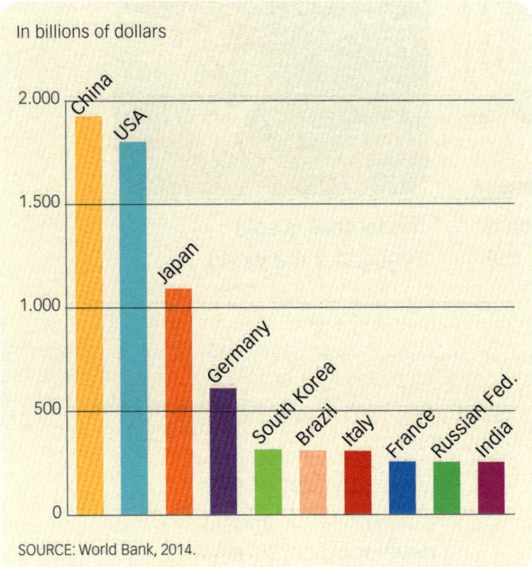

GDP BY SECTOR

	Primary	Secondary	Tertiary
DEVELOPED COUNTRIES			
USA	1.1%	19.2%	79.7%
Japan	1.1%	26.3%	72.6%
Germany	0.8%	28.0%	71.2%
BRICS			
Brazil	5.2%	26.3%	68.5%
Russian Federation	3.9%	36.0%	60.1%
India	17.4%	26.1%	56.5%
China	10.1%	45.3%	44.6%
South Africa	2.6%	29.3%	68.1%
OTHER ECONOMIES			
Mexico	4.1%	34.2%	61.7%
Pakistan	20.1%	25.5%	54.4%
Singapore	0%	26.8%	73.2%

Traditional and emerging powers

Industrial production is not evenly distributed throughout the world, and 85% of total production is concentrated in just a few countries. Traditionally, the industrial powers were the United States, Europe and Japan. However, several emerging countries are now catching up.

The **BRICS** take their name from the first letter of the countries Brazil, the Russian Federation, India, China and South Africa. Together, these five countries account for around 27% of the world's industrial production.

China, the world's leading industrial power

China has become the world's leading industrial power. Its growth has been based on:

- **Changes in economic policy**. Until the 1960s, China followed the Soviet model of giving priority to basic industries and capital goods. Then the economy was liberalised. This led to the privatisation of companies and the arrival of foreign businesses.
- **Abundant resources**. China has vast natural resources and an abundant source of cheap labour.

The following are China's **main industries**:

- **Electronics**. A third of the world's electronics production is based in China. Electronics is a **strategic industry** because electronic equipment and components are essential in many production processes.
- **Basic industries**. These include chemicals, steel, heavy machinery and cement.
- **Consumer goods**. China is a major producer of textiles, toys, cars and food.

Industry is mainly concentrated in **coastal areas** around Shanghai, Beijing, Hong Kong and Guangzhou.

WORK WITH THE GRAPH AND THE TABLE

1. Which countries are the world's main industrial powers?
2. What is the composition of the GDP by sector in the United States? Why is it still a great industrial power?
3. Compare India, Singapore and Pakistan. Analyse the role of industry. Are these countries at the same level of development?

The traditional industrial powers

The developed countries have a strong tradition of industry, a skilled labour force and highly developed technology.

- **The United States** has fallen to second place (after China) in industrial production. The main industrial areas are located near the Great Lakes, the Gulf of Mexico and the Pacific Coast. Production is diversified. It includes traditional industries like cars, chemicals, iron and steel and food, and high technology industries like the aerospace industry, electronics and telecommunications.
- **Japan** is the world's third largest industrial power. A strong work ethic, aggressive business policies and advanced technological development have all contributed. Japan is among the leading producers of cars, electronic equipment, ships and steel. The main industrial area is the Tokyo-Yokohama region.
- **Germany**, the **United Kingdom**, **France** and **Italy** are important industrial powers. They are members of the European Union, one of the world's greatest industrial regions.

Other economies

- **India** has been expanding rapidly since the 1990s. Textiles are the country's most important industry. India is also one of the world's main steel producers. The petrochemical, pharmaceutical, car and software industries are growing.
- **Brazil** has the biggest economy in Latin America with a large industrial region stretching from São Paulo to Rio de Janeiro and Belo Horizonte. The growth of its automobile, naval, aeronautic, and pharmaceutical industries indicates that Brazilian industry has a promising future.
- **South Africa** is Africa's main economic power, and has a vigorous mining economy. It is a leader in the manufacture of combustible synthetics and railway rolling stock.
- Other **second level emerging countries** have benefited from outsourcing in recent years. Most are in Latin America and Asia.
- The Pacific area is home to strong economies: the so-called **Asian Dragons** (like Singapore and South Korea), and also Australia.

THINK ABOUT IT

Look at a relief map of China. What is the terrain like on the coast? What is it like inland? Where do you think it would be easier to set up industries? Are there any disadvantages?

INDEX OF INDUSTRIAL PRODUCTION

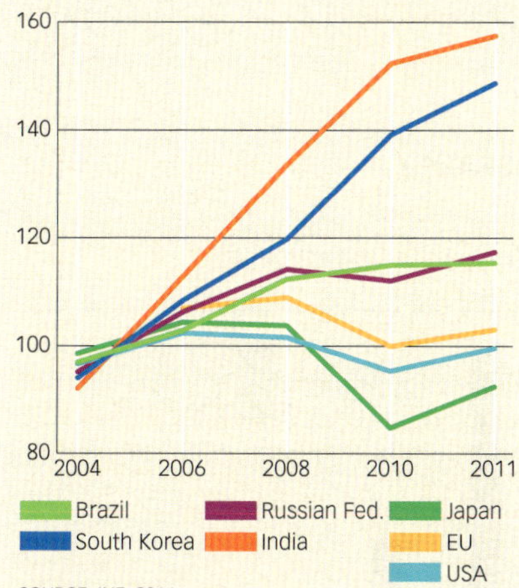

SOURCE: INE, 2011.

WORK WITH THE GRAPH

4 Industrial production has not evolved in the same way in all the countries shown on the graph. What are the differences? How do you explain them?

The industrial zone of São Paulo (Brazil).

8 Mining, energy and industry in Spain

ENERGY IN SPAIN

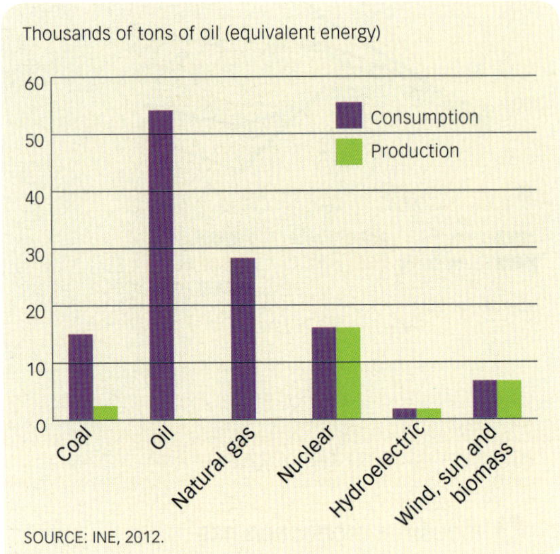

WORK WITH THE GRAPHS

1. From which sources does most of Spain's energy come? Is it produced in Spain?
2. Which are the most important renewable energies in Spain?
3. Explain the evolution of industrial production and employment in Spain.

PRODUCTION AND EMPLOYMENT IN SPANISH INDUSTRY

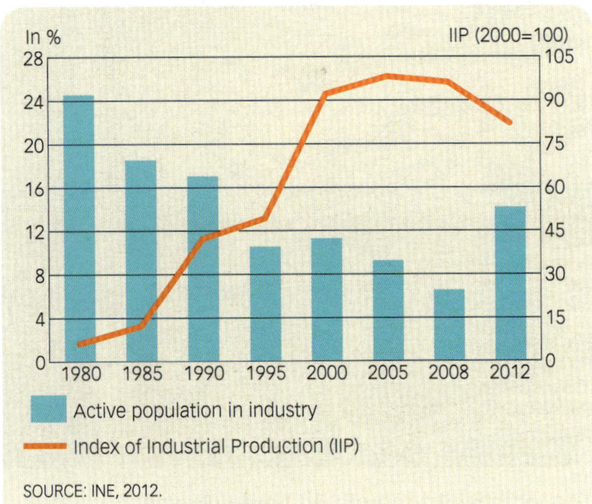

Mining and energy resources

In Spain, mining employs around 0.2% of the working population, and it contributes a similar percentage to the country's GDP.

- Spain is an important producer of **non-ferrous rocks** and **minerals**. It has the world's largest reserves of sepiolite (an industrial absorbent), and it is Europe's leading producer of gypsum.
- Spain has many different **ferrous metals** like copper, tin, zinc, nickel, lead, gold and silver. However, its production does not cover the country's needs. In recent years, production of copper, tin and wolfram has increased.
- **Energy minerals** are scarce. Coal mining was important in the past, but today it is cheaper to import coal from other countries.

Spain is strongly dependent on non-renewable energy. Attempts are being made to reduce this by developing renewable energies, especially wind and solar energy.

Industrialisation

Industrialisation in Spain began in the mid-19th century, later than in other European countries. The earliest industries were textiles in **Catalonia**, and coal and iron mining in the **Basque Country** and **Asturias**.

In the early 20th century, industry grew thanks to the capital brought back from Cuba and the Philippines after their independence, Spanish neutrality in the First World War, and the building of new infrastructures. Protectionist policies favoured the consumption of national products. In 1900, 15% of the working population was employed in industry, and this proportion had risen to 31% by 1930.

This growth was interrupted by the Civil War, but the Spanish economy expanded strongly between 1959 and 1975. Existing industrial areas (Catalonia, the Basque Country and Asturias) were joined by the coastal areas of Levant, Cantabria, Galicia and western Andalusia, as well as inland areas like Madrid, Zaragoza and Valladolid.

After 1975, many traditional Spanish industries became uncompetitive. In the 1980s, there was **restructuring** when many companies in traditional industries like iron, steel and ship-building were closed down. Industrial production reached its peak in the 1990s and the early 21st century.

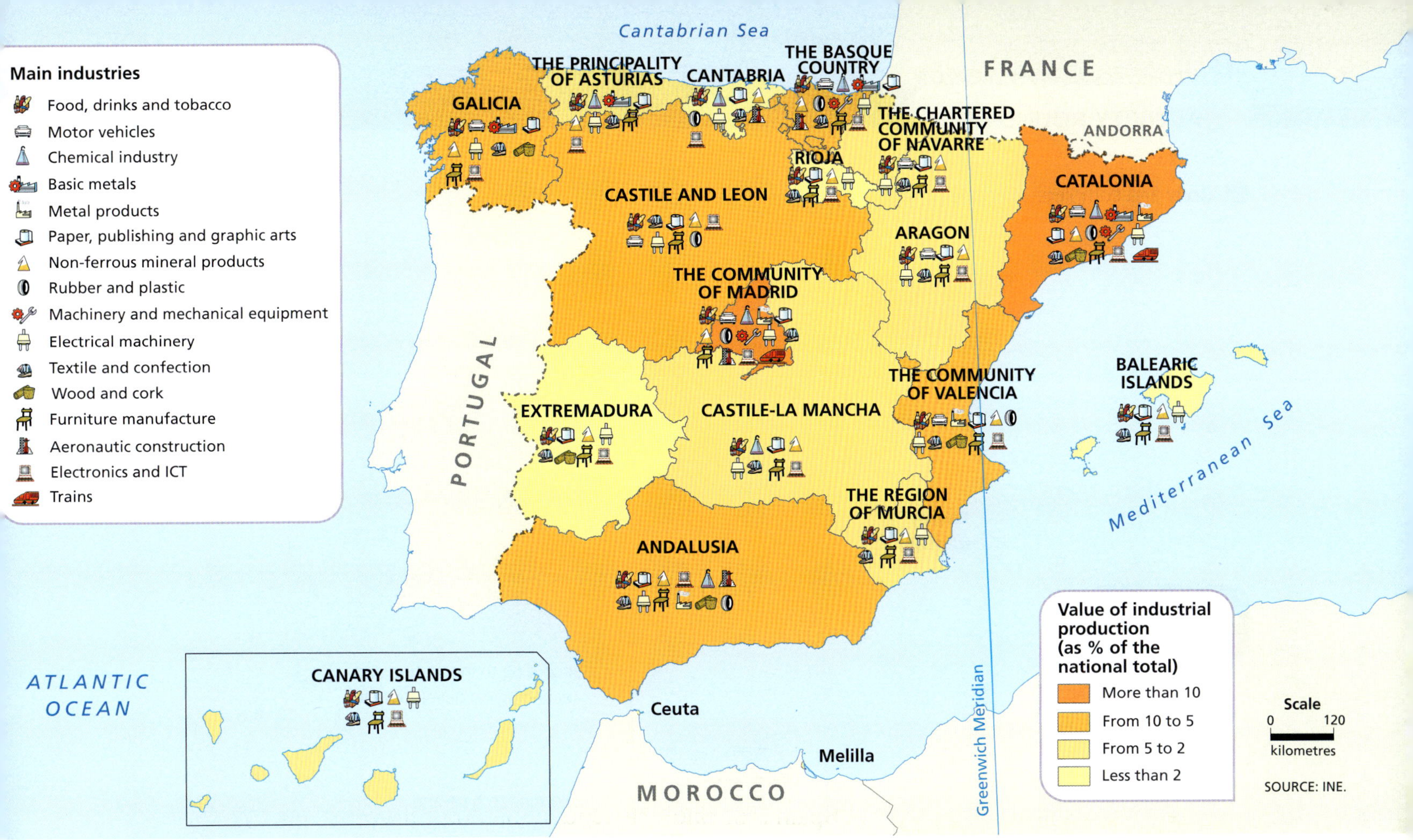

MAIN INDUSTRIES IN SPAIN

Industry in Spain today

The world economic crisis, beginning in 2007, had serious effects throughout the world. However, in Spain it coincided with the **property bubble**, when inflated property prices collapsed.

For years, a large part of secondary sector production was connected to the boom in building construction. When this slowed down, many other activities were also affected. Employment fell, and salaries and consumer buying went down. This was the cause of a fall in production and employment in industry.

- In 2013, 13.7% of the active population in Spain was employed in **industry**, and 6% in **construction**. In the same year, industry contributed 16.1% to **GDP**, while construction contributed 5.2%.
- The **food, drinks and tobacco** sector represents around 18% of production and industrial employment.
- Spain's **automobile manufacturing industry** is the second largest in Europe, and car exports make up 20% of total industrial exports. The **chemical industry** is next in importance.
- Most industrial manufacturing takes place in the Autonomous Communities of Catalonia, Madrid, the Basque Country and the Valencian Community.

WORK WITH THE MAP

4 In which Autonomous Communities is industry most important?

5 Which branches of industry are especially important in each Community?

ACTIVITIES

6 When did industrialisation begin in Spain? In which regions? Are they the same regions that are industrialised today?

7 What consequences do you think the restructuring of Spanish industry had?

8 Which sectors of Spanish industry are especially important today?

ACTIVITY ROUND-UP

1 Copy and complete the table.

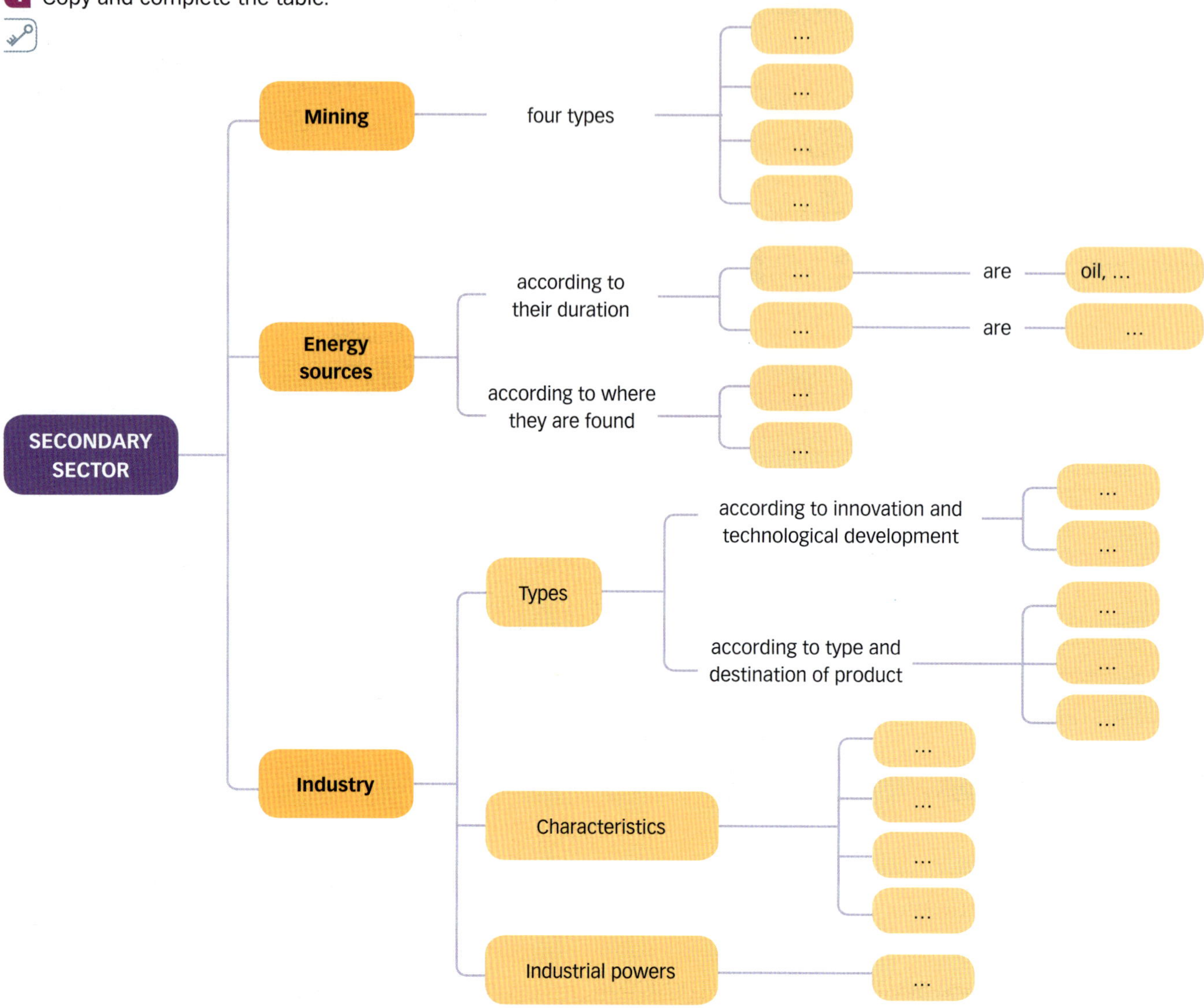

2 Match the words and phrases with their definitions.

- Renewable energy
- Industry
- Mining deposit
- Offshoring
- Multinational

- A place where minerals are concentrated.
- A natural source of energy like the Sun, wind or waves.
- Economic activities which involve manufacturing.
- A large company with subsidiaries in different countries.
- The relocation of business activities in a different country.

3 Explain the difference between these concepts and give an example of each.

- Primary energy and secondary energy
- Basic industry and consumer goods industry

4 Explain the meaning of the following terms:

- Decentralisation of production
- Tertiarisation of industry

5 What are the main challenges today in mining, energy exploitation and industry?

6 Copy and complete the table. Then find the producing countries on a world map.

Energy source	Main uses	Big producers
...	...	...

7 Compare the energy consumption of three countries.

a What type of energy is consumed most in each country?

b How do you explain each case?

c Which of the three countries has the most varied consumption?

d Is there any country where a single type of energy makes up more than half of consumption? Is it produced by the country itself? Is it renewable?

e What could happen in this country if energy supplies ran out?

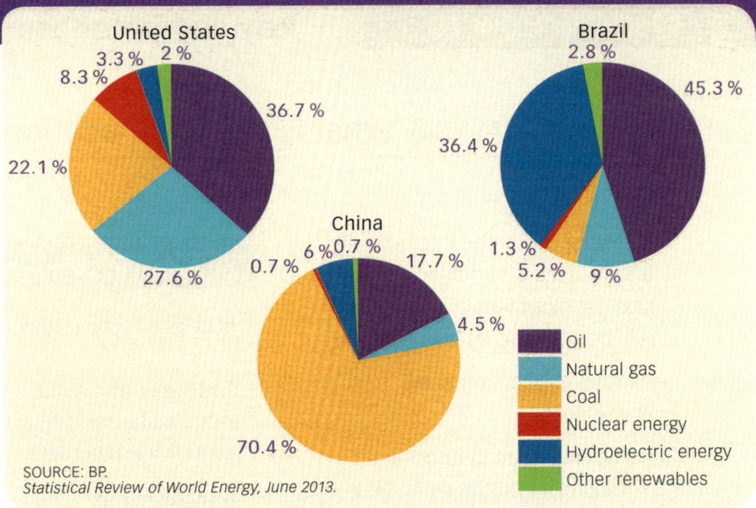

SOURCE: BP. *Statistical Review of World Energy, June 2013.*

8 Look at the photograph and answer.

a Classify the industry by its type of product and its technological development. Explain your answers.

b Which factors did the company take into account when it decided where the factory should be?

c What type of place would be best as a location for this type of industry?

ENERGY POVERTY

Energy poverty is the lack of access to modern energy services. It affects a fifth of the world's population. More than 1,300 million people have no electricity in their homes, and around 3,000 million depend on traditional types of energy like coal and wood for heating and cooking. Sub-Saharan Africa is the area with greatest energy deficiencies.

The lack of clean energy puts health at risk. It causes breathing problems and the death of two million people every year. Moreover, without energy, industry cannot develop, basic social services like hospital care are missing, and the water supply is not assured. Energy poverty prevents economic and social development.

Energy poverty also exists in developed countries although it means something different. In these cases, people are unable to cover their basic energy needs, for example to take a shower or heat their houses. In Spain, more than 10% of homes suffer from energy poverty.

9 Type "satellite image of the earth at night" into a search engine on the Internet.

a Which parts of the Earth are the darkest at night? Why?

b Look at Africa. Where are the most illuminated areas? Why?

10 Imagine spending a whole day without electricity.

- Which of your activities would be affected? Make a list.

11 Look at the picture. How have these people solved the problem of energy poverty? Can you suggest other solutions?

SCIENTIFIC ANALYSIS. What is the best location for an industry?

Deciding on the best location for an industry is complex. The growth and development of a company will be affected by this choice.

Various aspects must be considered:

- The main aim is to obtain maximum profits by finding a place where company costs are reduced to a minimum.
- The importance of each aspect depends on the type of industry. Some factors are decisive while others are not so important.

When evaluating the location of an industry, identify the most important factors from a list like that on the right. Think about the characteristics, needs and production of the industry involved. Answer questions like those asked here. Then draw your conclusions.

LOCATION FACTORS	NEEDS OF AN INDUSTRIAL PLANT
Access to innovation	Is there a need for developed technology, or technical and scientific assistance (universities)?
Existence of a solid industrial structure and good infrastructures	Is there a need for auxiliary companies to share services and infrastructure? Are there industrial parks and estates?
Transport and communications	Is there a need for a complete network of transport and communications?
Proximity to markets (sales)	Should the company be near the consumer markets?
Labour force	Is much labour needed? Does it require skilled workers?
Proximity to raw materials and energy sources	Does the company need to be near raw materials or energy sources?
Environmental impact	Does this industry affect the environment? Can it be near a town or city?
Availability of land	Is much land needed for its construction?
Political factors	Does it require political support?

Assess the location of a petrochemical plant using this information.

The petrochemical industry obtains chemical products from oil or natural gas. It needs a large area, a lot of energy and a specialised labour force. It also requires a large scale investment of capital and modern technology. Its environmental impact can be great, involving the dumping of toxic material and harmful gas emissions.

Petrochemical plant of the Reliance company in Jamnagar, India.

Location of Jamnagar.

1. Assess the importance of each factor. Which factors are decisive? Why?

2. Explain your conclusions. Do you think a petrochemical plant should be located:
   - In a developed, emerging or less developed country?
   - Near a port?
   - Near gas or oil pipelines?
   - Near a large city?

ENVIRONMENTAL ISSUES. **A sustainable energy project**

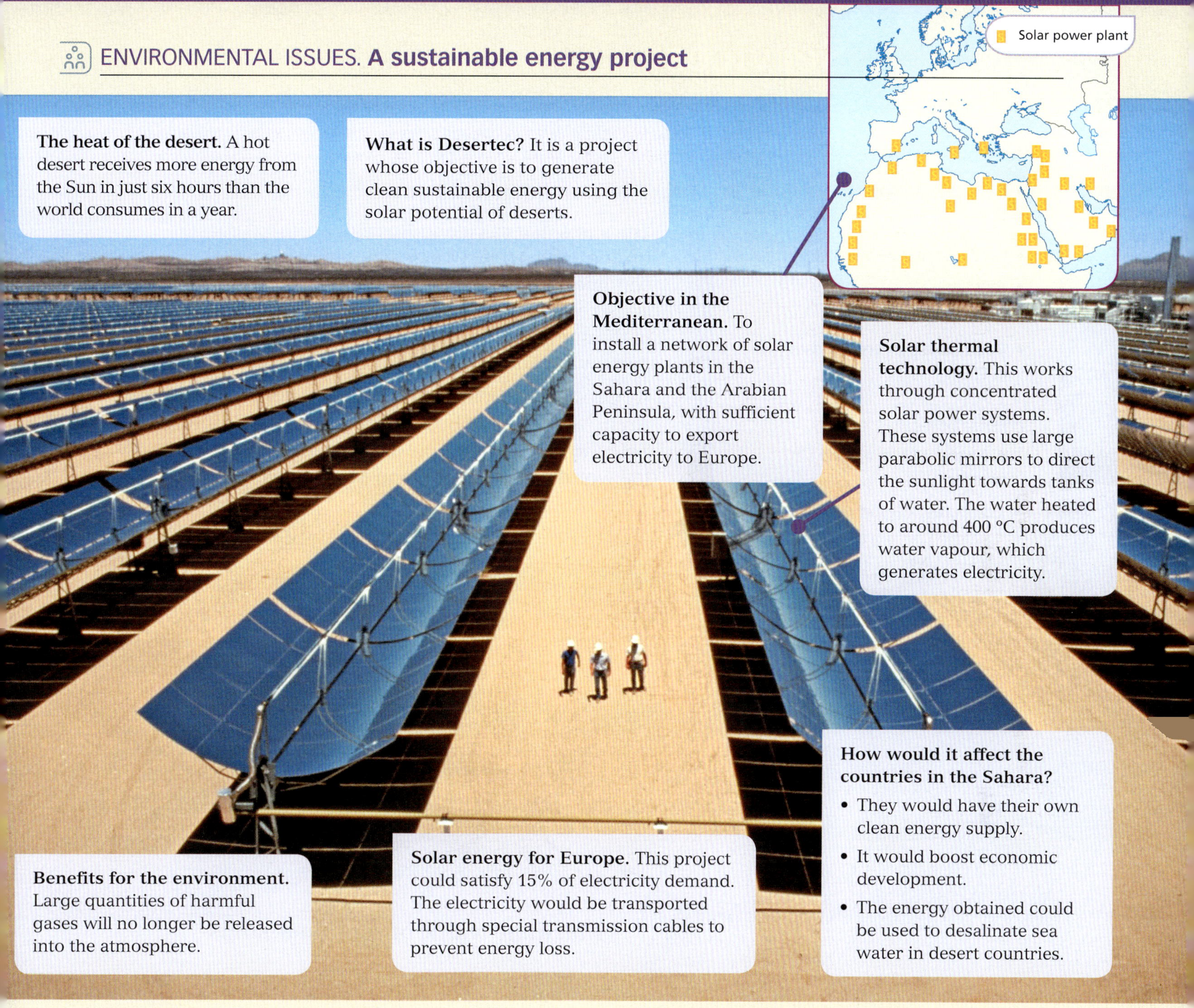

The heat of the desert. A hot desert receives more energy from the Sun in just six hours than the world consumes in a year.

What is Desertec? It is a project whose objective is to generate clean sustainable energy using the solar potential of deserts.

Objective in the Mediterranean. To install a network of solar energy plants in the Sahara and the Arabian Peninsula, with sufficient capacity to export electricity to Europe.

Solar thermal technology. This works through concentrated solar power systems. These systems use large parabolic mirrors to direct the sunlight towards tanks of water. The water heated to around 400 °C produces water vapour, which generates electricity.

How would it affect the countries in the Sahara?

- They would have their own clean energy supply.
- It would boost economic development.
- The energy obtained could be used to desalinate sea water in desert countries.

Benefits for the environment. Large quantities of harmful gases will no longer be released into the atmosphere.

Solar energy for Europe. This project could satisfy 15% of electricity demand. The electricity would be transported through special transmission cables to prevent energy loss.

1 Are the following statements true or false? Listen to the recording and check your answers.

a The world consumes more energy in a year than a desert receives in six hours.

b Desertec is a project aimed at using solar power to produce clean energy.

c The energy would go to solar plants in Europe.

d Mirrors are used in concentrated solar power systems.

e Thanks to the Desertec project, all the electricity demand in Europe would be covered by solar energy.

f The project would be detrimental to Saharan countries.

2 Investigate hydroelectric power stations on the Internet.

a When and where was the first hydroelectric power plant built?

b What resources do hydroelectric power plants need?

c How can the construction of hydroelectric power stations affect people and wildlife?

d How did it affect the people of Riaño in León?

Present your findings to the class.

6 Services

FIND OUT ABOUT:

- What are services?
- Types of trade
- Land, sea and air transport
- Tourism
- Other services
- Trade, transport and tourism in Spain

KNOW HOW TO:

- Analyse a tourist resort: Benidorm
- Compare transport networks

The northern sea routes

The ships that sail along Russia's northern coast through the Arctic Ocean experience extreme conditions. It is dark for much of the year, and there are strong winds, fog and abundant ice. However, in the summer enough ice melts to create a very narrow corridor through which ships can navigate.

As a result of climate change, many parts of the Arctic Ocean could become free of ice in the future. Ships would then be able to navigate between the world's main ports along new routes.

Maritime traffic of container ships

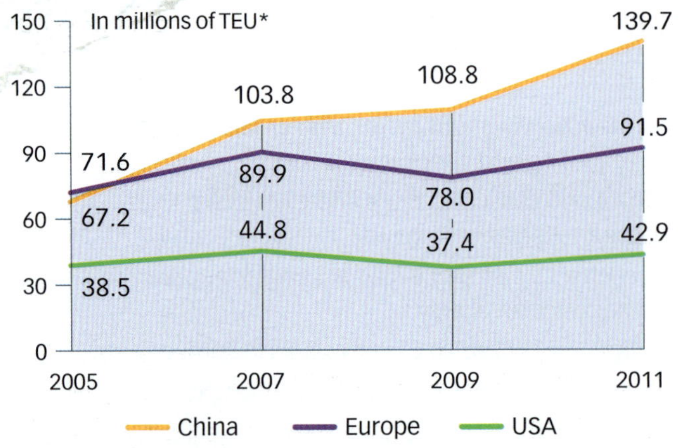

*TEU: Twenty-foot equivalent unit. Cargo capacity of a container ship of twenty feet (6.1 metres).

? WORK WITH THE IMAGE

- Where would the new northern shipping routes be?
- What would their advantages be?
- Would any present-day shipping routes lose traffic?
- How did maritime traffic evolve between 2005 and 2011?

Shipping routes

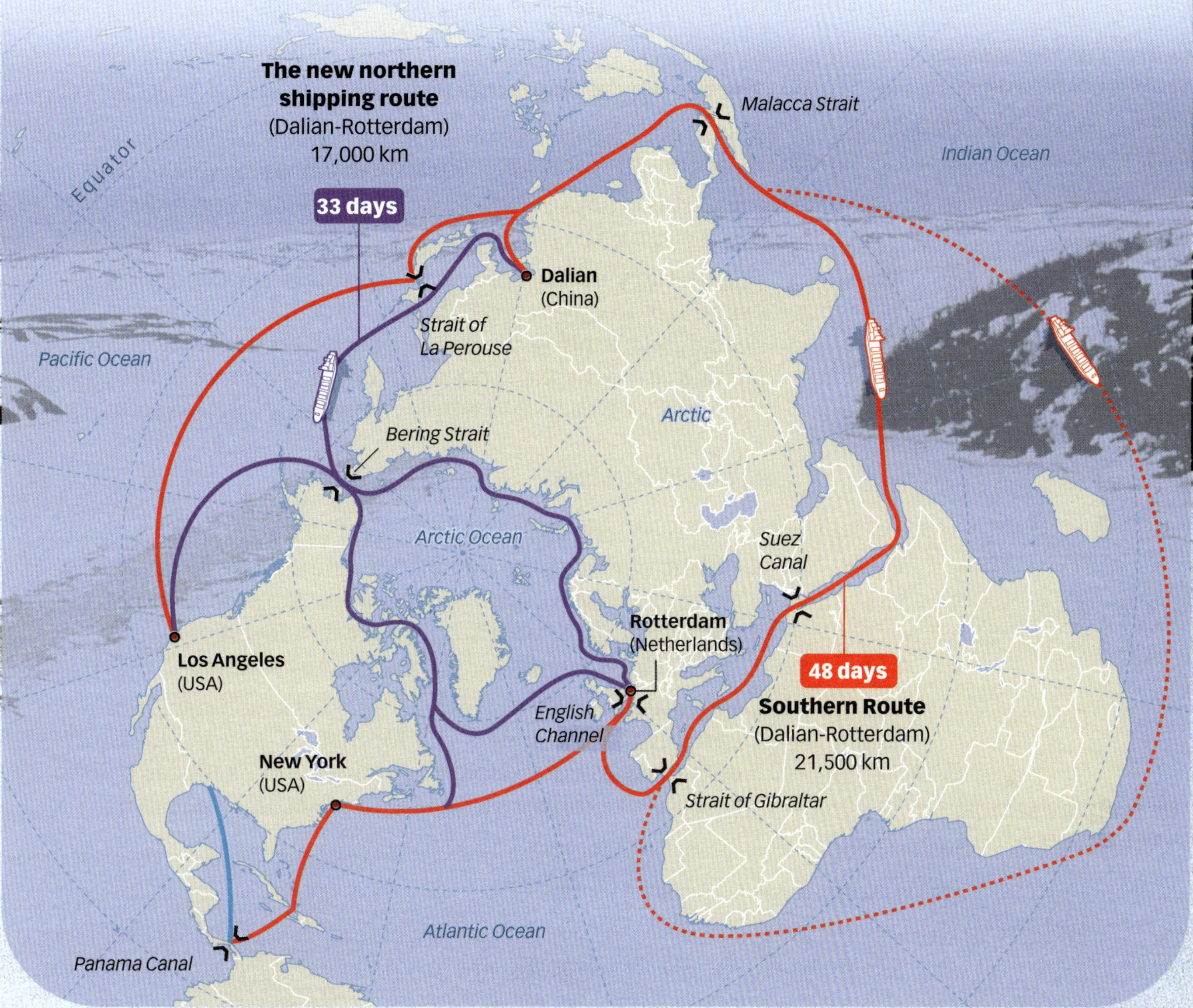

HOW DO WE KNOW?

In the light of climate change, many scientists are now studying the evolution of ice on the Earth's surface.

Satellite images provide information about changes in the ice caps. Radar equipment measures the thickness of the ice.

- How will the disappearance of the Arctic ice cap affect climate, landscape, and human life on Earth?

Satellite images of the North Pole.

1979-1981

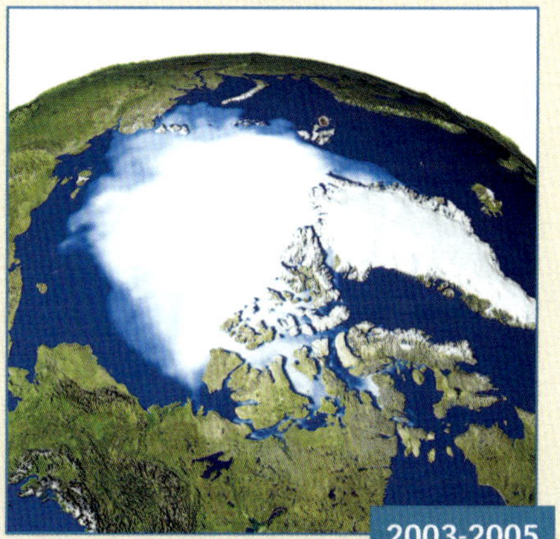
2003-2005

1 Services

GROWTH OF THE SERVICE SECTOR IN THE WORLD

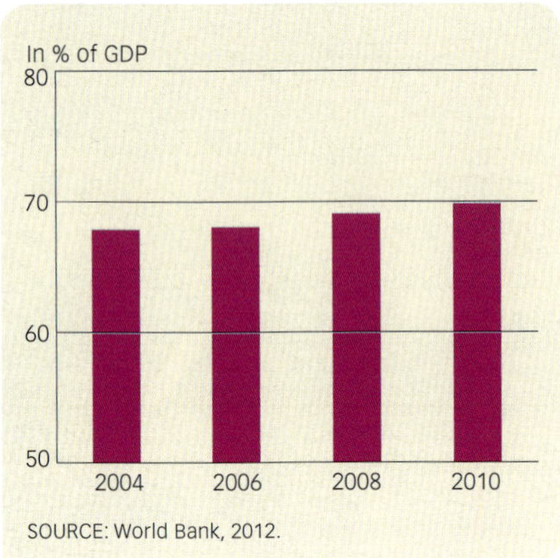

The growth of the tertiary sector

Services consist of all the activities which do not directly produce tangible goods, but help to satisfy people's wants and needs. They include the work of teachers, doctors, bus drivers or hairdressers.

The **tertiary sector** (or service sector) of the economy has expanded greatly in the 20th and 21st centuries. In particular, developed countries have experienced **tertiarisation**, which is a shift towards the service sector. Today, services provide over 70% of wealth and employment in developed countries.

Types of services

Services can be classified into two groups:

- **Public services** are financed by the state from taxes. They run public institutions in areas like administration, security and justice. They also meet many of the basic needs of the population in health or education.
- **Private services** are offered by private companies or individuals. Consumers pay for them. Some services are commercial, educational or cultural; other services belong to fields like transport, health and telecommunications.

THE TERTIARY SECTOR IN THE WORLD

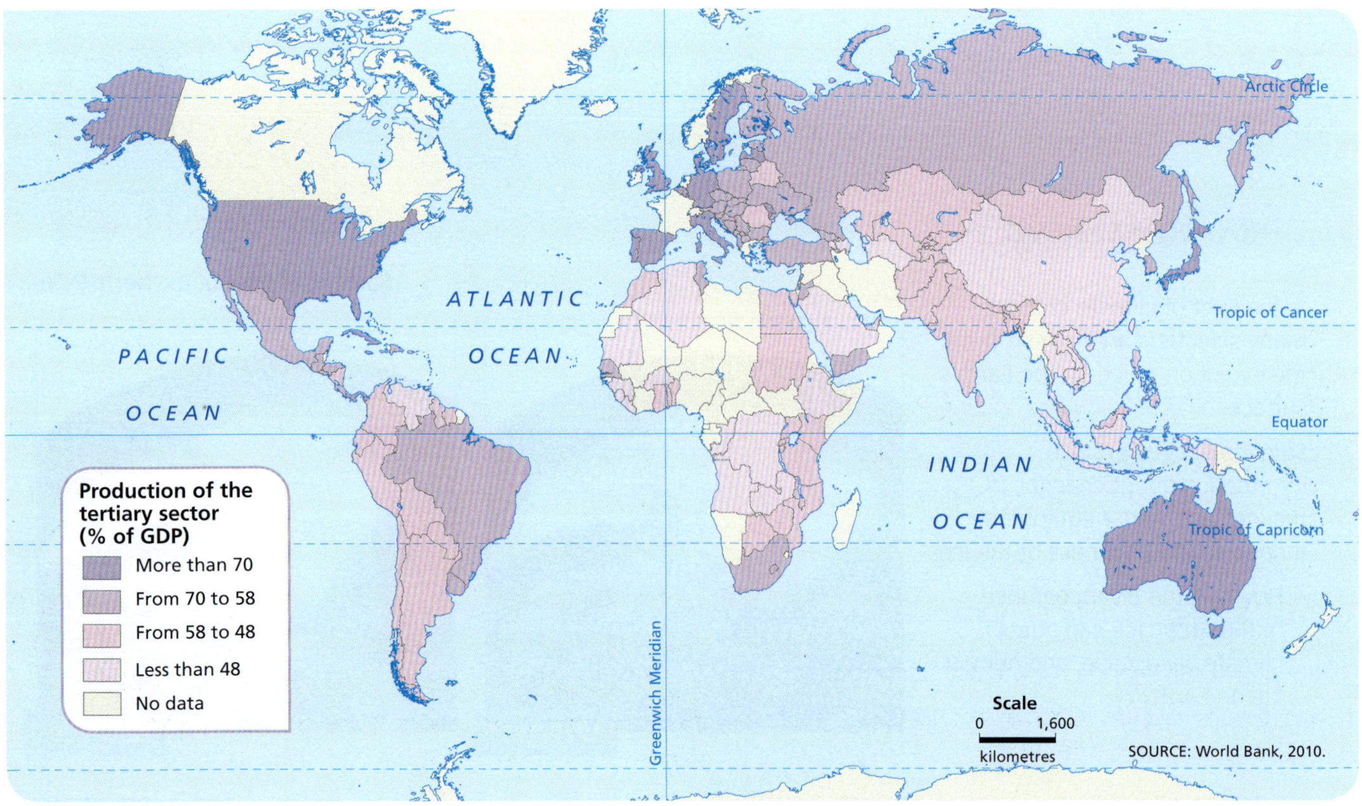

Where are services located?

Traditionally, services have mainly been located in urban areas, where there are many consumers, companies and official institutions.

Many **advanced services**, involving new technologies, are located in **business districts** in metropolitan areas. Services provided by public institutions and small-scale commercial activity are spread across a much wider area.

Improvements in international communications and transport have led to **outsourcing**. This process is the transfer of services like information technology from developed economies to places where labour is cheaper. Countries like India and Mexico provide a well-qualified work force with technical expertise as well as fluency in English.

Outsourcing. A call centre in Morocco.

WORK WITH THE MAP

1 Study the map of the service sector on the opposite page.

a Find five countries where the service sector is more than 70% of the value of GDP.

b Find five countries where it is less than 48%.

c Why is there such a marked difference between the service sector in developed and less developed countries?

d Does the tertiary sector predominate in any emerging countries? Why?

ACTIVITIES

2 Give examples of services that are provided to the populations of developed countries. Which of these services reflect high standards of living?

3 Explain if these services can or cannot be outsourced:

- Defining a company's business strategy.
- A customer service call centre.
- Swimming lessons.

4 Why is India an attractive destination for outsourcing?

THINK ABOUT IT

The importance of public services

Public services are provided in areas that are considered essential to all groups in society. In developed countries, the state (known as the "welfare state") guarantees many of these services. In less developed countries, many of these services fail to reach the whole population.

Defence, security and justice are always public services. In fields like education and health, some services are provided by the state, but others are financed by the private sector.

- In groups, discuss health and education. Should they be public services?

The British Museum, London. Many of the world's great museums are run as public services.

2 Trade

What is trade?

Trade is the exchange of goods and services between producers and consumers. In many countries, it is the most important type of service in terms of turnover and the number of workers employed.

Trade depends primarily on three factors:

- **Transport and infrastructures**. Places with good communications, such as cities, have the most trade.
- **Market size**. This is the number of consumers who can purchase goods or services. There is more trade if the market is bigger.
- **Wealth**. If people have greater purchasing power, they will consume more.

Domestic trade

Domestic trade takes place within a country's borders. It distributes domestic products (produced inside the country) and foreign products throughout the territory.

Wholesalers and retailers supply products to consumers:

- A **wholesaler** buys large quantities of a product directly from the producer to sell to retailers.
- A **retailer** sells products in small quantities to a customer.

Foreign trade

Foreign trade takes place with other countries.

- **Imports** are the purchases of goods and services from a foreign country, while **exports** are the sales of goods and services to a foreign country.
- The **balance of trade** is the difference between the value of a country's exports and the value of its imports.

MAIN EXCHANGES OF PRODUCTS IN WORLD TRADE

	2000	2011
Agricultural and livestock products	551.2	1,659.5
Fuel and mining products	852.63	4,000.83
Manufactured products	4,692.27	11,510.9

*Data in millions of dollars

BALANCE OF TRADE

If the value of a country's exports is greater than the value of its imports, its balance of trade is **positive** (favourable).

If the value of a country's imports is greater than the value of its exports, its balance of trade is **negative** (unfavourable).

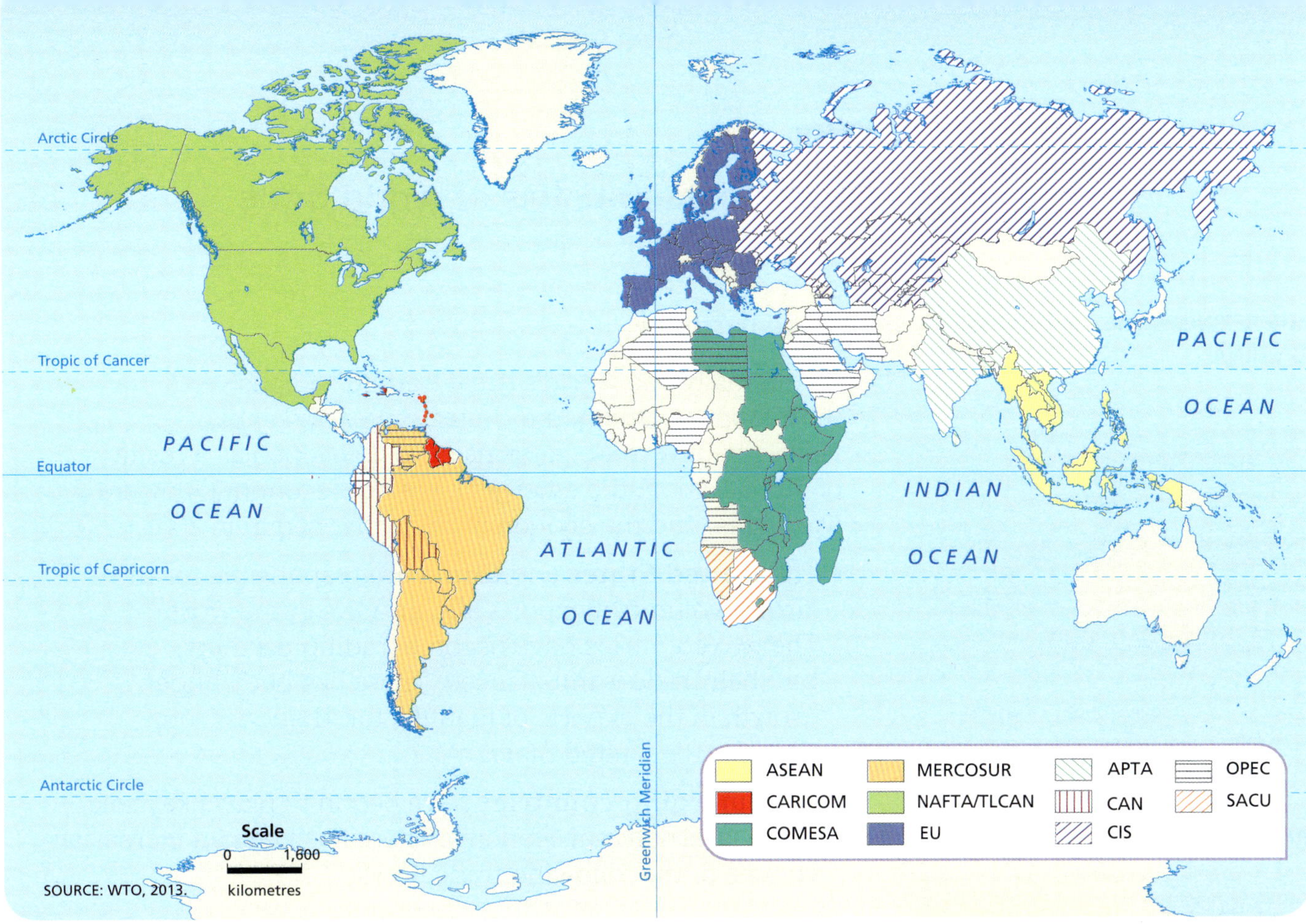

MAIN WORLD TRADE BLOCS

The organisation of world trade

World trade is regulated by agreements between countries and the **World Trade Organisation (WTO)**. Nearly all the countries in the world belong to this organisation. It aims to achieve agreements that promote world trade by reducing trade barriers, and eliminating unfair practices such as selling products at prices that are lower than the costs of production.

Many countries form **trade blocs** in order to promote trade with each other. There are two types:

- **Free trade area**. This is made up of a number of countries whose trade with each other is exempt from customs taxes. Each country has an independent trade policy with countries outside the area. For example, the North American Free Trade Agreement (NAFTA) is a free trade union between the United States, Mexico and Canada.
- **Customs union**. This is a free trade area which also establishes a common customs policy with regard to non-member countries. The Southern African Customs Union (SACU), and the European Union (EU) are customs unions.

There are also associations created for the trade of a specific product, like the Organisation of the Petroleum Exporting Countries (OPEC).

ACTIVITIES

1. Copy and complete the diagram in your notebook with explanations of each term.

PRODUCER
↓
WHOLESALER
↓
RETAILER
↓
CUSTOMER

2. What is trade? Why is it so important.

3. Look for information and complete this table about three trade blocs.

Trade blocs	Member countries
......	......

4. Is trade between Spain and other countries in the European Union foreign or internal trade? Explain your answer.

3 Characteristics of world trade

TRADE BETWEEN THE UNITED STATES AND CHINA

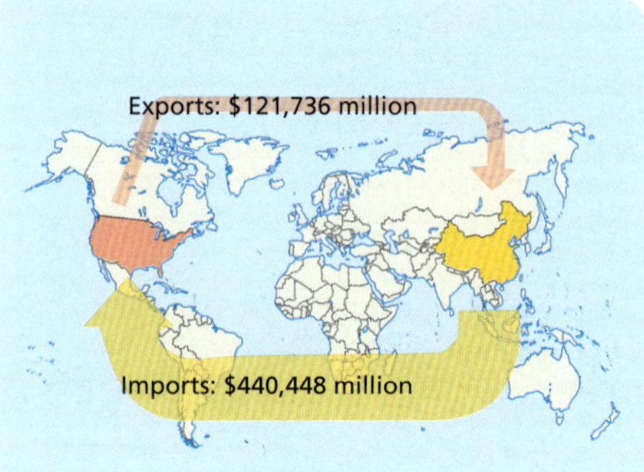

SOURCE: US Census Bureau, 2013.

WORLD TRADE IN GOODS

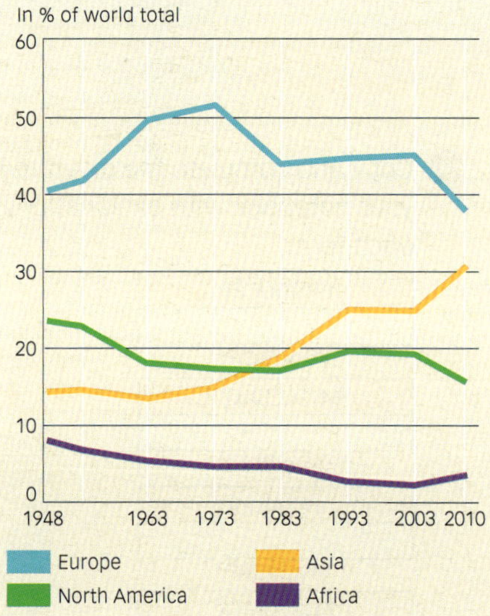

SOURCE: WTO, 2011.

WORK WITH THE GRAPH

1. Describe the evolution of trade in the different regions. What factors could explain their growth?
2. What proportion of world trade do you think Asia is likely to have in the year 2030?

Trade patterns

Traditionally, the world's great trading regions were the European Union, the United States and Japan. However, new trade patterns have developed as a result of globalisation, the emergence of new economic powers like China, and the economic crisis that began in 2007.

- The **United States** and **China** are responsible for the largest volume of exports and imports in the world. The two countries have become trading partners, but their trade is imbalanced. There has been huge growth in the exports of manufactured goods from China to the United States.
- The **emerging countries** have begun to play a key role in the international economy. There has been an increasing volume of the following trade flows:
  - Between the emerging economies in Asia, Africa and Latin America and the traditional powers (USA and EU).
  - Between the emerging countries themselves, especially between Asia and Latin America, and between Asia and Africa.

Trade in the European Union

If we group together all its member states, the **European Union (EU)** is the world's largest trading power. More than 20% of the world's exports are from the EU:

- 71% of the exports from EU member states are to other parts of the EU.
- Outside the EU, the main destinations for its exports are the United States, China, Switzerland, Russia and Turkey.
- The EU's imports come mainly from China, Russia, the United States, Norway and Switzerland.
- Germany is the leading exporting country in the EU.

86% of the EU's exports are **industrial goods**, such as machinery, transport vehicles, chemical products, and raw materials (iron and steel). The remaining 14% are mainly agricultural products like wine, cereals and meat products.

The EU negotiates **trade agreements** with other countries. This benefits growth and employment in Europe, and creates new markets in the rest of the world.

The USA and EU as trading partners

The United States and the European Union are important **trading partners**. Trade between them amounts to 2,000 million euros each day. It also creates around 15 million jobs, which are shared equally between the two areas. Trade between the USA and the EU represents around 31% of world trade, and 53% of world GDP.

Trade relations between the USA and the EU are influenced by the following factors:

- The close links between large industrial groups.
- Negotiations aimed at creating a free trade area and a common investment policy.

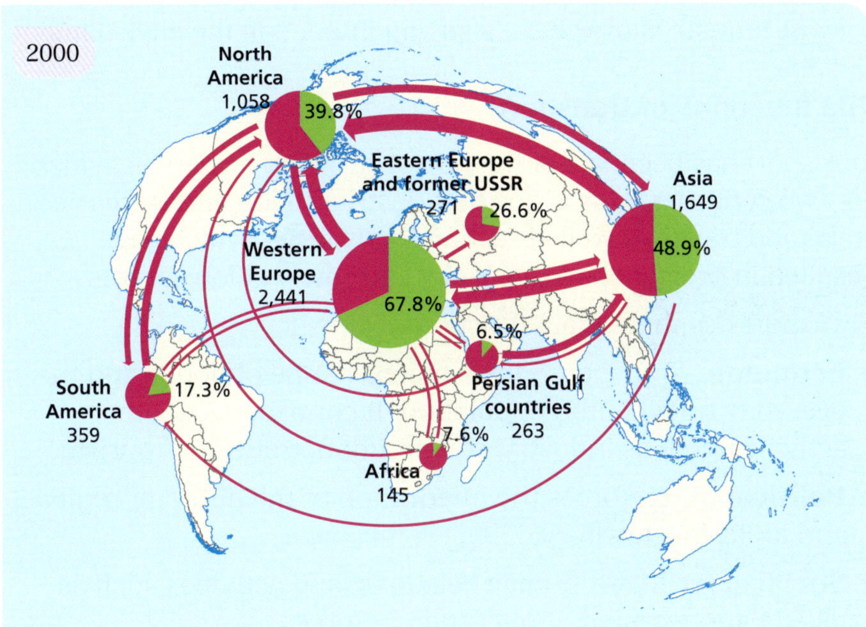

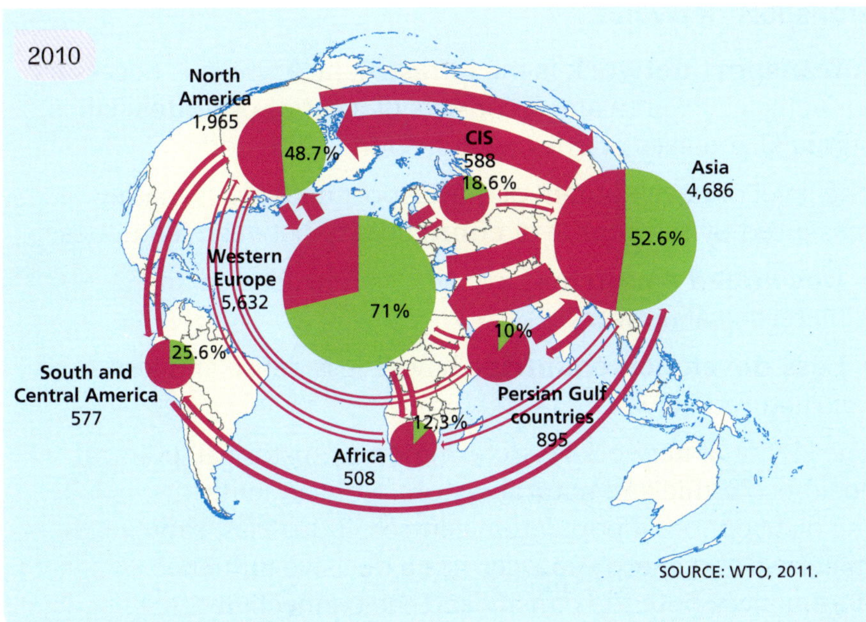

ACTIVITIES

3 Look on the Internet for information about the "Doha Development Agenda". What is it? What are its aims? Which countries are involved?

4 Why do you think there is so much trade within the European Union?

WORK WITH THE MAPS

5 Analyse the two maps and answer the questions.

a Did the volume of world trade increase between 2000 and 2010?

b What were the main trade flows in the world in 2010? Were they the same as in 2010?

c What were the most significant changes?

d Do you think these trade flows are likely to continue in the future?

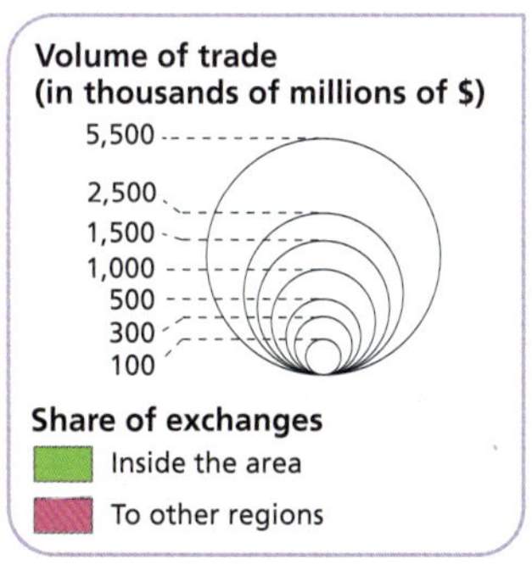

4 Transport networks and land transport

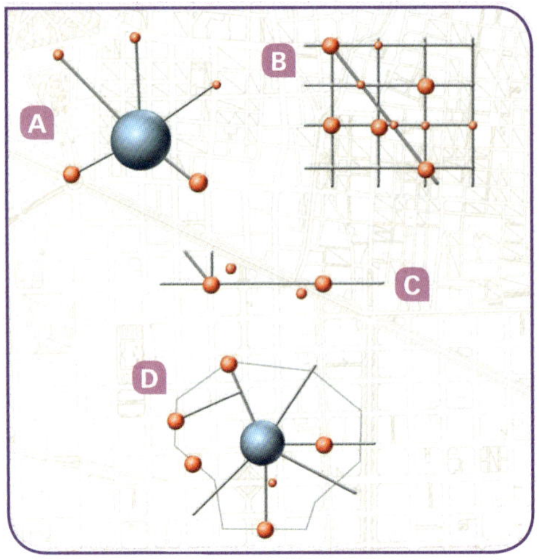

Transport networks. Centralised or radial (A). Grid-shaped (B). Linear (C). Anarchic (D).

Transport services

Transport is the movement of people and goods from one place to another:

- **Means of transport** are the different vehicles used for travel: car, train, plane or boat.
- **Infrastructures** are the elements necessary for transport such as roads for vehicles, railways for trains, airports for aircraft, and ports for ships. Pipelines, power lines and telephone lines are also considered transport infrastructures. Most infrastructures have a strong impact on the environment.

The functions of transport

Transport has been one of the fastest-growing tertiary activities for two centuries. This is due to industrialisation and increasing trade and travel, as well as technical advances, which have resulted in greater speed, capacity and safety at lower costs.

Transport has three main functions:

- **Economic**. Transport allows the exchange of goods and the everyday travel of the population. It contributes to other activities such as industry, construction, trade and tourism.
- **Political**. It reinforces the integration of the different regions and areas in a territory.
- **Social**. It facilitates human relations, and activities such as leisure and learning about other cultures.

Transport networks

A **transport network** is the infrastructure which is necessary for vehicles to circulate. It consists of a series of **nodes** (cities, stations) and axes or lines, which connect the nodes.

The level of development of a transport network is mainly measured by the degree of **connectivity** between the nodes:

- **Developed countries** have a dense, structured and hierarchical network.
- **Less developed countries** have a less dense and less structured network.

The layout of networks is determined by historical, political, socio-economic and natural factors. Consequently, the distribution of population centres, industries, raw materials and energy sources has a decisive influence. It influences both the density and the connectivity of the transport axes.

THINK ABOUT IT

Intermodal transport

Many transport networks are intermodal – this means that they involve more than one means of transport. For example, trains, buses and taxis all circulate at transportation hubs. In international trade, goods are often shipped in containers using several means of transport, like ships and lorries.

- Think about your own daily life. Do you and other members of your family use an intermodal network?

ACTIVITIES

1. What is the difference between 'transport' and 'infrastructures'?

Land transport

- **Road**. Cars, buses and lorries travel by road. This type of transport is very flexible and very useful for moving people and goods. However, it has some disadvantages:
  - It is **expensive** to build and maintain the roads.
  - There are more **accidents** in this type of transport, and it also creates **pollution**.
- **Rail**. Rail transport includes trains, metros and trams. Large cities need good rail infrastructures so that people can travel by local trains and metro. Rail transport has several advantages:
  - **High-speed trains** are very useful to travel rapidly to distant places.
  - Trains can transport **goods** in containers refrigerator compartments, etc.
  - Rail transport is **cheap**, **fast** and **safe** and does not contaminate as much as transport by road.

The high-speed rail network in Japan.

WORK WITH THE MAP

2 Answer the questions.

a Which countries have the most land transport networks? Explain your answer.

b Why do you think rail transport is more important than roads in Russia?

c Look at Australia. Why do you think that land traffic density is very low?

WORLD LAND TRANSPORT

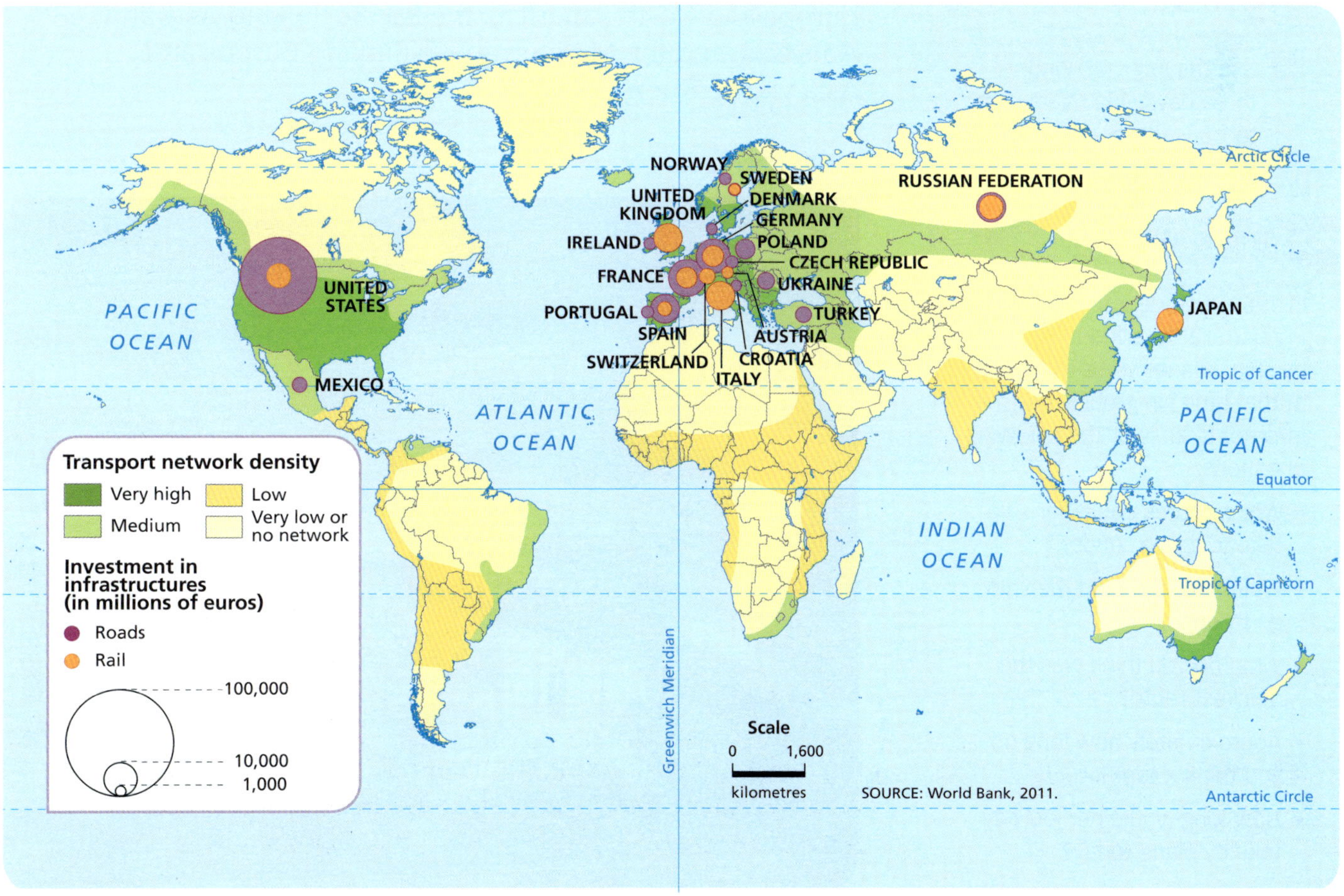

5 Ship and air transport

SHIPPING OF GOODS

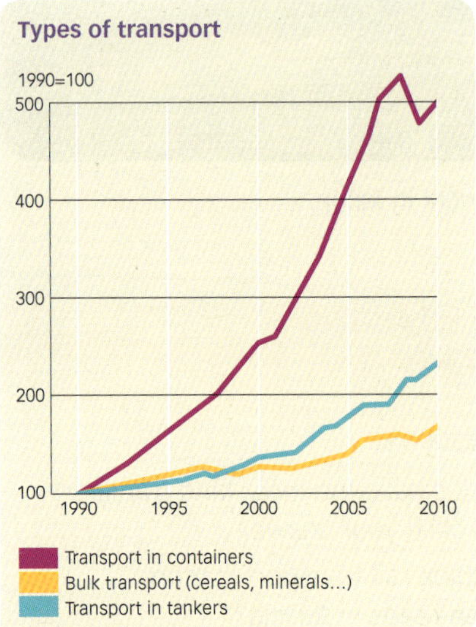

ACTIVITIES

1. What are the advantages and disadvantages of ship transport?
2. Use the Internet. Find three navigable rivers used to transport goods.

THINK ABOUT IT

In 1969, a student from Sydney in Australia travelled to London by ship. (Planes were expensive at that time.) In groups, use the Internet to answer the following questions:

- Which routes could her ship take to get to London?
- Which canals could the ship go through?
- What do you think was the quickest route?
- Approximately how long do you think it took?
- How long would her journey take by plane today?

Ship transport

Ship transport includes transport by sea and on navigable rivers. It is mainly used to transport big heavy goods, and it is used for over 90% of world trade. Its main characteristics are:

- It is **cheap** but **slow**. That is why today people travel long distances by plane.
- It needs **port infrastructures** to handle large ships. These require huge investments. Consequently, states finance ports, although ships belong to large shipping companies, which can be public or private.
- It is a very **polluting** means of transport. Fuel is lost when the tanks are cleaned at sea or when ships have accidents and produce oil slicks.
- It cannot take goods directly from the producer to the consumer. Goods are taken to consumers by road or rail.

Huge **containers** are used to carry goods. These are easy to load onto the ships and then place on lorries or special trains. Modern ports have large container terminals.

The main **shipping routes** are the North Atlantic (linking Europe and North America), the Pacific (linking Asia and America), and the Mediterranean (linking Europe and Asia via the **Suez Canal**).

The Panama Canal.

Air transport

Air transport is mainly used to move people over **long distances**. Some goods are also transported by air, but these are usually only small, urgent or valuable products.

- Air travel is a **quick** and **safe**, but also an **expensive** means of transport.
- Planes cause **noise pollution**, and their fuel causes **air pollution**. New airports are therefore built at some distance from cities. Consequently, roads and rail lines have to be built to connect the cities and the airports.
- The **infrastructures** required for air travel are complex, technical and expensive. That is why many airports are managed by state companies.

In the last ten years, traditional airlines have been faced with competition from **low cost companies**. As a result, Governments have privatised or merged their companies to make them more competitive.

Airports attract businesses from other economic sectors because of the volume of passengers who use them. These sectors include catering, banking, retail trade and travel agencies.

Main air routes

International airports are connected by **air routes**. These have grown rapidly in recent decades.

- The largest number of flights is between North America and Europe. Air traffic between these regions represents 29.3% of the world total.
- Next, air routes linking Europe with Asia represent 13.2% of world air traffic.
- In third place, routes from Asia to America are 12.2% of the total.

The Middle East has become a key air hub. Airports like Doha, Abu Dhabi and Dubai manage around 15% of all the traffic connecting Europe and Asia.

ACTIVITIES

3 What are the advantages and disadvantages of air transport?

4 Which regions had the world's main airports in 2000? Did the situation change between 2000 and 2010? Why?

PASSENGERS IN THE WORLD'S MAIN AIRPORTS

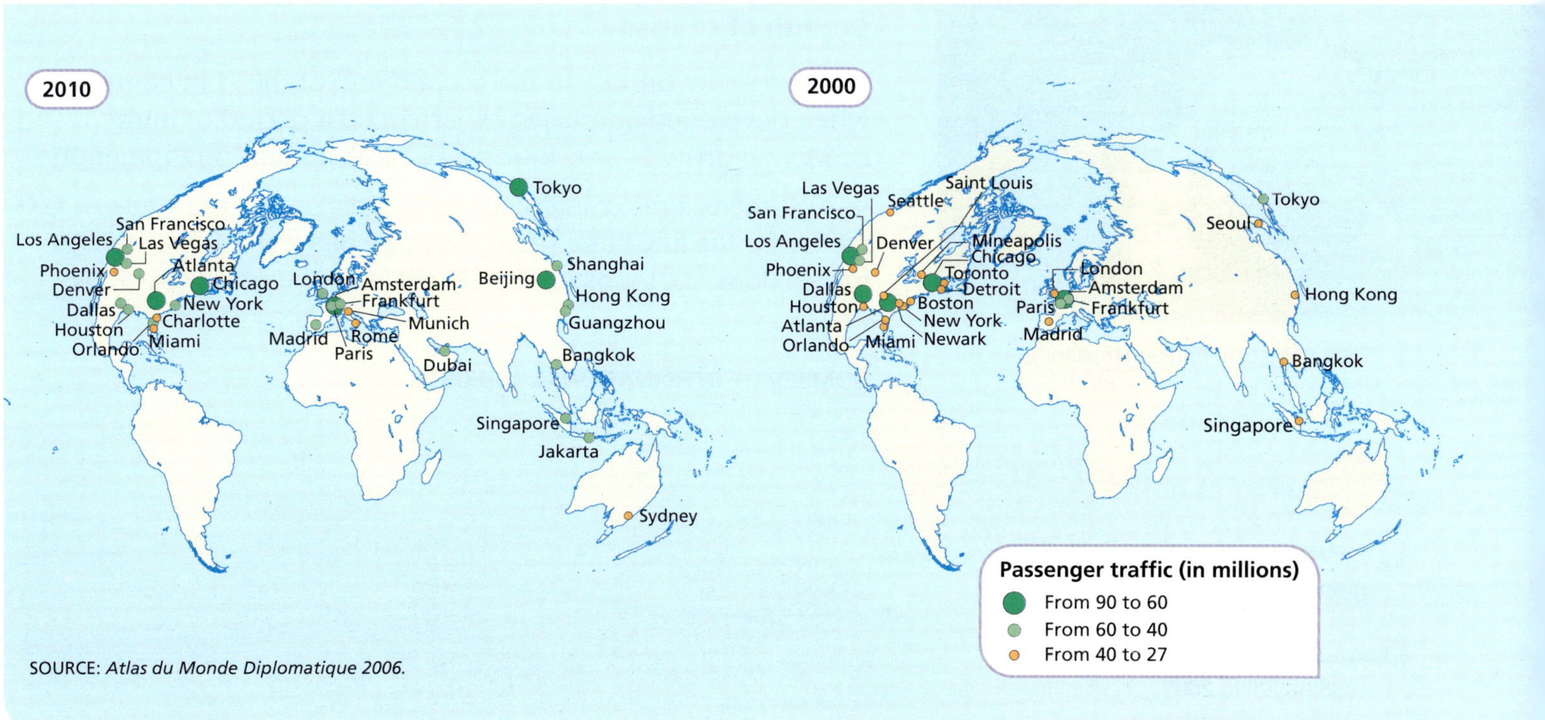

SOURCE: *Atlas du Monde Diplomatique 2006.*

6 Tourism

What is tourism?

Tourism consists of the activities that people do for pleasure during their travels. Many services satisfy tourists' needs for travel, accommodation, food and leisure.

The importance of tourism for the wealth and prosperity of countries is increasing. In 2013, tourism represented 9% of world GDP, and 6% of world exports.

Tourism has important **benefits**:

- It creates **business** opportunities and **jobs**, both directly (in hotels, restaurants and shops) and indirectly (in construction, transport and trade.)
- It adds economic value to a region's natural and cultural **heritage**, which may contribute to its conservation.
- It leads to **cultural** exchanges between different countries.

At the same time, tourism has **negative effects**. For example:

- Seasonal and **precarious employment**. Jobs are often temporary and poorly paid.
- **Coastal tourism** has a strong impact on the environment and requires the construction of many new buildings. This causes the degradation of fragile ecosystems, pollution and increased pressure on natural resources such as water.
- There is sometimes a threat to a region's **local traditions**.

Growth of tourism

Tourism grew rapidly in the second half of the 20th century, when developed countries experienced a period of high economic growth, and tourism became a mass phenomenon.

In 2012, the volume of international tourists passed the figure of 1,000 million for the first time. There were also between 5,000 and 6,000 million domestic tourists, who did not leave their own countries.

ACTIVITIES

1 Find out about different types of tourism. Classify your information in your notebook.

Types of tourism	
Leisure tourism	......
Tourism for personal reasons	......
Tourism for professional reasons	......

2 What is "sustainable tourism"? Look for information on the Internet.

Ecotourism in Phuket (Thailand).

NUMBER OF INTERNATIONAL TOURISTS

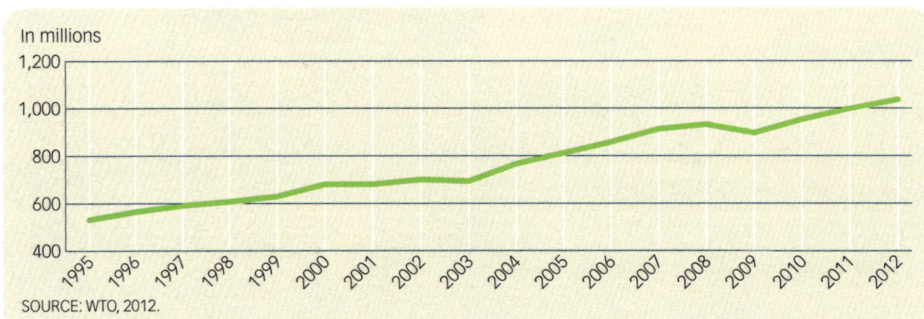

WORK WITH THE GRAPH

3 Describe the change shown on the graph. How do you explain it?

4 Why do you think the number of tourists fell in 2009?

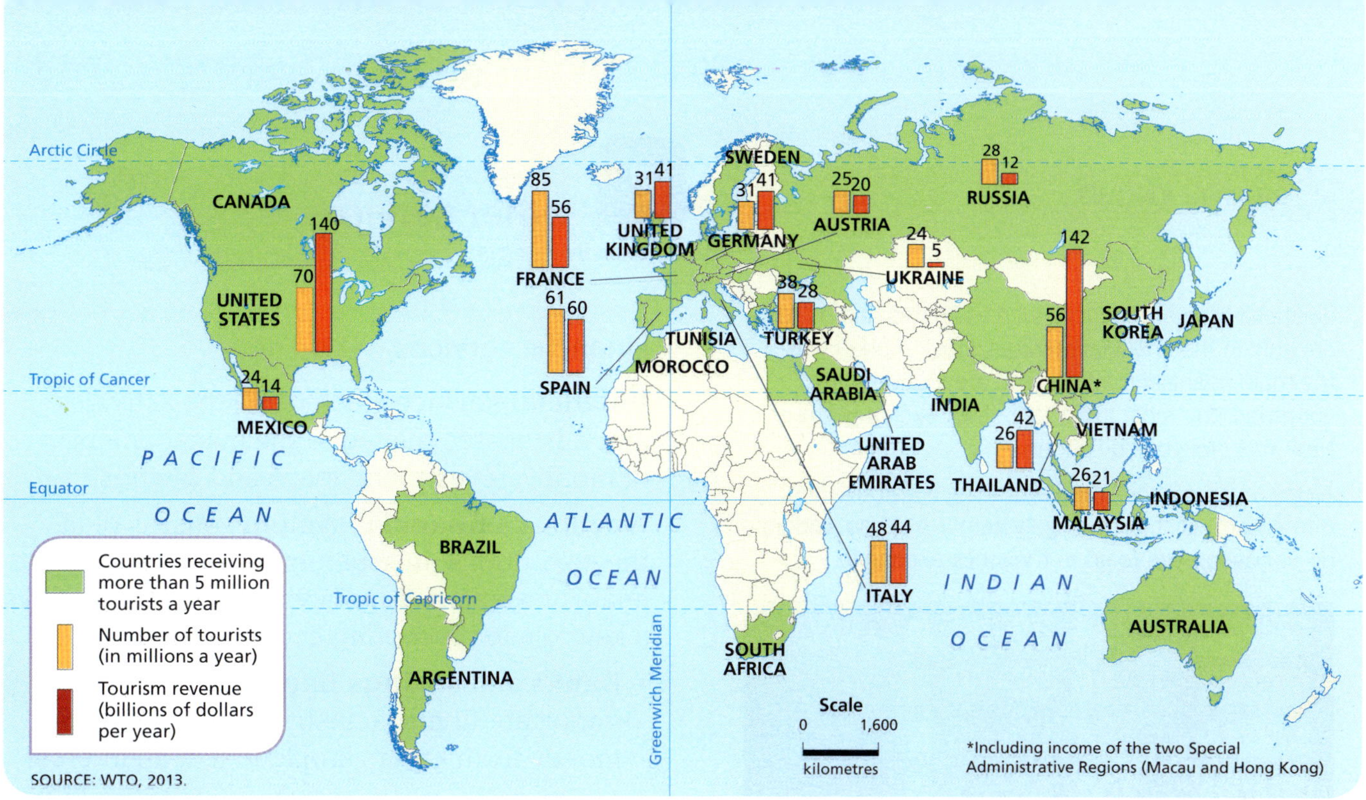

WORLD TOURISM

International tourist flows

Until recently, nearly all international tourists came from developed countries. But today, the inhabitants of emerging countries also practise international tourism.

Europe is the main tourist destination in the world. France, USA, Spain and China are the countries which receive most foreign visitors and income.

- **Europe** received 52% of all international tourists in 2013, when it received 42% of world tourist revenues. Europe's artistic and cultural heritage, and its Mediterranean climate and beaches, were among its attractions. France and Spain were the main destinations.
- **The Asia-Pacific region** grew the fastest in 2013, when it received 23% of all tourists as well as 31% of world revenues. Asian culture is a great tourist attraction for western visitors. China, Thailand and Malaysia are the leading destinations.
- **America**. The United States has many attractions, such as its cities and natural spaces. Mexico's cultural heritage and beaches also attracts tourists.
- **Africa**. The natural environment (rainforests and deserts) and lack of infrastructure limits the growth of tourism. Morocco, South Africa and Tunisia are leading destinations.
- **Middle East**. Tourism has declined as a result of instability, which has affected countries like Egypt and Syria.

WORK WITH THE MAP

5 In terms of the number of visitors, which countries are the main tourist destinations in each region?

6 Why do you think France receives more tourists than China, but visitors spend more money in China than in France?

ACTIVITIES

7 Look on the Internet for more information about why tourism to the Middle East has fallen. How long do you think it will take countries like Syria to recover?

8 Use the Internet (or go to your local Tourist Information Office) and find out about your own region. What are the main types of tourism?

DID YOU KNOW?

The Stock Exchange

Financial assets (products such as shares and bonds) are sold in the financial markets. Stock markets are the main financial markets, and are found in most countries. The most important stock exchanges are New York, Tokyo and London.

Today much of stock market trading is through the Computer Assisted Trading System (CATS), so traders do not even have to go to the stock exchange.

Tokyo Stock Exchange.

INTERNATIONAL FINANCIAL MOVEMENTS

SOURCE: World Bank, 2012.

WORK WITH THE CHART

1. How did financial movements evolve between 1995 and 2012?
2. In your opinion, why did financial movements drop suddenly around 2007?

7 Other services

Financial services

Financial services include all the activities and businesses related to the management of money: investment, loans and savings.

A country's **financial system** consists of all the institutions, instruments and practices that enable transfers of money to take place. It plays a key role in promoting economic activity:

- **Banks** and **savings banks** are the most important financial institutions. They attract money from savers and lend it at an interest.
- **Insurance companies**. In exchange for regular payments, these companies cover risks to people and goods.
- **Investment funds**. A large number of people make contributions to these funds, which a management company administers.

Every country has **supervisory bodies** to control these institutions.

Information and communication services

We use different means of communication:

- **Person-to-person**. One person contacts another, for example by telephone.
- **Social communication or mass media**. Information is communicated to the general public, for example through the press.
- **The Internet** is both a person-to-person and a social means of communication.

Today, communication networks enable us to connect with the rest of the world in real time. **Information and Communication Technologies (ICT)** have led to many changes:

- **Business**. New ways of doing business include e-commerce.
- **Work**. People use video conferencing and do teleworking from home.
- **Culture** and **leisure** activities have changed.

These technologies continue to grow but their coverage and the quality of their service vary considerably from one country to another.

INFORMATION AND COMMUNICATION TECHNOLOGIES IN THE WORLD

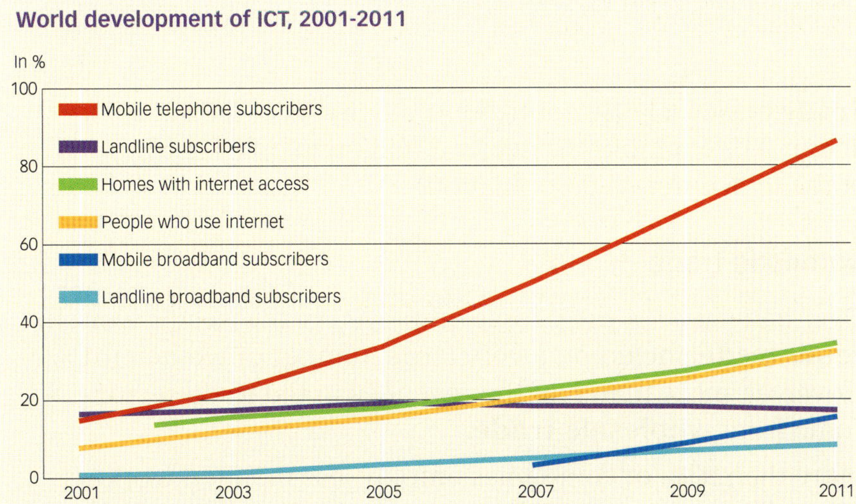

SOURCE: ITU (International Telecommunication Union).

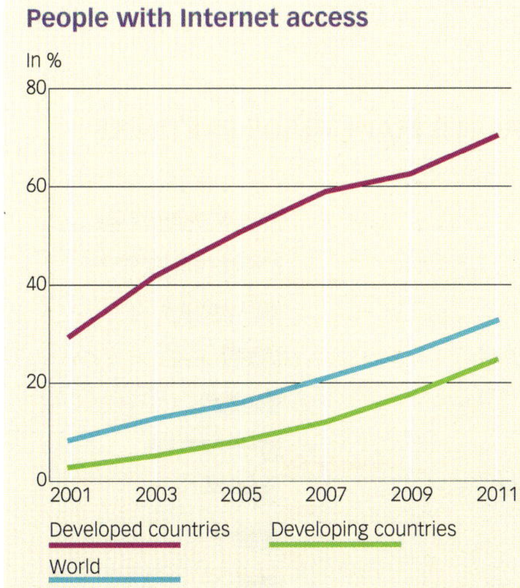

SOURCE: ITU (International Telecommunication Union).

Business services

These services are designed to meet the needs of businesses. They replace or complement services that used to be provided within companies:

- **Knowledge-intensive services**. Highly qualified people provide legal and fiscal advisory services, marketing, design, IT and training.
- **Operating services**. These are services like cleaning, security and transport.

These services make company operations easier and contribute to general economic growth. They improve the **competitiveness** of the business, as the companies can concentrate on their core business and commission other tasks from specialised companies or self-employed workers, who have the resources to perform them. This is reflected in higher quality work.

Personal services

Personal services are services which are provided directly for one person or a small group of people. They include:

- Domestic staff who work in people's homes.
- Assistance to the elderly at home or in centres.
- Child care at nursery schools.
- Maintenance and repair, for example by painters and electricians.
- Independent professional services from lawyers and architects.
- Other services, such as those provided by hairdressers, etc.

WORK WITH THE GRAPHS

3 Answer the questions.

a How has the use of mobile phones evolved?

b What about Internet access?

c Which variable has fallen? How can you explain this?

d What percentage of the population has Internet access in developed countries? And in less developed countries?

ACTIVITIES

4 Define the following:

- Financial services
- Financial system
- Mass media
- Business services

5 How have information and communication technologies revolutionised the world?

6 What personal services do you or your family use?

8 Trade, transport and tourism in Spain

SPAIN'S MAIN TRADE PARTNERS

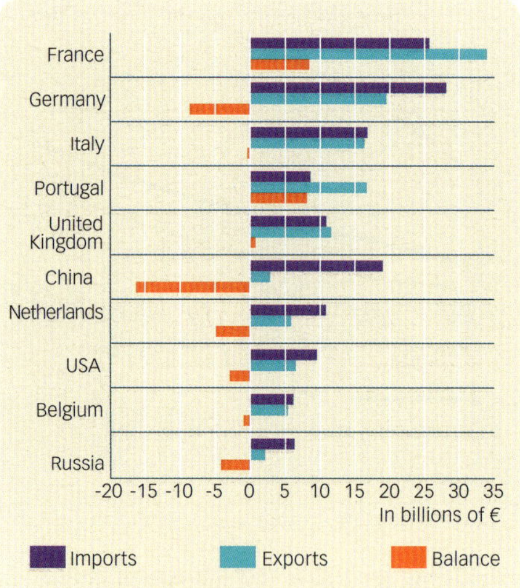

SOURCE: Tax authorities, 2010.

Trade makes a large contribution to Spain's GDP (32.3%), and in 2014 it employed nearly 3 million people (16.5% of the total employed population).

Domestic trade

Over half (56%) of the people employed in this sector work in **retail trade**, including a very large number of women (61%). However, more than half the revenues in the trade sector come from **wholesale trade**.

More than 95% of companies employ less than ten workers. Although small companies employ nearly half the workers, they only generate a third of total revenues in the trade sector.

In addition to traditional commercial activity and self-service, the following systems have grown rapidly:

- **E-commerce**. People shop on the Internet and through mobile devices.
- **Franchises**. A franchise is an agreement between a brand and an individual. The individual opens a shop under the brand's name and sells its products, in exchange for part of the revenues.

The trade sector was hit particularly hard by the economic crisis that started in 2007. People's spending power was affected by rising unemployment and falling salaries.

Foreign trade

Spain used to have a negative balance of trade or a trade deficit. This means that it imported more than it exported. In recent years, this trend has been reversed because exports have increased and imports have fallen.

Spain's main trading partners are France and Germany, and its largest trade deficit is with China. Spain imports petroleum, agricultural, metal and mineral products. It exports manufactured goods and agricultural, metal and mining products.

Transport in Spain

Spanish road and rail networks are better and denser in the areas with the largest populations and greatest economic activity.

- **Land transport** is radial with its centre in Madrid. Better roads and high-speed rail lines (AVE) have been developed.
- **Ship transport**. Spanish ports play a key role in trade between the EU and other parts of the world.
- **Air travel**. The leading airports are in Madrid, Barcelona and Palma.

WORK WITH THE CHART

1. With which region does Spain do most trade?
2. With which countries does it have a trade deficit? What does this mean?
3. With which country is Spain's balance of trade most positive?

Tourism

Tourism is a key economic activity in Spain, as it accounts for nearly 11% of the GDP and 12% of employment. Spain ranks third in the world for the number of foreign visitors.

This leading position can be explained by a number of factors:

- Excellent **beaches**. Spain is the country with the most blue flag beaches of the 49 countries involved in the blue flag project. Blue flags are given to beaches which meet high quality standards.
- **Government policies** promoting tourism.
- Large **investments** and improvements in infrastructures to meet tourists' needs.
- Spain's rich **cultural heritage**.

In the past, tourism in Spain was mainly beach tourism. However, since the 1990s great efforts have been made in other areas, such as business, or cultural and rural tourism.

FOREIGN TOURISM IN SPAIN

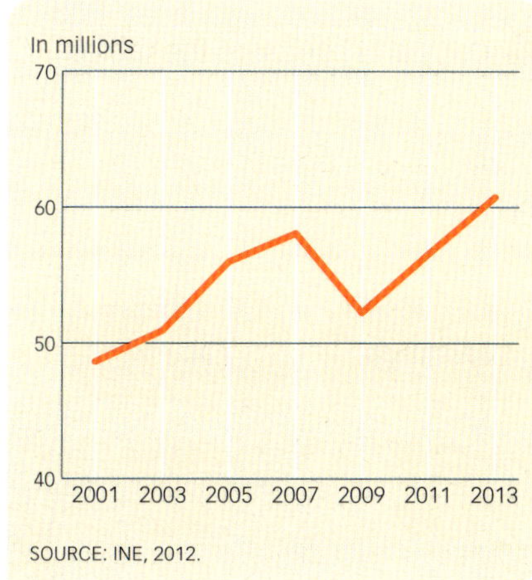

National and foreign tourism

National tourism includes travel in Spain (domestic tourism) and Spanish people's trips abroad (outbound tourism). Overall, around 160 million trips are made every year, although the figure fell after the crisis in 2007. The largest volume corresponds to **domestic tourism**, with 144.5 million trips in 2013.

As regards **outbound tourism**, Spanish people made 11.2 million trips abroad in 2013. Their main destination is Europe (77%), especially countries in the European Union and, amongst these, the nearest countries: France, Portugal and Italy.

As regards **foreign tourism**, Spain received **60.6 million foreign visitors** in 2013. They spent 59,082 million euros on tourism.

The main reasons for their visits are holidays and leisure (86%). The majority of tourists are from the United Kingdom, Germany and France. However, tourism from emerging markets, especially Russia and China, is increasing.

The **main destinations** for foreign tourists are Catalonia (nearly 26% in 2013), the Balearic Islands (18.5%), the Canary Islands and Andalusia.

WORK WITH THE MAP

4 Study the map with a partner and discuss why some regions receive more tourists than others.

MAIN TOURIST AREAS IN SPAIN

ACTIVITY ROUND-UP

1 Copy and complete the tables.

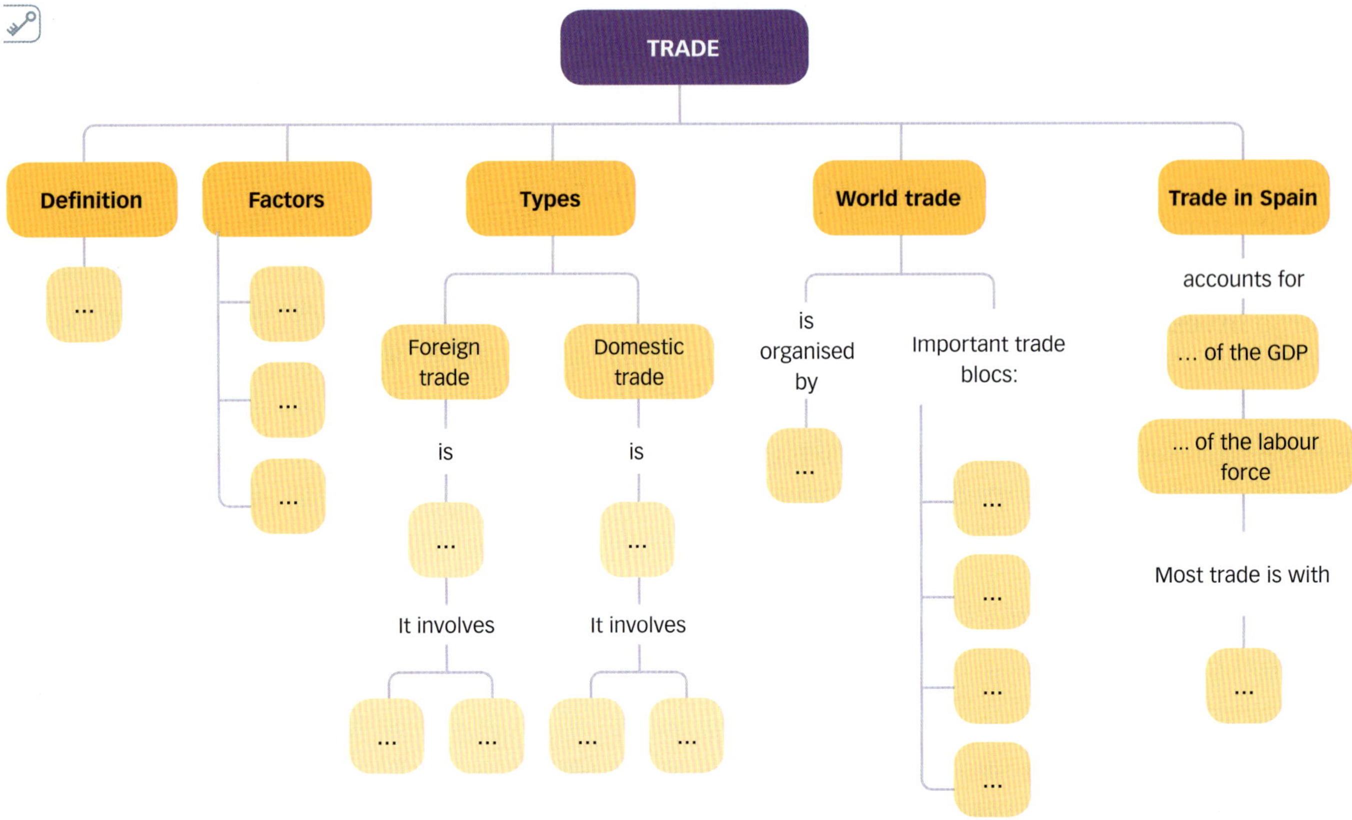

	TRANSPORT					
	Means of transport	**Infrastructures**	**Types**	**Advantages**	**Problems**	**Main networks or routes**
Land	...	...	...	...	...	...
Ship	...	...	...	...	...	...
Air	...	...	...	...	...	...

2 Answer the questions. Explain your answers.

a Are only goods exchanged in trade?
b Are hospitals private or public services?
c What is a trade bloc?
d What is a trade surplus?
e What are the advantages of having different means of transport?
f What is the relocation of services?

3 Explain the difference between the following:

- A free trade area and a customs union
- Person-to-person communication and mass media

4 Form small groups and investigate on the Internet.

a Choose three countries from three different continents.
b For each country, find out about:
- the size of the population
- the means of transport
- the country's per capita GDP

c Compare the three countries' transport networks. In your opinion, how does the wealth of each country affect its transport networks?
d Present your findings to the class.

5 Look at the photo and answer the questions.

a Why are most service activities concentrated in cities?

b In your opinion, how has the fact that Madrid is the capital of Spain influenced the development of the service sector?

c What services are there in large cities, such as Madrid and Barcelona, which are not found in smaller towns? Why?

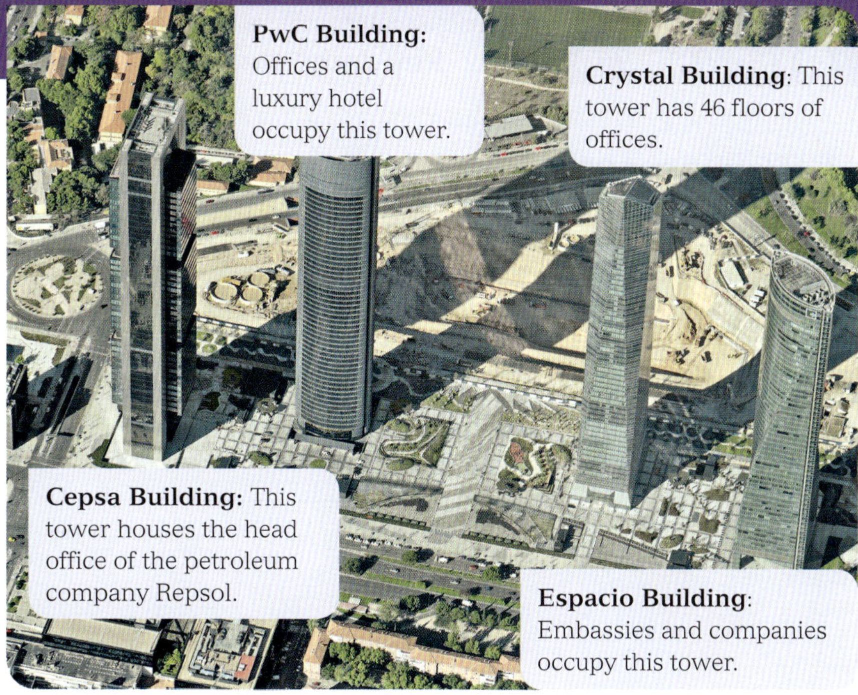

Four Towers Business Area (Madrid).

FAIR INTERNATIONAL TRADE RELATIONS

One of the challenges of the 21st century is to achieve fairer trade relations. The production of goods and raw materials has moved to developing countries. However, many of these countries are not progressing and extreme poverty still exists there. Why?

A

Unfair trade

- Large companies, generally multinational firms, dictate conditions to the workers and the local companies they subcontract.
- Some companies buy raw materials as cheaply as possible. Occasionally, they speculate with the price of raw materials.
- They reduce labour costs, and their workers live in very poor conditions. They try to prevent any attempts to draw up union laws in the countries. They sometimes employ children.

6 Form groups. Read texts A and B and answer the questions. Then present your answers to the class.

a Why is it cheaper for large multinational companies to produce their goods in developing countries?

b How does this affect the inhabitants of the developing country?

c Is there a way of making a product, such as trainers, in a developing country but in fair conditions? If so, how can it be done?

B

Fair trade

- Would it be possible to establish other types of companies (either family or local) with different objectives?
- Prices are determined by the international markets – is there an alternative?
- Why are the workers in less developed countries treated differently from workers in developed countries?
- What are the minimum conditions that we should demand for every worker?

A Fair Trade shop sells products from less developed countries at prices that allow the local producers to make a profit.

A CASE STUDY. **Analyse a tourist resort: Benidorm (Alicante)**

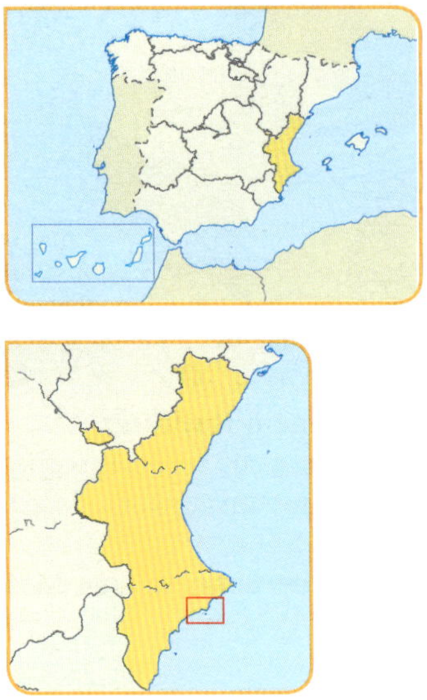

Localisation of Benidorm.

POPULATION

- **1883:** 3,150 inhabitants
- **1960:** 6,259 inhabitants
- **2014:** over 69,000 inhabitants; 400,000 inhabitants in summer

TRANSPORT

- **Motorway:** AP-7
- **Main road:** N-332
- **Airport:** Alicante-Elche (El Altet), 20 km from Benidorm

Timeline of the growth of tourism in Benidorm.

Climate	J	F	M	A	M	J	J	A	S	O	N	D
Average temperature (° C)	12	12	14	16	19	23	25	26	24	20	15	12
Rainfall (mm)	21	20	20	27	28	10	4	5	40	46	34	22
Hours of sunshine	181	180	227	247	277	302	330	304	250	217	173	164
Sea temperature (° C)	14	13	14	15	17	21	21	25	24	21	18	15

1 Location.

- Where is Benidorm?
- What province is it in?
- Find out the name given to this coast.

2 Positive factors for tourism.

- Look at the photographs and analyse the information about Benidorm. Make a list of the natural and human factors that favour tourism in Benidorm.

3 Are the following statements true or false? Listen to the recording and check your answers.

a Some tourists arrived in Benidorm in the 19th century.
b The population then was 7,000 people.
c The population rose by 69,000 people between 1883 and 1960.
d The tourist boom started in 1970.
e Today Benidorm has over 400,000 inhabitants in the summer.

4 The impact of tourism on the population.

a Is there a relationship between the development of tourism and the number of inhabitants? Why is this?
b How do you think tourism has affected the culture and customs of the local population?

5 The impact of tourism on the economy.

a Which economic activities were important in Benidorm before the tourist boom?
b What activities and economic sectors have benefited from tourism in Benidorm?

6 The impact of tourism on the environment.

- Describe the landscape in the two photographs. How has it changed?
- What type of buildings predominate?
- How do you think the increase in population has affected natural resources such as water?

1967

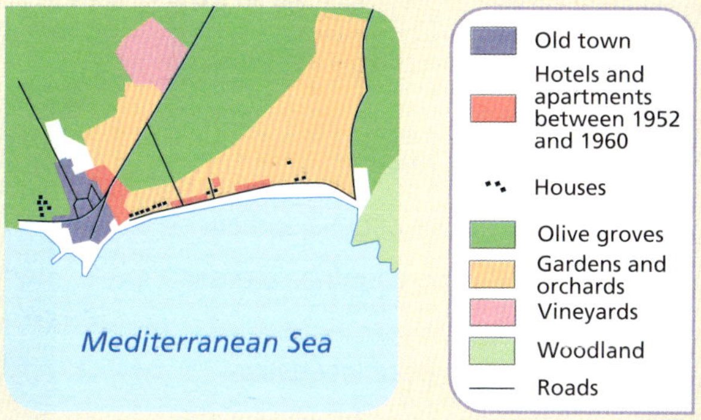
Mediterranean Sea
Old town
Hotels and apartments between 1952 and 1960
Houses
Olive groves
Gardens and orchards
Vineyards
Woodland
Roads

2013
Mediterranean Sea
Old town
Hotels and apartments
Other buildings
Abandoned fields
Empty space
Olive groves
Roads

7 Development

FIND OUT ABOUT:

- The nature of development
- Why some countries are less developed
- The causes of inequality
- How to promote international development
- Poverty and social exclusion in Spain

KNOW HOW TO:

- Classify a country in terms of its human development
- Compare the development of two countries
- Assess development in Africa

Is education a universal right?

Education is an essential means to fight poverty and inequality. However, in 2011, 57 million children did not go to primary school. Another 70 million young people, especially girls, did not continue their studies at secondary level.

COUNTRIES WITH THE HIGHEST ILLITERACY RATES

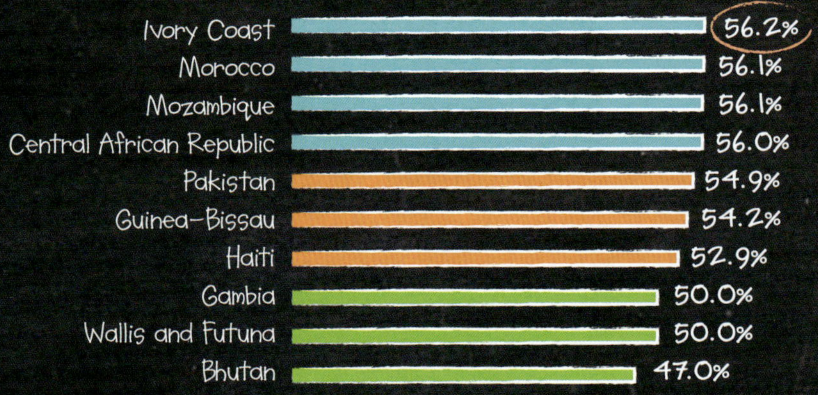

SOURCE: www.indexmundi.com

COUNTRIES WITH THE HIGHEST INVESTMENT IN EDUCATION (In % of GDP)

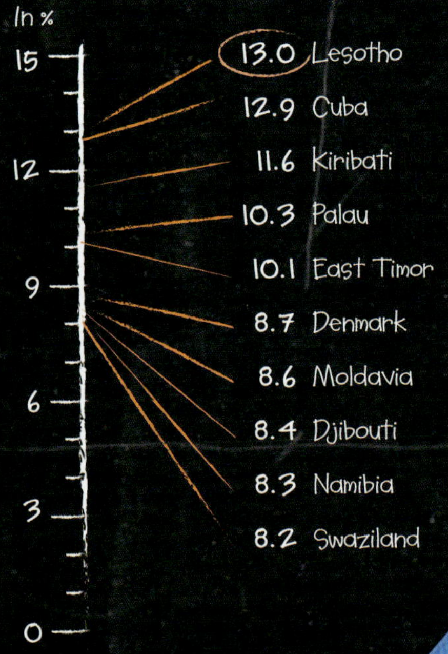

SOURCE: CIA, *The World Factbook*.

CHILDREN WHO DO NOT GO TO SCHOOL 1999–2010

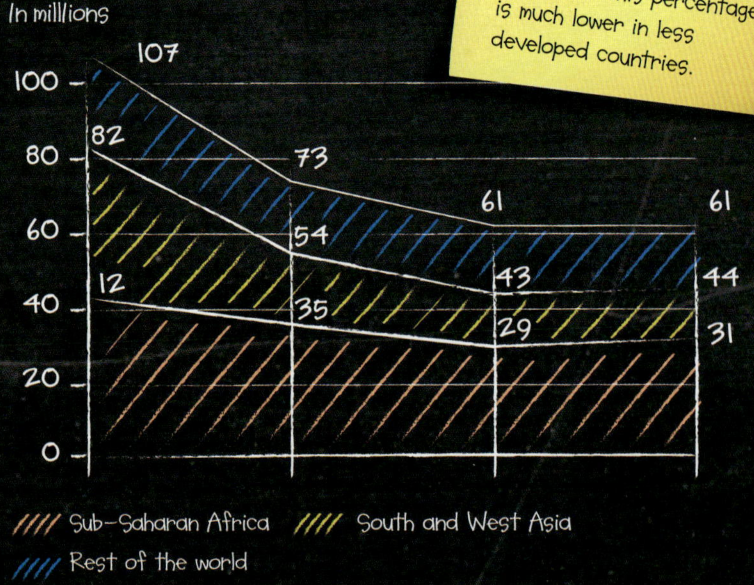

84% of the world's population can read and write. But this percentage is much lower in less developed countries.

SOURCE: Unesco, *Global Monitoring Report on Education (EFA)*, 2012.

China

Mexico

Cameroon

Equality of the sexes is a basic human right.

? WORK WITH THE IMAGE

- Why is the literacy rate higher in Europe and America than in Africa?
- Do you agree that education reduces poverty and inequality? Why?
- Is the universal right to education a real possibility in the near future?

HOW DO WE KNOW?

International organisations, as well as some NGOs, publish annual reports on aspects of life around the world.

- Use an Internet search engine to find "Teaching and learning: EFA global monitoring report". Which organisation prepared this report? What are its objectives in the field of education?

1 What is development?

COUNTRIES WITH HIGH AND LOW HDI

THE TEN COUNTRIES WITH HIGHEST HDI			
1. Norway	0.95	6. New Zealand	0.91
2. Australia	0.93	7. Ireland	0.91
3. United States	0.93	8. Sweden	0.91
4. Netherlands	0.92	9. Switzerland	0.91
5. Germany	0.92	10. Japan	0.91

THE TEN COUNTRIES WITH LOWEST HDI			
1. Niger	0.30	6. Mali	0.34
2. Dem. Rep. of Congo	0.30	7. Eritrea	0.35
3. Mozambique	0.32	8. Central African Republic	0.35
4. Chad	0.34	9. Guinea	0.35
5. Burkina Faso	0.34	10. Burundi	0.35

WORK WITH THE TABLE

1 Analyse the table.

a Classify the ten countries with the highest HDI by continent. Then classify those with the lowest HDI.

b Which continents are most represented in each category?

How do we measure human development?

The development of a country is calculated by taking into account its wealth, and other factors such as life expectancy, the population's access to health care and the literacy rate. It can be measured in different ways:

- **Human Development Index (HDI)**. This was developed by the United Nations and is published annually. It evaluates social and economic aspects on a scale from 0 (lowest human development) to 1 (highest human development).
- **The Human Poverty Index (HPI)** measures social deprivation in three areas:
  - **Life expectancy**. This measures the likelihood at birth of not surviving to a given age. In less developed countries, this age is 40 years but it is much higher (60 years) in developed countries.
  - **Adult illiteracy rate**.
  - **A decent standard of living**. This takes into account the percentage of the population without access to drinking water, and the percentage of underweight children.

Countries according to their development

The United Nations report of 2013 indicated an improvement in human development, especially in less developed countries. However, there are still great economic and social differences:

- **Developed countries** have a high per capita income which favours consumption and economic growth. They have advanced industries and modern infrastructures. Quality services are available.
- **Emerging countries** are acquiring a level of production and wealth which will enable them to join the group of developed countries. They are modernising their economies through industrialisation, agricultural transformation and development of services.

  In emerging countries, infrastructure and services are still deficient and educational levels are low. They depend on more developed countries for technology and consumer goods.
- **Less developed** countries have low per capita income and low levels of consumption. Infrastructure and services are scarce and unavailable to most of the inhabitants. Life expectancy and educational levels are low. The prospects for economic development in the near future are poor. Most of these countries are in Sub-Saharan Africa.

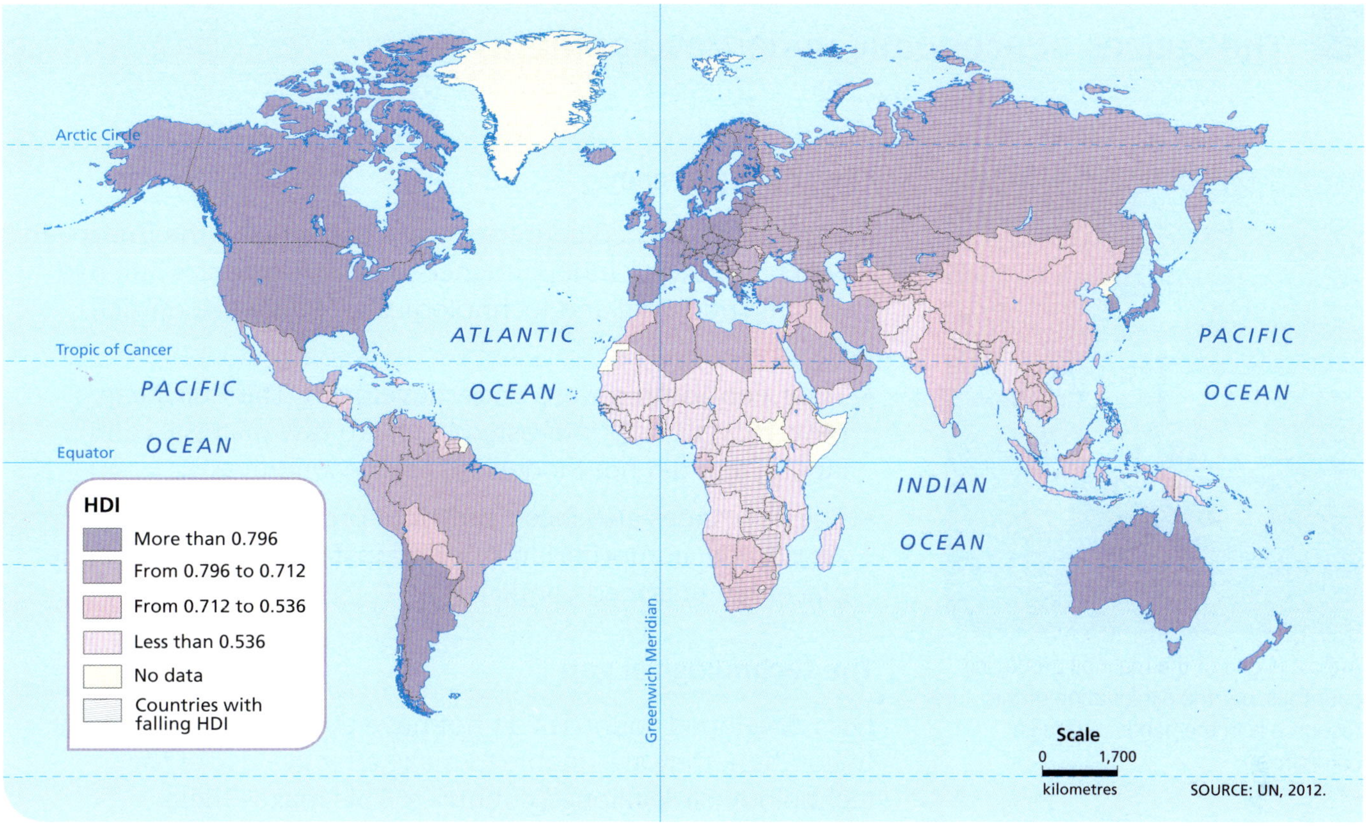

COUNTRIES BY HUMAN DEVELOPMENT INDEX (HDI)

Classification by HDI

The United Nations Development Programme (UNDP) classifies countries into four groups according to their HDI:

- Countries with **very high** human development. Their HDI is above 0.796.
- Countries with **high** human development. Their HDI is between 0.796 and 0.712.
- Countries with **medium** human development. Their HDI is between 0.712 and 0.536.
- Countries with **low** human development. Their HDI is below 0.536.

ACTIVITIES

2 How can we measure a country's level of development?

3 How are countries grouped according to their development? Describe each group.

4 The country of Qatar is fourth in the world in wealth and production, but it is 36th according to its HDI. Is the wealth of a country reflected in the quality of life of its inhabitants? Explain your answer.

WORK WITH THE MAP

5 The United States and China have the highest GDP in the world. Do they both have high HDI?

6 How do you explain the difference in their HDI?

7 Are the countries with the highest HDI also developed countries? Why?

8 Which countries have a lower HDI today than ten years ago? Why?

9 Look on the Internet for statistics about three countries with high, medium and low HDI. Look at their GDP per capita, life expectancy and literacy rate. Compare your findings with a partner's, and draw conclusions together.

2 The causes of economic underdevelopment

Angola is one of the main oil producing countries, but the exploitation of this resource is in the hands of foreign companies.

The colonial legacy

Many less developed countries were **colonies** of the European powers. Following independence, many ex-colonies still had financial, industrial and technological dependence on rich countries.

- Their **economies** were underdeveloped. The European powers were only interested in cheap raw materials and labour, and did not modernise their economies.
- The new states also faced **political** problems. There was corruption, authoritarian rule and instability which in some places led to armed conflicts.

The technological gap

Less developed countries do not have their own **technology**. In addition, they are unable to make effective use of the technology of developed countries. This makes them uncompetitive in the world market.

Foreign companies usually control the activities which generate economic growth. These companies exploit natural resources and carry out some of their economic activities in these countries because it is cheaper.

CORRUPTION AROUND THE WORLD

Population

In less developed countries, the **population grows** at a faster rate than resources. This means that income is insufficient to satisfy the basic needs of the population.

In addition, **disease** and **famine** have a serious long-term impact. Undernourished and sick adults may have to leave their jobs, and children sometimes leave school in order to do the work of their parents. An **unskilled labour** force is an obstacle to development.

External debt

External debt is the money that a country owes to other countries or international institutions, such as the World Bank or the International Monetary Fund (IMF).

- Some countries do not use the money that they have borrowed to promote their development.
- They sometimes find it difficult to pay the money back. If their economies do not grow, they remain in debt.

Less developed countries have less external debt than developed countries, but they find it more difficult to repay their debt. In consequence:

- They ask for new credits, which increases their debt.
- They reduce investment in the modernisation of their economies.
- They overexploit their natural resources in order to increase exports.
- They reduce public spending on basic services like health and education.

The World Bank and the IMF have created programmes to alleviate the debt of **Heavily Indebted Poor Countries (HIPC)**. Some developed countries have cancelled all or part of this debt.

Armed conflicts

Economic interests (such as the control of mineral resources) and ethnic and religious differences have led to armed conflicts. Some of them have lasted for many years. They can be devastating.

EXTERNAL DEBT

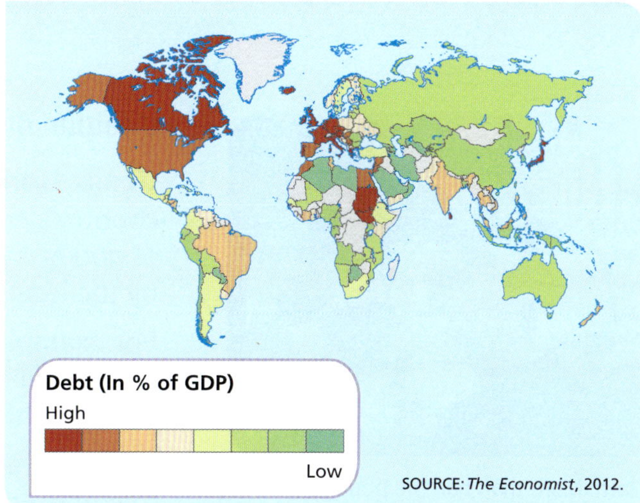

SOURCE: *The Economist*, 2012.

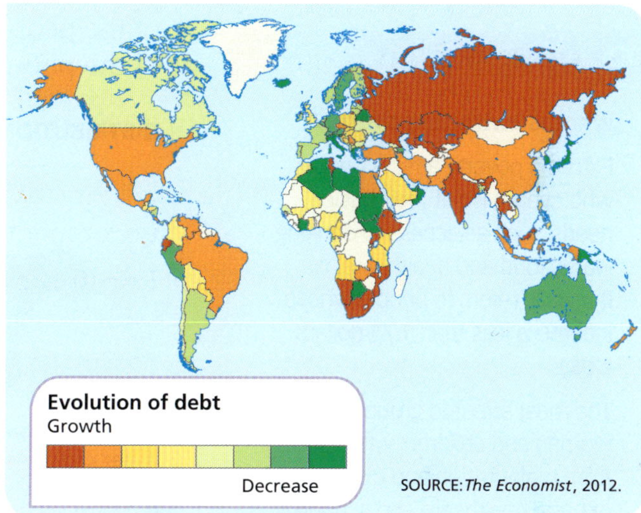

SOURCE: *The Economist*, 2012.

WORK WITH THE MAPS

1. Which countries have most debt? Which have least? Find them on an atlas.
2. Which countries have the highest growth in their external debt?

ACTIVITIES

3. With a partner, discuss how each of the following factors prevents the development of less developed countries:
   - High external debt
   - High population growth
   - Low level of education
   - Lack of technology
   - AIDS
   - Corruption

3 Types of inequality

Haiti. More than half the population lives below the national poverty line.

THINK ABOUT IT

What is extreme poverty?

Extreme poverty affects people who cannot meet their basic needs, such as access to drinking water, food and healthcare. It usually refers to people whose income is less than 1.25 dollars a day.

The most affected groups are women and children, who suffer from malnutrition and disease, lack of basic healthcare and schooling.

- In groups, investigate poverty in one of the less developed countries in Central or South America, Africa or Asia. Think of measures for overcoming extreme poverty, and explain your proposals to the rest of the class.

WORK WITH THE MAP

1. With a partner, discuss which parts of the world have high GDP, and which regions have low GDP per capita. Are there any exceptions or surprising differences? How do you explain them?

Unequal distribution of wealth

Gross Domestic Product (GDP) is often used to measure a country's economic development. **GDP per capita** is the value of goods and services produced in a country in a year divided by the number of inhabitants.

For example, the inhabitants of France have a GDP of 40,000 dollars per capita and are therefore wealthier than the inhabitants of Haiti, whose GDP is under 800 dollars per capita.

GDP per capita shows the wealth of a country's people but not the differences among them. These differences are greater in countries which do not redistribute their wealth. In Brazil, 10% of the people own 75% of the country's wealth, while 8% of the population lives in extreme poverty.

DISTRIBUTION OF GDP PER CAPITA AROUND THE WORLD

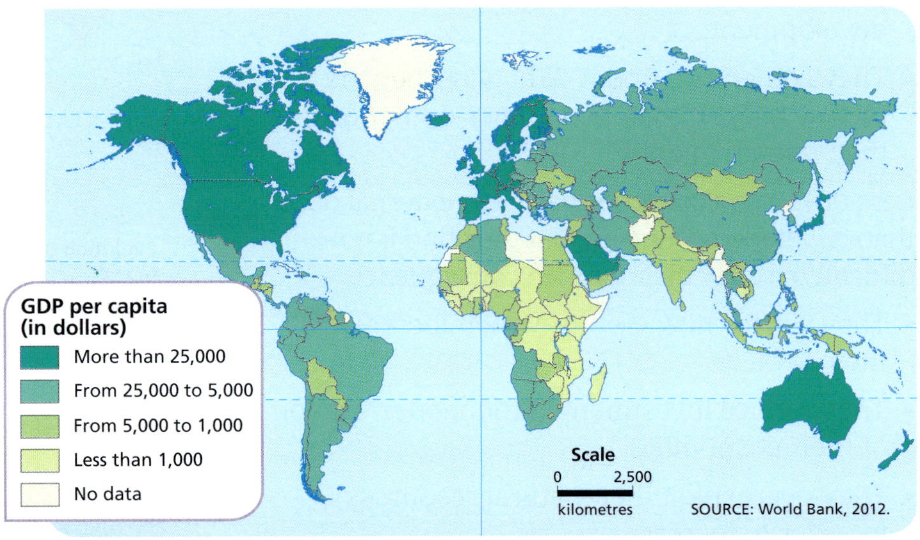

Contrasts in consumption

While there is excessive consumerism in rich societies, poor societies live near subsistence levels. This is reflected in consumption of food, consumer goods and energy.

- **Food consumption**: in developed countries, people consume many more calories, and some countries suffer from an 'epidemic' of obesity. In contrast, in many less developed countries a high proportion of the population suffers from **malnutrition**.
- These differences are even greater in the consumption of expensive **goods and services**.
- **Energy consumption** also varies considerably. A small number of countries consumes most of the world's energy.

Inequality in health care

- **Illnesses**. In less developed countries, millions of people die every year from illnesses (like AIDS or malaria) which could be prevented or cured. Almost 70% of the world's AIDS victims live in Sub-Saharan Africa, and about 15% of these are children. However, the number of new cases has dropped in recent years.
- **Maternal mortality**. In less developed countries, many women die from complications during pregnancy and birth. Only half of pregnant women receive prenatal care, and a third of mothers give birth without medical aid. South Asia and Sub-Saharan Africa are the areas with the highest level of maternal mortality. However, the global rate of maternal mortality has fallen by 47% in the last twenty years.
- **Infant mortality**. Since 1990, the rate of global infant mortality has fallen from 59 deaths per thousand inhabitants to 42 in 2011. Even so, 7 million children under 5 die every year. Many of them die from preventable illnesses such as pneumonia, diarrhoea and malaria.

Social inequality

- **Gender inequality**. Inequality between men and women has major consequences in emerging and less developed countries. In some parts of Asia and Africa, **discrimination** against women deprives them of legal rights and basic human rights. Women often have badly paid jobs.
- **Educational opportunities**. 84% of the world's population of 15 years and over is **literate**, that is, able to read and write. However, in less developed regions, there are countries with much lower literacy rates. For example, in Mali only 31% of people over 15 can read or write. Today, 57 million children in the world do not go to school, and more than half of these are girls.
- **Child exploitation**. There are 215 million children in the world who work instead of going to school. Most of these children do unpaid agricultural work to help their families.

EXPENDITURE ON HEALTH

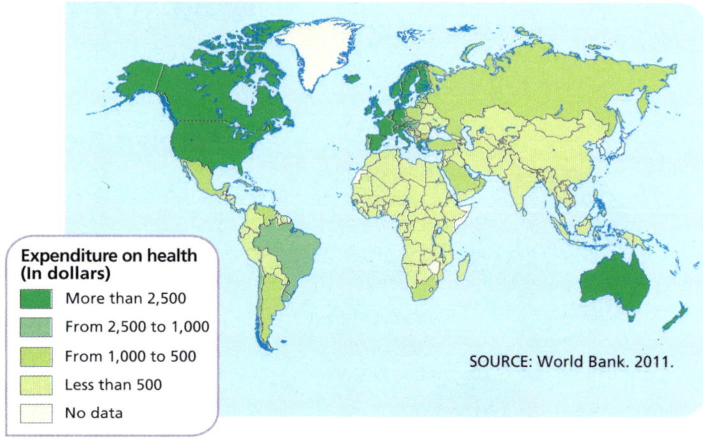

WORK WITH THE MAPS

2 Look on the Internet for information about the GDP of five countries on different continents. Does it match their expenditure on health?

3 Think about children in a country where health expenditure is low. What services do they fail to receive?

DID YOU KNOW?

Child soldiers

The Convention on the Rights of the Child (2000) prohibits the participation of children in armed conflicts.

However, Unicef estimates that around 30,000 children worldwide are recruited as soldiers by governments and guerilla groups. A third of these child soldiers are found in Africa, in countries such as the Central African Republic, Mali, Chad and the Democratic Republic of Congo.

Many of these children are kidnapped. If they resist or try to escape they are beaten or killed. There are also children, often orphans, who volunteer to join armies to escape extreme poverty.

4 Overcoming underdevelopment

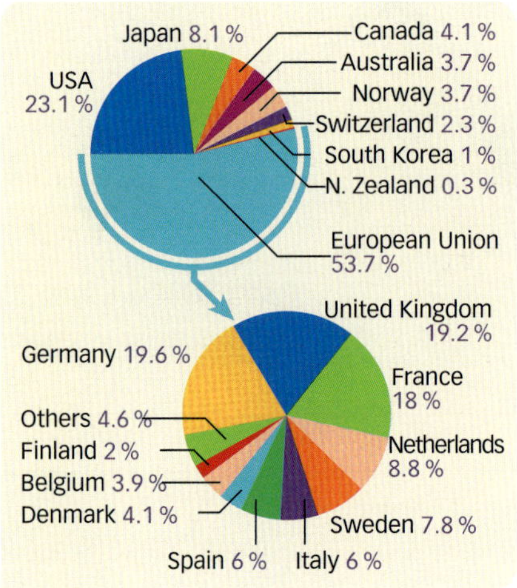

Contributions by the Development Assistance Committee (DAC) to Official Development Assistance (ODA).

WORK WITH THE GRAPHS

1. Which five countries contribute the most? Are they the same countries as those which contribute 0.7% of their GDP to development?
2. What position does Spain occupy?

Red Cross. This organisation has provided humanitarian aid to people affected by armed conflict for over 150 years.

Foreign Aid

Less developed countries receive foreign aid from three main sources: international organisations, developed countries and NGOs.

- **International organisations**.
  Organisations like the **United Nations (UN)** and the **Organisation for Economic Cooperation and Development (OECD)**, contribute to world development. They provide credits at low interest, donate money after natural disasters and promote commercial treaties.

  The **International Monetary Fund (IMF)** provides loans to countries in economic difficulties. However, countries are often required to reduce government spending as a condition for receiving loans. This forces them to reduce social services.

- **Direct aid from developed countries**.
  Some developed countries contribute directly to less developed countries. In 1980, the most advanced countries agreed to donate 0.7% of their GDP to foreign aid. So far, only Denmark, the Netherlands, Luxembourg, Norway and Sweden have met this target.

  Some countries have pardoned all or part of their foreign debt to promote development. For example, in 2007 Spain pardoned external debt in eleven countries.

- **Aid through NGOs**.
  An NGO (Non-governmental organisation) is a private, non-profit organisation.

  Many NGOs organise humanitarian aid and development projects in the poorest countries. They rely on volunteers to carry out projects to cover needs which are not met by international institutions.

  The Red Cross, Oxfam and Doctors without Borders are some of the most important NGOs. Their projects are financed with the economic aid of private individuals and companies.

ACTIVITIES

3. Who provides foreign aid to less developed countries? How is it provided?

Government management

Government management should be transparent and accountable. Effective governmental administration attracts foreign aid and investments. This promotes economic growth.

The **World Bank** and the **European Union (EU)** base their aid on reforms to government. They grant aid if a country reduces corruption, ensures free competition, respects private property and freedom of information and establishes effective administration. The EU also assesses these aspects before it establishes economic and trade agreements with other countries.

PERCEPTION OF WORLD CORRUPTION	
Countries with greatest corruption	**Countries with least corruption**
Somalia	Denmark
North Korea	New Zealand
Sudan	Finland
Afghanistan	Sweden

Source: *Corruption Perceptions Index (CPI)*, by Transparency International, 2014.

Fair trade

The **World Trade Organisation (WTO)** establishes the rules for international trade. Wealthy countries have great influence in the organisation. The globalisation of the economy has led to the liberalisation of international markets, which has had a negative effect on less developed countries.

- The developed and emerging countries sell more **raw materials** and **agricultural products** than less developed countries because their production systems are more efficient. Consequently, salaries stay low in less developed countries, causing deprivation.
- Many of the exports of less developed countries are in the hands of **multinational companies** based in developed countries. They often control the exports of raw materials and agricultural products. As a result, less developed countries fail to benefit directly from their own natural resources.
- International agreements to protect the **environment**, such as the Kyoto Protocol, commit countries to reducing pollution, respecting biodiversity, and exploiting resources in a sustainable way. However, this affects the ability of less developed countries to compete because production costs are increased through the application of "clean" technology.

Developed countries need to **change the rules** to make trade fairer, and allow poorer countries to achieve effective development. They need to:

- Eliminate **high customs duties** on products from less developed countries.
- **Reduce subsidies** on their own agricultural products, and so make it possible for less developed countries to compete.
- Pay a **fair price** for basic products, like coffee, cotton and sugar, which workers in less developed countries depend on.

WORK WITH THE TABLE

4 With a partner, study the table on this page and the map on page 132. Is there a connection between corruption and economic underdevelopment? If so, why are they related?

ACTIVITIES

5 Why is efficient government in less developed countries essential for their development?

6 What changes should there be in the organisation of world trade in order to promote development?

5 Poverty and exclusion in Spain

THINK ABOUT IT

What is poverty?

Poverty is a relative term. It is not measured in the same way in a developed country as in a less developed country.

In EU countries, a person is considered to be poor if their income is less than 60% of the average income in their country.

Poverty and social exclusion

Poverty is not found exclusively in less developed and emerging countries. In wealthy countries, too, many people live in poverty and **social exclusion**, which means they are deprived of basic rights like housing, employment and healthcare.

In Spain, 20.4% of the population live below the poverty line, with incomes which put them at risk of social exclusion.

Poverty is worse in some areas than others. In 2011, the Autonomous Communities of the Canary Islands (33.8%), Extremadura (31.9%), Andalusia (31.7%) and Castile-La Mancha (31.7%) had the highest percentage of people at risk of poverty.

However, extreme poverty is not widespread in Spain. The state provides many people with the means to meet their basic needs. Thus, people receive retirement pensions, disability or unemployment allowances, and public education and healthcare.

PEOPLE AT RISK OF POVERTY

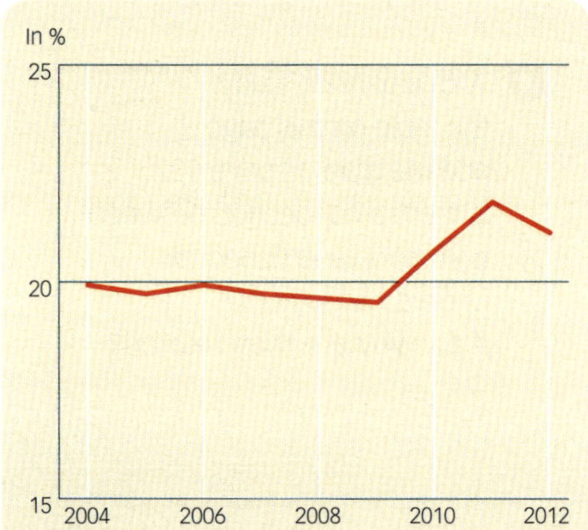

SOURCE: INE, 2012.

THE AT-RISK-OF-POVERTY RATE BY AUTONOMOUS COMMUNITY

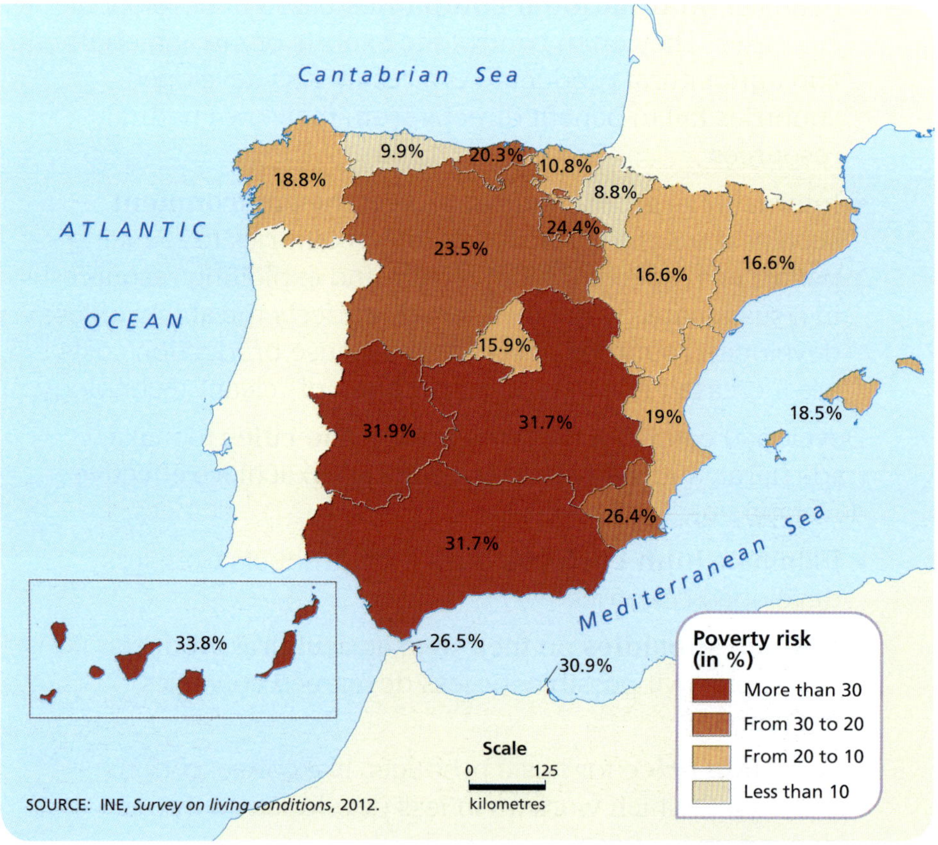

SOURCE: INE, *Survey on living conditions*, 2012.

ACTIVITIES

1. In groups, prepare a report on poverty in Spain using the map and graphs. Examine:
   - The evolution of poverty risk.
   - The differences in the degree of poverty between Autonomous Communities, age groups and nationalities.

The effects of the economic crisis

There was an increase in poverty in Spain following the beginning of the economic crisis in 2007:

- **Unemployment** grew and salaries fell or stayed at the same level. As a result, the income of Spanish families fell.
- The state **reduced spending** in order to meet its debts. This led to a marked fall in expenditure on social services.

Who is at risk of poverty?

In Spain, **unemployed** people and **immigrants** from outside the European Union are especially at risk of poverty. The at-risk-of-poverty rate is more than 47% in the latter group.

The at-risk-of-poverty rate among people over 65 fell from 21.7% in 2011 to 12.7% in 2013. On the other hand, the at-risk-of-poverty rate grew from 19.3% to 20.8% among people between 16 and 64.

The at-risk-of-poverty rate is also linked to people's educational level. It is 10% for people who have completed secondary education or higher studies, but rises to 28.9% for people who have only had a primary education.

DID YOU KNOW?

The key role of grandparents

In Spain, there are more than 8 million people aged over 65. When the economic crisis began in 2007, their pensions became vital to many Spanish families. Today, pensions are the main source of income for one fifth of Spanish households. For more than 300,000 families, they are the only source of income.

In homes where everyone is out of work, there are often three generations living together – grandchildren, parents and grandparents. By sharing their resources (pensions and unemployment subsidies), they are able to make better use of their money.

EVOLUTION OF AVERAGE INCOME

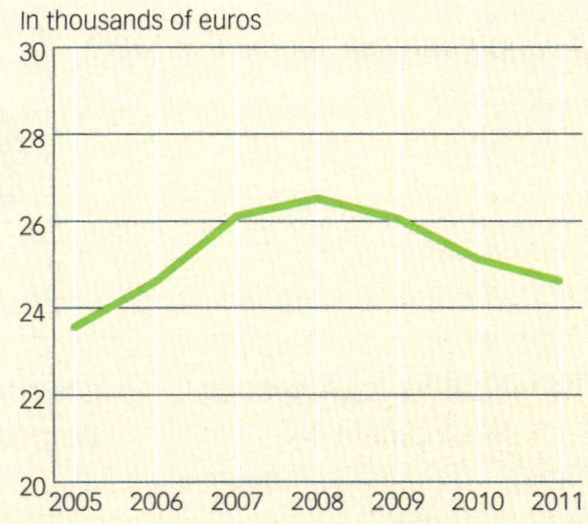

SOURCE: INE, *Survey on living conditions, 2012.*

STATE INCOME AND EXPENDITURE ON SOCIAL PROTECTION

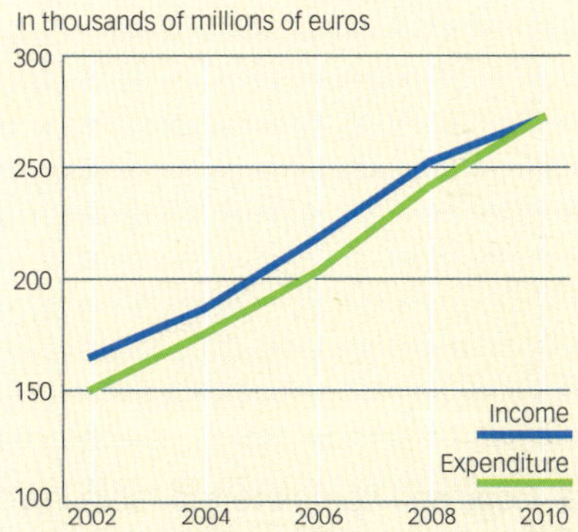

SOURCE: Ministry of Employment and Social Security, 2012.

ACTIVITIES

2 What is social exclusion?

3 Which of the following people could be considered poor or at risk of poverty in Spain?

- A 75-year old widow with no pension.
- An unemployed single person receiving unemployment benefit.
- A single mother with two children who works and receives an average salary.

Discuss your ideas in groups and think of two more examples. Then share your ideas with the rest of the class.

ACTIVITY ROUND-UP

1 Copy and complete the table.

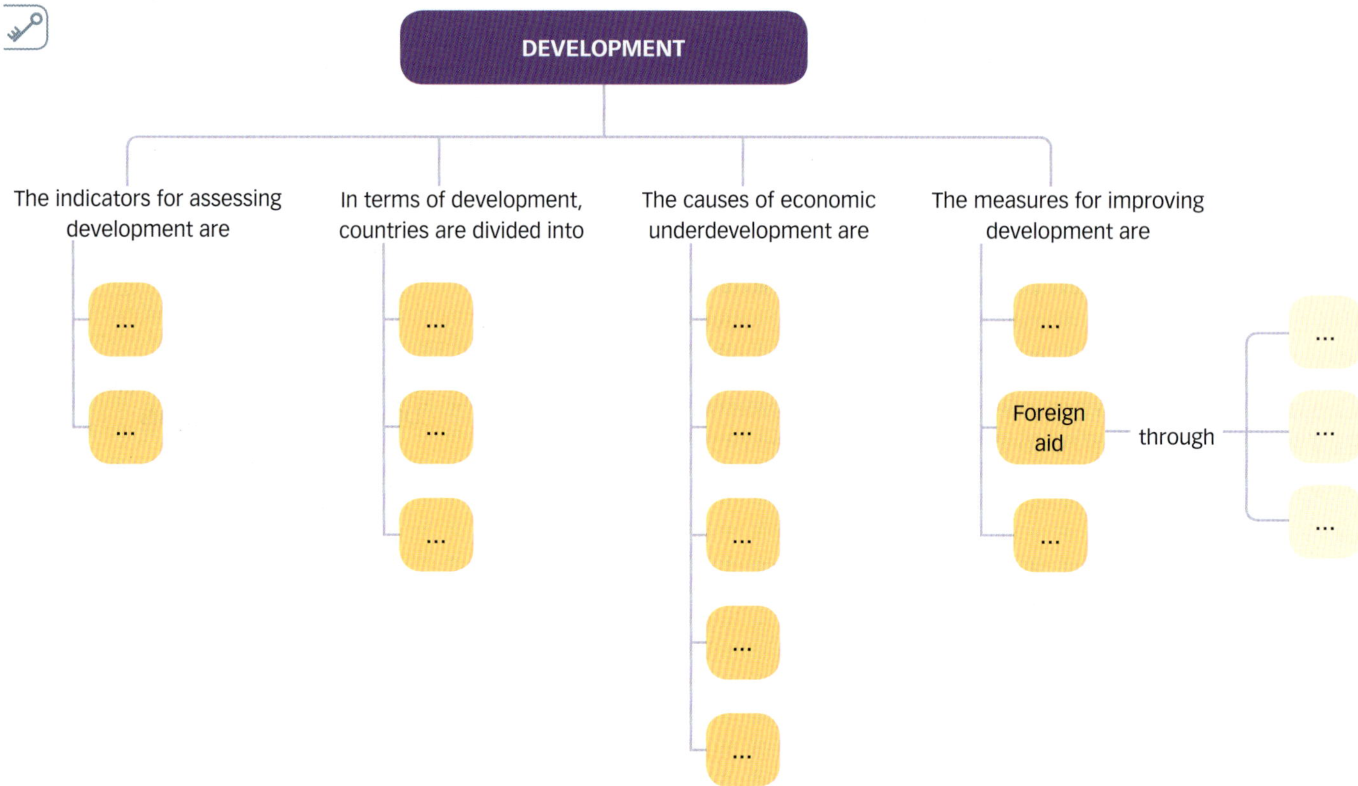

2 Copy and complete the table in your notebook.

COUNTRIES ACCORDING TO THEIR DEVELOPMENT INDEX			
	Developed countries	**Emerging countries**	**Less developed countries**
Characteristics	......	......	......
Examples	......	......	......

3 Explain the meaning of these concepts:

- Development
- Extreme poverty
- External debt
- Foreign aid

4 Think about the following issues, and then discuss them in groups. Write a short summary of your conclusions.

- Is the elimination of international trade barriers a problem for poorer countries?
- Is there a connection between malnutrition, illiteracy and child exploitation?
- Why do former colonies continue to depend on wealthy countries?

5 Analyse the Gini index.

Country	Gini
Norway	23.6
Slovenia	23.8
Hungary	24.1
Sweden	24.1

Country	Gini
Panama	51.9
Brazil	52.7
Colombia	55.9
Namibia	61

The Gini index measures inequality in a society:

- 0 means that wealth is equally distributed.
- 100 means that all the wealth is in the hands of one person.

What conclusions do you draw from these figures?

6 Look for information about Denmark and Bolivia on the Internet. (Type "CIA The World Factbook" in a search engine and then look for "Country Profiles"). Complete the table in your notebook with as much information as possible.

	DENMARK	BOLIVIA
GDP per capita	......	......
Population below poverty line	......	......
Illiterate population	......	......
Doctors per 1,000 inhabitants	......	......
Birth rate (‰)	......	......
Internet users	......	......

a Which is a developed country? Why?

b What are the greatest differences between the two countries?

7 Think about the barriers to development in the case of Bolivia.

a Why does the existence of malnutrition make it difficult for this country to develop?

b Does the high illiteracy rate affect development?

c Are there enough doctors to attend to the population?

d Is the birth rate in Bolivia high or low? How does this affect development?

e Describe the population of Bolivia in terms of its age groups. How might this affect the country's development?

f Is the number of people who use Internet high or low? What does this tell us about the level of technology in the country?

g What effects will the use of technology have on the development of the country? Why?

THE MILLENNIUM DEVELOPMENT GOALS

The **Millennium Development Goals** were established by member states of the United Nations in 2000. They have become an important part of the international fight against world poverty.

Significant progress has been made in many areas. However, there is much to be done: one in eight people in the world today suffers from hunger.

8 What are the eight Millennium Development Goals?

9 Form eight groups. Choose a single goal to study as a group.

- Look on the Internet for the official United Nations web page by typing "UNDP Millennium Development Goals" in a search engine. Click on the icon of the goal corresponding to your group.
- Assess the progress made in this field and say what remains to be done.

10 Discuss your conclusions with the rest of the class.

- In which area has most progress been made?
- In your opinion, which goal is the most important?

To eradicate extreme poverty and hunger.

To achieve universal primary education.

To promote gender equality.

To reduce child mortality.

To improve maternal health.

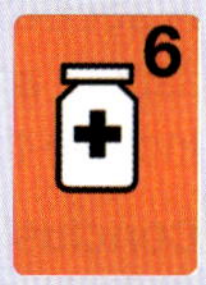

To combat HIV/AIDS, malaria and other diseases.

To ensure environmental sustainability.

To develop a global partnership for development.

TECNIQUES. Compare the development of two countries

Follow these steps in order to compare the development of two countries:

1 **Write a list** of the aspects that you are going to compare.

2 **Make a table** on which you can record the information about each country.

ASPECTS	COUNTRY 1	COUNTRY 2
......	......	......
......	......	......
......	......	......
......	......	......

3 **Assess the factors** that explain the situation in each country.

4 **Explain the consequences.**

5 **Write a report with your conclusions.** The report should include the following:

- **Introduction**. Explain the aspects that you are going to consider in the report. Begin like this: *I am going to analyse... To do this, I will compare these aspects...*
- **A section on each aspect**. Describe each aspect and how it is reflected in each country. Include their causes and consequences.
- **Conclusion**. Write your conclusions about your findings.

Spain

Population: 46,217,961 inhabitants

GDP: 1,349 thousand million

GDP per capita: $32,320

Cameroon

Population: 21,699,631 inhabitants

GDP: 25 thousand million

GDP per capita: $2,320

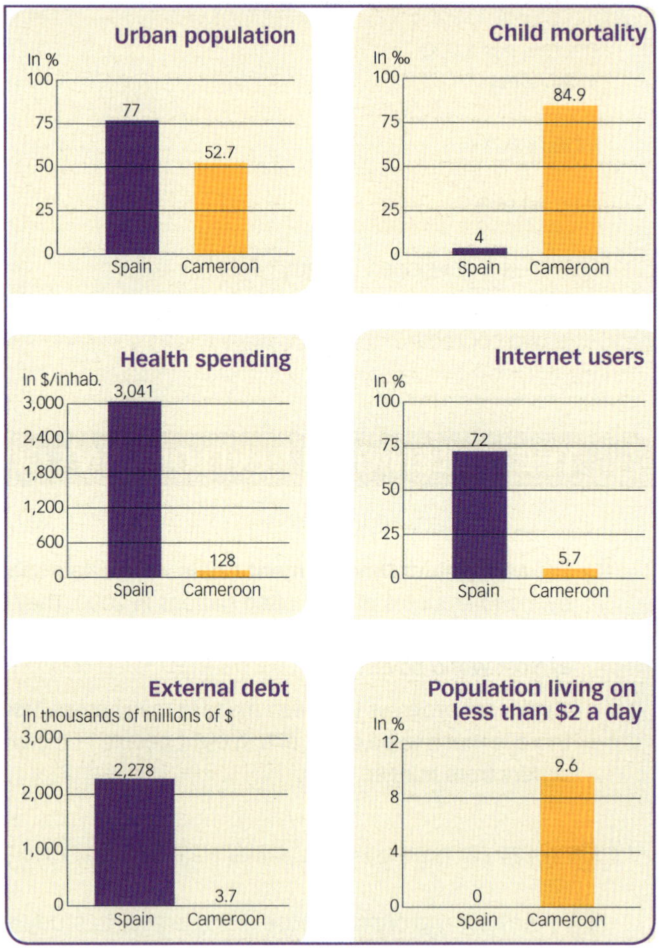

1 Write a list of aspects that you want to compare. Use them to complete a table.

2 There are great differences between the development of Spain and Cameroon. Write a report, including the following aspects.

a Economic indicators. Which country generates more wealth? Which has the higher per capita income? Which has the higher level of extreme poverty? Which has greater external debt? Which country has more Internet users?

b Demographic and social indicators. Which country has the larger urban population? Which has higher child mortality? Which spends more per capita on health? Which has greater social development?

3 Draw conclusions.

a Spain is number 15 in HDI classification, while Cameroon is number 153. Which data reflect this difference?

b In your opinion, which are the most significant differences between the two countries?

A CASE STUDY. **Is Africa taking off?**

We live at a time of optimism with regard to Africa. It is the region of greatest growth in the world, and is an El Dorado for investors. It represents a market of 1,000 million consumers, with a growing middle class.

Since the beginning of the decade of 2000, the real GDP of the African continent has grown on average by 1.5% per year. This is double that of the previous decade and higher than the rest of the world, especially Europe, since the 2008–2009 economic crisis. However, it has not been an easy journey. This positive trend has seen two stumbling blocks. In 2009, demand fell from the OECD countries which were affected by the world economic and financial crisis. In 2011, the Arab Spring suddenly put a stop to growth in Tunisia, Egypt and Libya.

The positive trend in Africa is due to various circumstances, beginning with the number of Africans. Today, Africa is the youngest continent in the world, with 200 million people between 15 and 24. This will become 400 million in 2045. Internal demand was one of the main driving forces behind the growth of African countries in 2012.

Violent conflict has become less common. Moreover, many African countries have benefited from partial or total cancellation of their debt to Western countries.

However, the sustained rise in demand, and in world prices of raw materials, is surely the essential factor in explaining the positive results in African economies. Africa has exceptional natural resources, and they are the least exploited in the world.

If wealth is accumulating more quickly in Africa, can we talk of an acceleration in development? The creation of employment has not been noticeably faster than in the "lost decade" of the nineties. Young people especially are victims of unemployment, underemployment and job insecurity.

In short, much of Africa has improved, but life has yet to improve for the majority.

If it is really to take off, Africa needs to transform the structure of its economies. Economic resources must be reassigned away from less productive activities, such as family farming or small family businesses, towards more productive activities, like the manufacturing industry.

Henri-Bernard SOLIGNAC-LECOMPTE, *African development, myth or reality?* (2013).

1 Check the meaning. With a partner, check that you understand any difficult words. What do the following terms mean?

- El Dorado (paragraph 1)
- Arab Spring (paragraph 2)
- Stumbling block (paragraph 2)
- Lost decade (paragraph 6)

2 Analyse the text.

a How did real GDP in Africa evolve between 2000 and 2012?

b Which factors explain the economic growth of Africa?

c Did employment evolve at the same rate as economic growth? What were the consequences?

3 Give your opinion.

a Is economic growth the same as development? Explain your answer.

b According to the article, what conditions are necessary for Africa to develop? Do you agree?

c According to this text (written in 2013), "we live at a time of optimism with regard to Africa"? Do you agree? Has the situation changed since 2013?

8 Environmental sustainability

FIND OUT ABOUT:

- Environmental awareness
- The main problems affecting the environment
- The reasons for these problems
- Solutions for protecting the environment

KNOW HOW TO:

- Compare the ecological footprint of two countries
- Analyse environmental impact using satellite images

Local decision, global effects

In 1984 Monaco Oceanographic Museum imported a dangerous invasive plant, a tropical alga called *Caulerpa*.

The museum submerged some of its aquariums in the sea to clean them. This had very serious environmental consequences. The alga *Caulerpa* spread throughout the Mediterranean and killed many of the *Posidonia* fields in Italy, France, Spain, Monaco and Croatia.

? WORK WITH THE IMAGE

- What does Posidonia look like?
- Why is it important to the environment?
- What is a danger to it? Where did this come from?
- Do you think individual actions can affect the environment?

WHY IS *POSIDONIA* IMPORTANT TO THE MEDITERRANEAN?

Posidonia is a Mediterranean marine plant. It is the habitat for over 400 species of plants and 1,000 species of animals. Its long leaves protect the coast from erosion caused by the waves. Illegal trawling, the contamination of sea water and the construction of infrastructures on the coast have a very adverse effect on its growth.

AGGRESSIVE ALGAE: *CAULERPA*

The *Caulerpa* is not toxic for human beings but it is very aggressive with flora and fauna. This invasive plant is currently colonising over 3,000 hectares of the Mediterranean Sea.

WE NEED TO TAKE ACTION TO PREVENT IT FROM SPREADING MORE!

- Divers should clean their material before and after each dive.
- Sailors should check their anchors and chains before and after anchoring.
- Fishermen should clean their nets before and after throwing them into the sea.

HOW DO WE KNOW?

Many non-governmental organisation (NGOs) are working to contain environmental problems. They make annual reports on the state of the environment.

- Find information on the Internet about an NGO which focuses on the environment, like Greenpeace UK, Friends of the Earth or Earthwatch.
- Make a list of some of the ecological disasters they mention on their websites.

1 Environmental degradation and green awareness

Human impact on the natural environment

Humans use the **natural environment** as a resource. It provides us with our water and food, and the raw materials that we need for our economic activities.

Several factors influence human impact on the environment:

- **Population**. Large populations need space for housing, raw materials, land to grow food and transport infrastructures.
- **Technology.** The use of advanced technologies makes it possible to modify vast areas of land.
- **Energy.** Increased use of energy causes changes in the environment.
- **Consumption.** The production of goods leads to an increased demand for raw materials and energy.

Many people argue that we need to take measures to conserve and protect our environment. We call this concern **green** (or **environmental**) **awareness**.

Threats to the environment

The main threats to the environment are:

- **Air pollution**, which is so severe that scientists believe that it is causing climate change.
- **Water pollution** and its excessive use.
- **Deforestation**, which affects the quality of the soil and alters ecosystems in many regions.
- **Endangered species:** the survival of many species of animals and plants is threatened.
- **Erosion** and **desertification**: the soil becomes less fertile.

ACTIVITIES

1 What do the following expressions mean?

- Natural environment
- Ecological awareness
- Sustainable development

WORK WITH THE DIAGRAMS

2 Discuss the diagrams with a partner.

a Put the environmental problems in order of importance.

b Which solutions will be most difficult to achieve?

THINK ABOUT IT

What do you think is the main environmental problem where you live? What can we do to prevent it?

FACTORS CAUSING ENVIRONMENTAL PROBLEMS

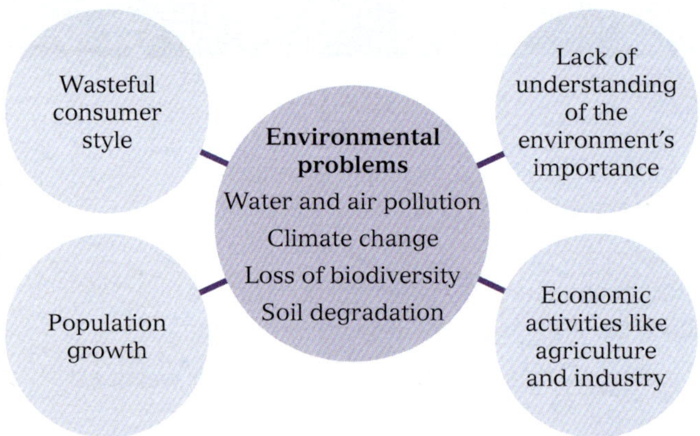

POSSIBLE SOLUTIONS

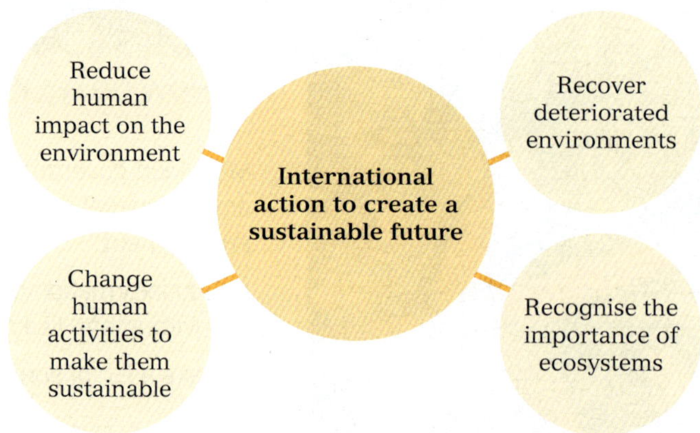

ECOLOGICAL FOOTPRINT PER CAPITA OF PRODUCTION

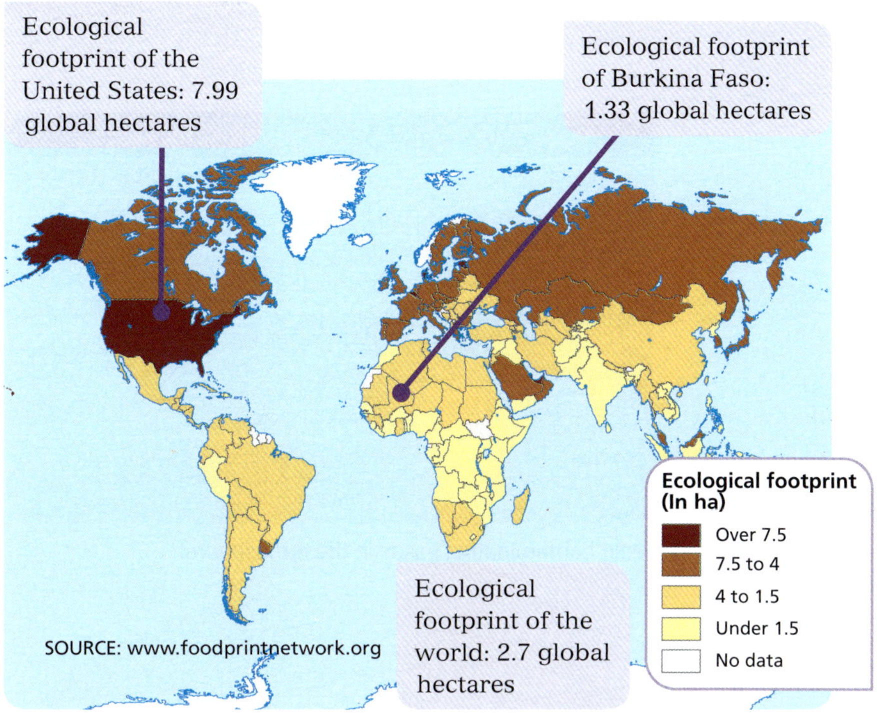

DID YOU KNOW?

Compare ecological footprints

An ecological footprint is a measurement of human impact on the planet. It is calculated by adding up the farmland, pastures, woods and fishing areas that a territory or a person needs in order to produce the food and elements they consume and to absorb the waste they produce.

How do we prevent environmental degradation?

There are alternative approaches to preventing the degradation of the environment:

- Some people argue that it is possible to achieve **sustainable development**, which is a balance between economic growth and the conservation of natural resources.
- Many ecological movements focus on the need to reduce consumption levels. They argue that strong economic growth is not compatible with respect for the environment.

This is a global problem that affects everybody. The United Nations has held several conferences to promote **international environmental policies**. However, its agreements have not been signed by all the countries in the world.

At a national level, policies have been implemented with three goals:

- **Protection** of spaces with great ecological value.
- **Prevention** of damage by evaluating the environmental impact of human activities.
- **Recovery** of degraded spaces by recycling waste, treating water, etc.

Many countries have created environmental agencies to deal with these problems.

ACTIVITIES

3 Use the internet to identify important international agreements on the environment.
   a Have all the countries in the world signed each agreement?
   b What has the attitude of the United States been to these agreements? Why?

WORK WITH THE MAP

4 Does Burkina Faso have a larger or smaller ecological footprint than the United States? Is it a developed country?

5 Which regions have a smaller footprint? Is there a connection between economic development and the ecological footprint?

6 What would happen if all the countries in the world left an ecological footprint like the United States?

2 Air pollution and climate change

Why does air pollution occur?

Air pollution is caused by the emission of gases into the air as a result of human activity. These gases cannot be recycled naturally because of their volume or toxicity.

Air pollution is bad for the health of living beings and causes serious environmental problems.

Smog

Smog is formed by a chemical reaction between sunlight and nitrogen oxides at ground level. This forms a thick mist that irritates people's eyes and lungs, and is also harmful to plants.

Smog is common in big cities in winter, when there are emissions from both cars and heating.

Acid rain

Acid rain occurs when nitrogen oxides and carbon dioxide mix with rain to produce acids. It is very harmful:

- It attacks plants and destroys them.
- The increased acidity of rivers and lakes affects their wildlife and vegetation.

GASES	HUMAN ACTIVITIES
Carbon dioxide	Burning fossil fuels; gas, oil and coal.
Sulphuric acid	Burning coal and heavy fuel oils.
Nitrogen oxides	Burning of fossil fuels. Use of fertilisers.
Methane	Rice cultivation. Livestock farming.
CFCs	Aerosols. Refrigerator coolants. Air conditioners.
Ozone	Chemical reactions of sunlight with gases from vehicles, industry and heating systems.

Main contaminating gases in the atmosphere.

Smog sometimes affects large cities.

WORK WITH THE IMAGE

1. Describe the effects of acid rain on the environment.

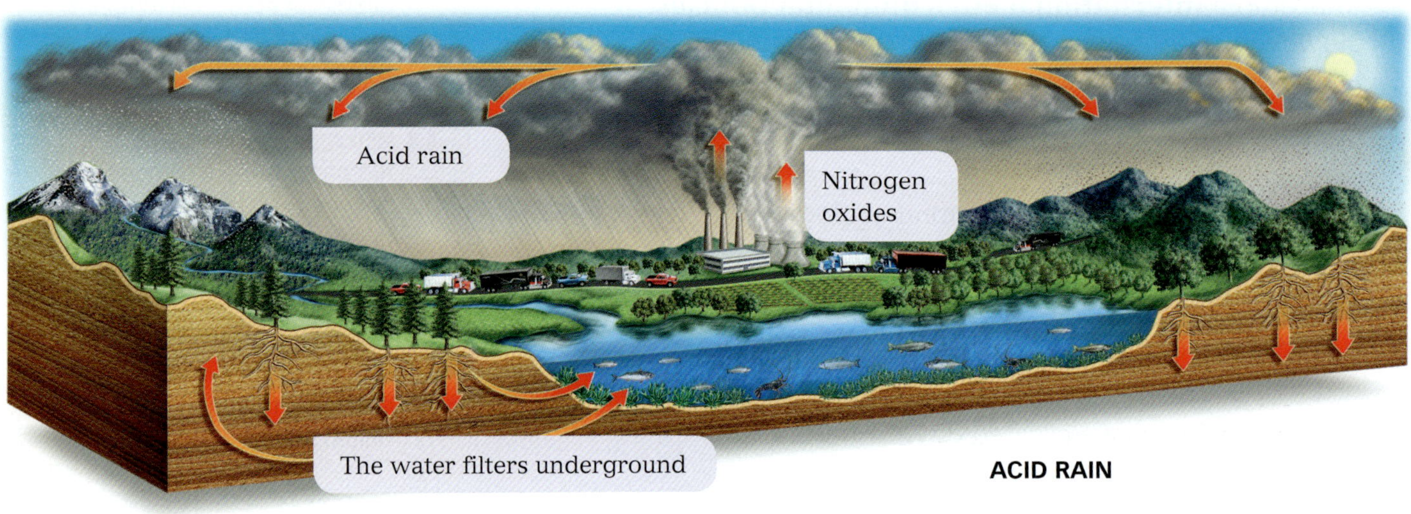

ACID RAIN

The hole in the ozone layer

Ozone is a gas found in the atmosphere. It is essential for life because it absorbs 90% of the ultraviolet radiation from the sun, which is bad for our health. The ozone layer becomes thinner in the polar regions in autumn and winter and thicker again in spring and summer.

However, the emission of CFC gases into the atmosphere upsets this natural balance. They dissolve the ozone. The ozone layer becomes thinner than normal in autumn and winter, and the gases prevent it from recovering in spring and summer.

The hole in the ozone layer started above Antarctica because ozone is destroyed more easily at very low temperatures.

1979

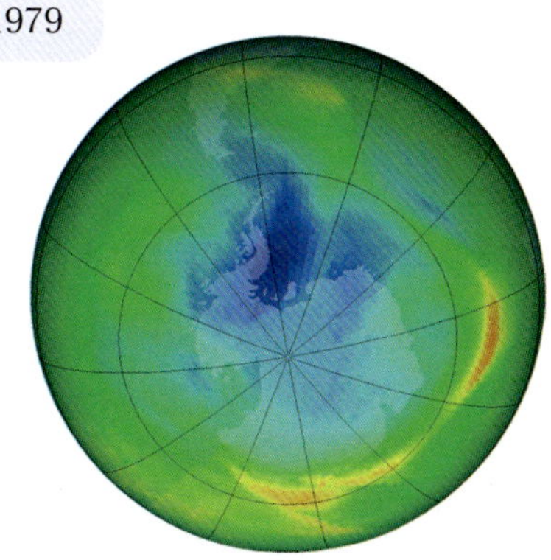

2011

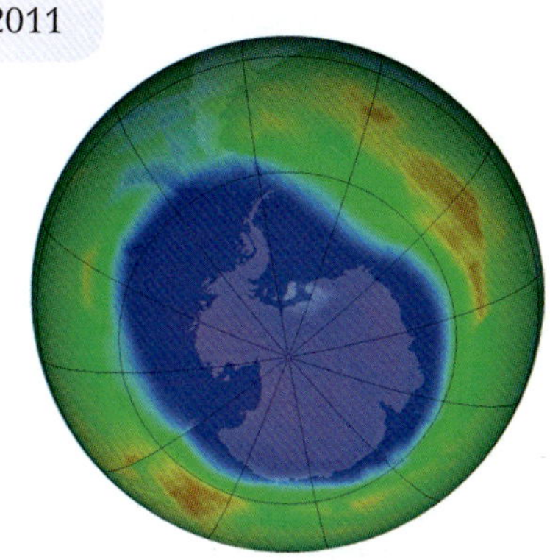

Evolution of the hole in the ozone layer.

The greenhouse effect and climate change

Some gases, such as methane, nitrous oxide and carbon dioxide, are known as **greenhouse gases**. They allow the heat from the Sun to reach the Earth and stop some of it travelling back into the atmosphere. In this way, temperatures are regulated.

Some human activities produce greenhouse gases. The increase in these gases in the atmosphere is leading to **global warming**: an increase in the planet's average temperature. In the 20th century, average temperatures rose by 0.6 °C. In only one hundred years, temperatures increased by the same amount as in the previous 18,000 years.

The consequences of global warming are very serious:

- The **polar ice caps** are melting faster. In the last hundred years, the sea level has risen by between 15 and 20 cm. If this trend continues, some coastal regions could disappear.
- According to scientists, **climate change** is taking place. If this continues, climate will become more extreme and unpredictable.
- Climate change will lead to the destruction of some **ecosystems**. If temperatures continue to rise quickly, many plants and animals will not be able to adapt and will die.

How can we prevent air pollution?

The problems caused by air pollution affect the entire planet, regardless of the place where most CFC gases are produced. The hole in the ozone layer over Antarctica was caused thousands of kilometres away in the most populated regions, where there are the most CFC emissions. Consequently, international solutions are the only way of reducing air pollution.

ACTIVITIES

2 Read the sentences. What do they refer to? Listen to the recording and check your answers.

a It is the increase in the Earth's average temperature.
b It is caused by gases that cannot be naturally recycled.
c It is a mixture of smoke and fog.
d It absorbs most of the radiation from the Sun.
e It increases the acidity of lake and river water.
f In excess, they stop heat leaving the planet.

3 List the effects of air pollution.

4 Look on the Internet for information about the following international agreements:

- The Montreal Protocol
- The Kyoto Protocol

3 Water supply

Water is a scarce resource

Water is a **natural resource** that is essential for all living things. It is also used in many economic activities, such as farming, stock-breeding, industry and leisure services, such as golf courses. Farming accounts for 75% of all the water used worldwide.

Fresh water is a **renewable resource** because evaporation and rainfall are continuous. However, it is still a **scarce resource**:

- Only 3% of the world's water is **fresh water**. Most of this is frozen at the poles and in glaciers.
- In addition, water is not evenly **distributed** throughout the world. In some regions there is insufficient water because rainfall is low. Moreover, there are not always the necessary infrastructures, such as reservoirs, pipes and water treatment plants.

Today, the excessive use of water is a problem for nearly half of the world's population. In many places, consumption is greater than the amount of water available. Consumption is much greater in developed countries than in less developed countries.

Treatment plants clean water in cities.

WORK WITH THE MAP AND THE CHART

1. What is physical water scarcity? What is economic water scarcity?
2. Which regions suffer from physical water scarcity? Look at the climate map in Unit 1. What is the relationship between climate and the lack of water in these regions?
3. In your opinion, are water scarcity and wealth related?

PHYSICAL AND ECONOMIC WATER SCARCITY IN THE WORLD

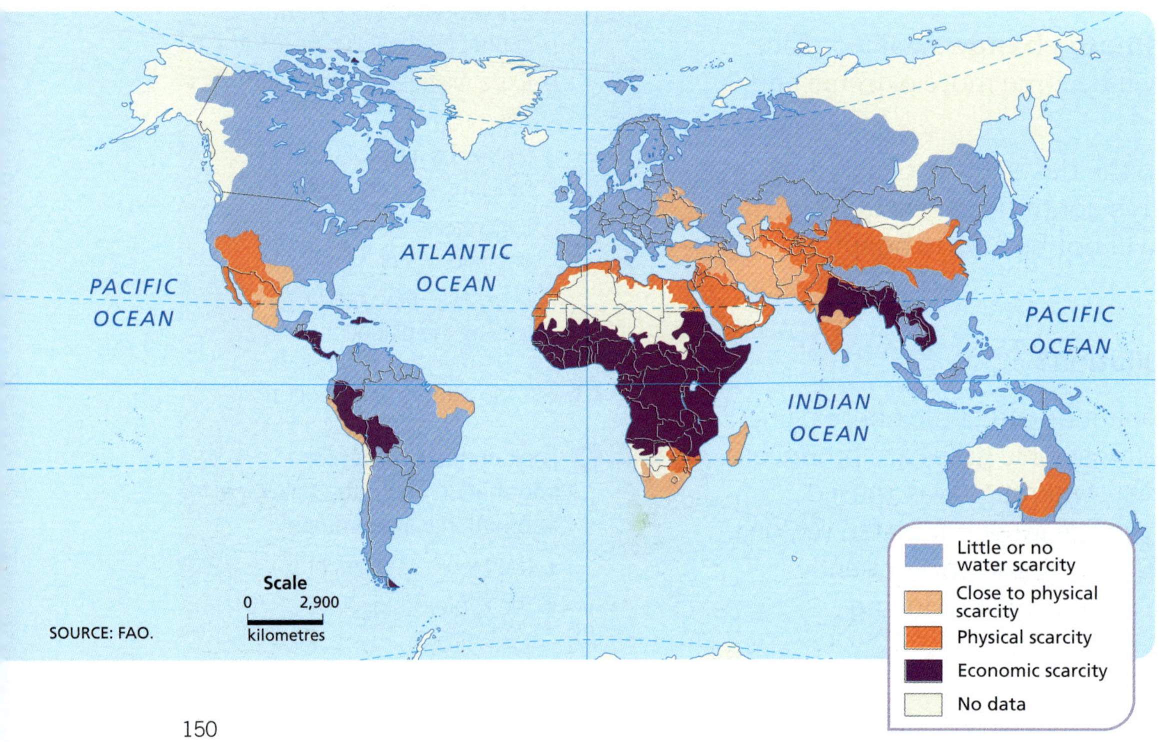

How can we increase available water?

We need to implement measures to increase the amount of water available in regions where it is scarce. At the same time, there are a number of problems:

WAYS TO INCREASE AVAILABLE WATER	
MEASURE	**PROBLEM**
Build hydraulic works, such as reservoirs, to store water when there is abundant supply.	This is expensive and has a negative effect on the environment.
Create desalination plants to convert sea water into drinking water.	This is expensive but in some places it is the only source of fresh water. In Spain there are desalination plants in the Canary Islands.
Treat waste water to use the treated water in public fountains, for watering parks, and in farming.	Treatment plants can be installed in towns and industrial zones, as the water is channelled through pipes. But this is impossible in farming.
Use transfers to take water from one place to another.	This is expensive and has a negative effect on the environment.
Use aquifers to increase the available water.	This is a very short-term solution. If we take water from aquifers, less groundwater flows into rivers, springs and wetlands.

DID YOU KNOW?

Water consumption in Spain

The Water Use Index relates the water resources available in a territory to the amount of water used there. Spain has one of the highest indices in the EU.

The water used in farming accounts for 75% of total consumption. Some irrigated regions, such as the Júcar and Segura basins, are extremely short of water and use transfers from the Tajo River to mitigate the shortage.

This situation is aggravated by urban consumption and the use of water for leisure activities, such as the creation of golf courses in the Canary Islands and on the Mediterranean coast.

How can we reduce consumption?

Farming is the human activity that uses most water. To reduce this consumption, we should use flood irrigation less. Flood irrigation uses a lot of water and loses it because of evaporation and infiltration. Instead, we should use more efficient **low water irrigation systems**, such as drip irrigation. In industry, water should be **recycled**.

The network of water supply pipes also needs to be in good condition to **prevent water loss**. The pipe systems lose as much as 50% of the water they distribute.

It is easy to reduce water consumption in the home if we use water-saving electrical appliances or if there are special prices to encourage saving water.

ACTIVITIES

4 Copy and complete the paragraph. Then listen and check your answers.

Water is a .... resource which is essential to life. Fresh water is a .... resource thanks to evaporation and .... Unfortunately, most fresh water is frozen at the poles or in .... Water consumption is much higher in .... countries than in .... countries. There is an economic water scarcity in some .... countries because they do not have the necessary ....

5 Investigate. Go to www.coloradoriverbasin.org and learn about the problems affecting the Colorado River. Then discuss the following questions:

a What has happened to the river?
b What are the reasons for this?
c How many people does the problem affect?
d What economic activities does it affect?

This golf course is on an island where water is a scarce resource.

4 Deforestation and soil degradation

What is deforestation?

Forests are of great economic importance but, above all, they are vital for the environment.

- Forests absorb **carbon dioxide** and help to fight climate change.
- They are the **habitat** for thousands of species of plants and animals.
- They cover the **soil**, protecting it from erosion and desertification.
- They lower **temperatures** at the Earth's surface.
- The trees play a role in the **water cycle** by helping water to evaporate into the atmosphere.

Deforestation is the loss of the area covered by forests as a result of human activity. Many forests are cut down to obtain wood, make farmland, create pastures for livestock or build housing, roads and railways.

In the last 5,000 years, around 30% of the world's forests have been lost. Most deforestation has occurred in the last forty years. However, deforestation has slowed down since 2000.

Deforestation is especially serious in **tropical rainforests**: between 1% and 2% are lost every year. Tropical rainforests are not the largest forests in the world, but they are extremely important because more than half of the Earth's animal and plant species live there.

Consequences of deforestation in tropical rainforests

One of the most serious consequences of deforestation is the **loss of biodiversity**. In the rainforests there are up to six ecosystems at different levels or heights. If trees are cut down, only one level remains: the ground.

New trees are sometimes planted, but most animals and plants disappear when a forest is cut down. They are used to ecosystems where there are hundreds different species of trees.

Another serious effect of cutting down many trees is **soil degradation**. The soil in tropical rainforests is very fragile. It becomes exposed to sunlight, which degrades the organic compounds needed by plants.

Tropical rainforests absorb 18% of the carbon dioxide generated by human activity. If these forests are lost, there will be a higher proportion of carbon dioxide in the atmosphere.

DEFORESTATION IN THE WORLD

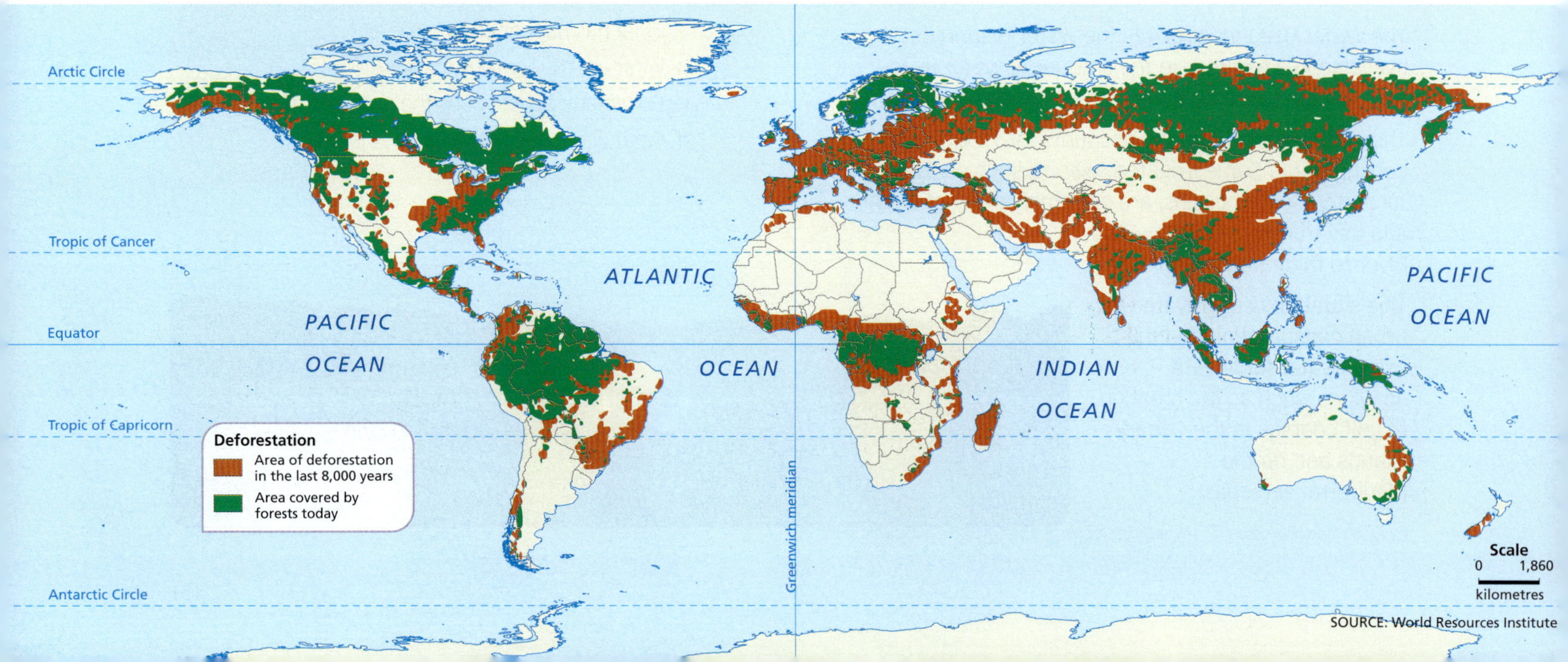

Soil and its role in the environment

Soil is the top layer of the Earth's crust, between bedrock and the atmosphere. The richness of the soil depends on four basic factors:

- **Vegetation** obtains minerals and water from the soil and enriches it with the nutrients from leaves and branches.
- **Water** filters into the soil and deposits dissolved substances.
- **Wind** brings mineral and vegetable particles from other places.
- **Bedrock** (the rocks under the soil) gives the soil chemical elements.

Soil degradation

Soil degradation is the loss of soil fertility due to the following factors:

- **Soil contamination** by toxic waste, pesticides used in agriculture and acid rain.
- **Soil erosion**. Vegetation is removed, so there are no plant roots to hold the soil in place. Consequently, soil can be washed away by rain.
- **Deforestation**. The land is exposed to sunlight, which breaks down the organic compounds needed by plants.
- **Loss of nutrients** caused by repeated use of land to cultivate the same crops.

Some **measures** can be taken to prevent soil pollution, such as reforestation, waste recycling, crop rotation, building safe landfills and carrying out decontamination activities.

Erosion and desertification

Soil **erosion** is the loss of the top layers of soil. This is a natural process caused by the action of water and wind. However, the problem is greater if human activities upset the natural balance and speed up erosion.

- The areas most exposed to **natural erosion** have low rainfall, are exposed to torrential rain or strong winds, or are on steep slopes.
- The areas most exposed to erosion caused by **human activities** are heavily populated, are in built-up areas, or have intensive farming.

When soil fertility and vegetation loss reach a certain level, the result can be desertification. Drought and climate change accelerate this process. Desert, or the risk of desertification, now affects 22% of the Earth's land surface.

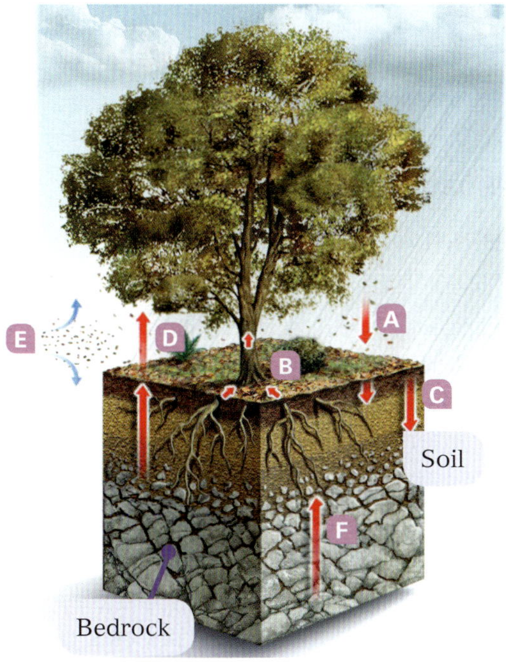

The formation of soil. Vegetation provides nutrients (A) and obtains substances (B). Water filters through (C). Evaporation takes substances to the surface (D). The wind brings and takes particles (E). The bedrock gives the soil chemical elements (F).

ACTIVITIES

1 Answer the questions.

a What is soil degradation?

b What makes soil erosion happen faster?

c Why do you think Africa is the continent most threatened by desertification?

DID YOU KNOW?

The threat of desertification

According to the United Nations Environment Programme, desertification threatens a quarter of the planet and directly affects over 250 million people.

Africa is the most threatened continent. Over 40% of its area is in danger of desertification. Considering that 43% of the continent is already occupied by deserts, only 17% of Africa has no risk of desertification.

5 The loss of biodiversity

What is biodiversity?

Biodiversity is the variety of species and ecosystems on the Earth.

Biodiversity is present on the whole planet, including deserts and regions covered with ice. However, in some regions there is a much greater variety of species. Tropical rainforests are the places with the greatest biodiversity. They only occupy 6% of the Earth's surface but they are home to over 50% of all species.

Biodiversity is essential for the environment and for life, since every species plays a specific role in its ecosystem. Losing a species has serious effects on other living things.

Threats to biodiversity

Human activity is the major threat to biodiversity for various reasons:

- **Changes in soil use.** Natural uses of the soil have been replaced by artificial and agricultural uses on 35% of the surface of the Earth. Deforestation, mining and the extraction of hydrocarbons threaten plant and animal life.
- **Overexploitation of animal species.** About 30% of fishing grounds have been depleted and another 50% are being fished to the limit of their capacity. On land, some animals are endangered because of the loss of their habitat.
- **The introduction of foreign species.** Native species are not prepared for the competition for food, or the arrival of new diseases.
- **Air, water and soil contamination.** This changes the environment by reducing food sources and modifying habitats.
- **Climate change.** The increase in temperatures and precipitation, the rise of sea levels and the desertification of some regions all modify the environment and affect many species.

BIODIVERSITY IN THE WORLD

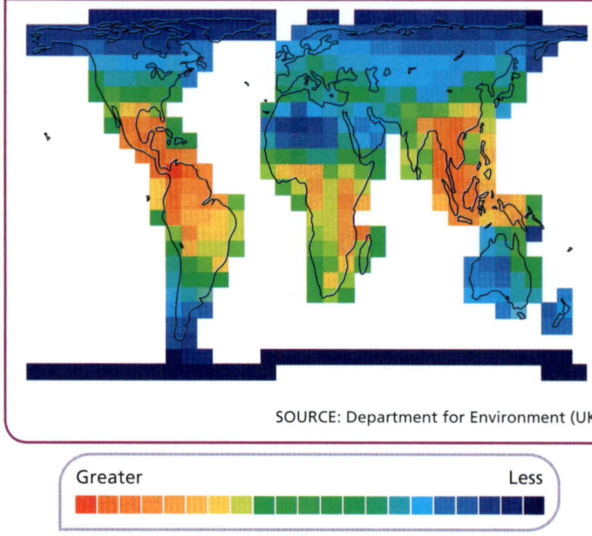

SOURCE: Department for Environment (UK).

MAIN REGIONS THREATENED BY LOSS OF BIODIVERSITY

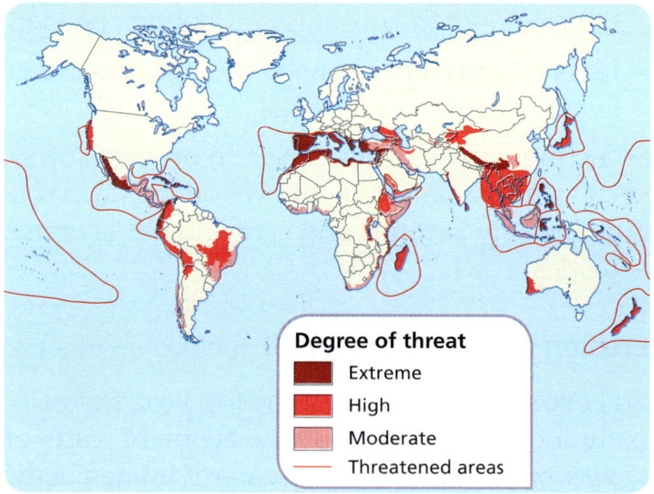

WORK WITH THE MAPS

1 Answer the questions.

a Which regions of the world have the greatest biodiversity?

b Are they in tropical, temperate or cold climate zones?

c In which zones is biodiversity most threatened?

Possible solutions

Solutions to prevent the loss of biodiversity need to be global and carried out by every country.

The **Earth Summit**, held in Rio de Janeiro in 1992, brought countries together for the first time in the fight to conserve biodiversity. The following strategies were proposed:

- Make **inventories** of vulnerable and endangered species so that we understand their situation and can adopt appropriate conservation measures.
- Develop **protected areas** to preserve habitats and ecosystems which are endangered or fundamental to the survival of vulnerable species.
- **Restore** damaged habitats.
- Pass **laws** that limit the negative impacts of economic activities on the natural environment and prevent the overexploitation of resources.

DID YOU KNOW?

Biodiversity in the EU and Spain

The EU has developed the United Nations' initiatives and has taken new measures to prevent the loss of diversity.

For instance, it created the Natura 2000 network, which includes all the protected areas in EU countries. In these areas, human activity and nature are compatible.

Spain has very rich biodiversity. More than 2,000 species living here are native to Spain, i.e. they are not found in other places. Nearly 30% of Spanish territory forms part of the Natura 2000 network.

ACTIVITIES

2 Work in a group with other classmates.

a You have inherited a large piece of land. At the moment, there is only a small house and a few trees on it.

b Plan what to do with it and explain your answers to the class.

c Listen to the other groups' plans. Object if any of their plans will have an adverse effect on biodiversity.

d Vote for the group with the best plan.

WORK WITH THE MAP

3 Look on the Internet to find out the meaning of Special Protection Area (SPA) and Site of Community Importance (SCI) on the map below.

NATURA 2000 NETWORK, SPAIN 2013

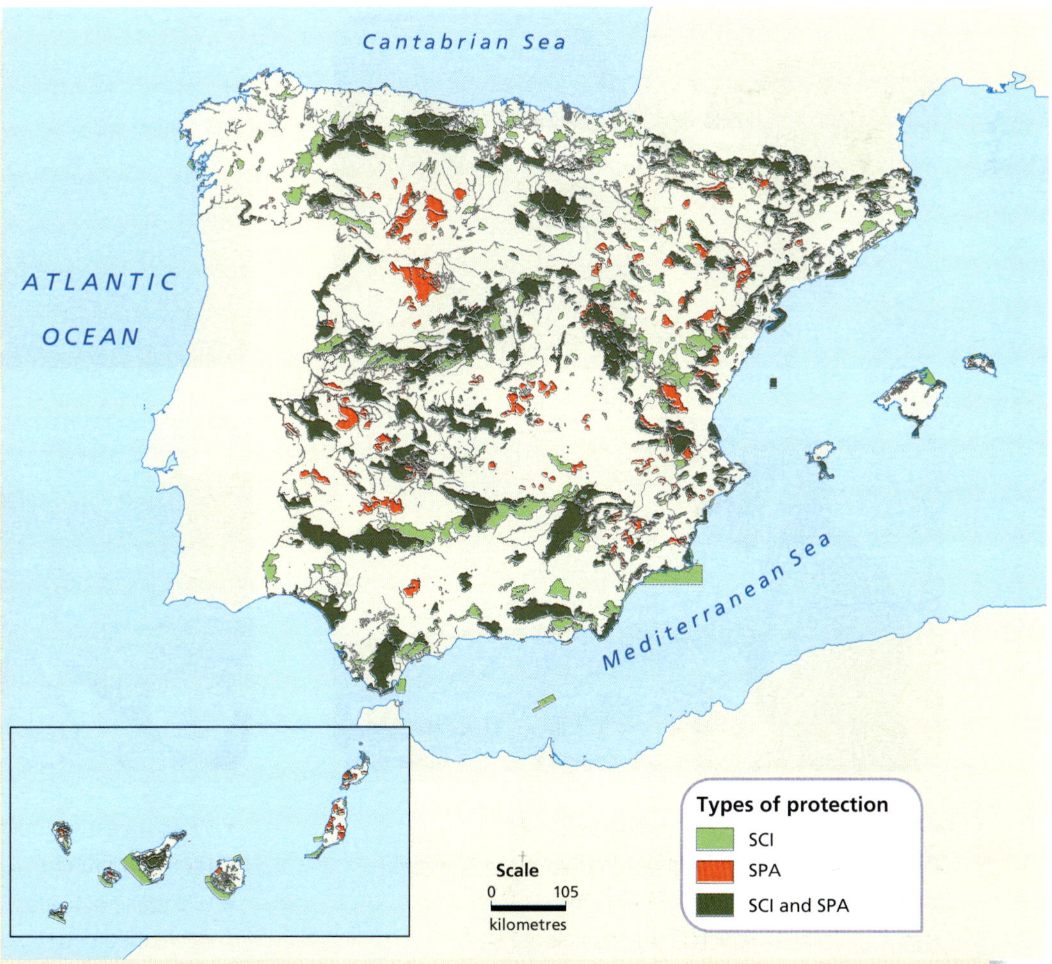

ACTIVITY ROUND-UP

1 **Explain each of the terms below.** Then copy and complete the table.

THE ENVIRONMENTAL CHALLENGE			
Main environmental problems	**Causes**	**Consequences**	**Preventive actions**
.....	.....	.....	.....
.....	.....	.....	.....
.....	.....	.....	.....
.....	.....	.....	.....
.....	.....	.....	.....

- Environment
- Environmental awareness
- Biodiversity
- Sustainable development
- Desertification
- Smog
- Greenhouse effect
- Acid rain

2 Analyse the photographs.

A

B

C

D

- Think of a title for each photograph.
- What environmental problems could these human activities cause?
- In your opinion, which is the most serious environmental problem? Why?
- What measures could be taken to prevent each of these problems?
- Will these activities have the same effects in a developed country and in a less developed country? Explain your answer.

3 Analyse the graph and answer the questions.

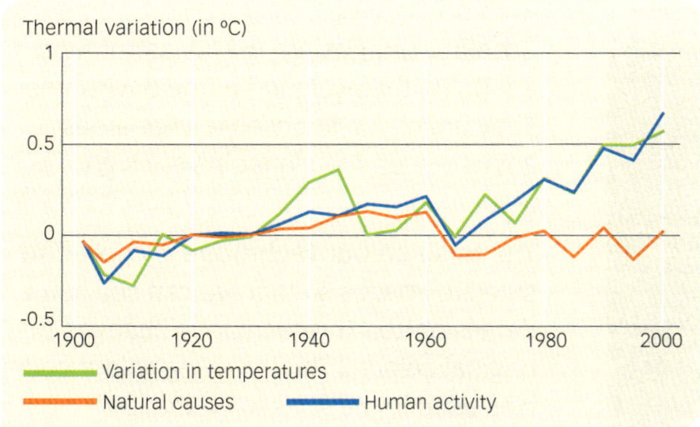

a What does the chart show?

b How much has the Earth's temperature increased in the last hundred years? Do you think this is a moderate or a rapid increase?

c Which factors had more to do with this increase – natural or human factors?

d What could the consequences of global warming be?

4 Compare the images of the Aral Sea and investigate.

1989

2008

- What changes have occurred?
- Why did they occur?
- What could be the consequences of this?

SUSTAINABLE CONSUMPTION

Today we produce and consume more than ever before. We consume resources so quickly that we do not give the planet time to recover.

The massive use of resources has very serious consequences for the environment. It leads to a loss of biodiversity, but it also affects the poorest people in the world because some essential resources are diminished.

5 The demand generated by very low prices leads to the unsustainable use of resources, such as overfishing. Look at the map. What will the environmental consequences be for this area?

6 Analyse the chart opposite and list the measures proposed for sustainable consumption. Are these measures fulfilled today?

PRACTICAL KNOWLEDGE. **Analyse the environment**

Rondônia State (Brazil), 1986

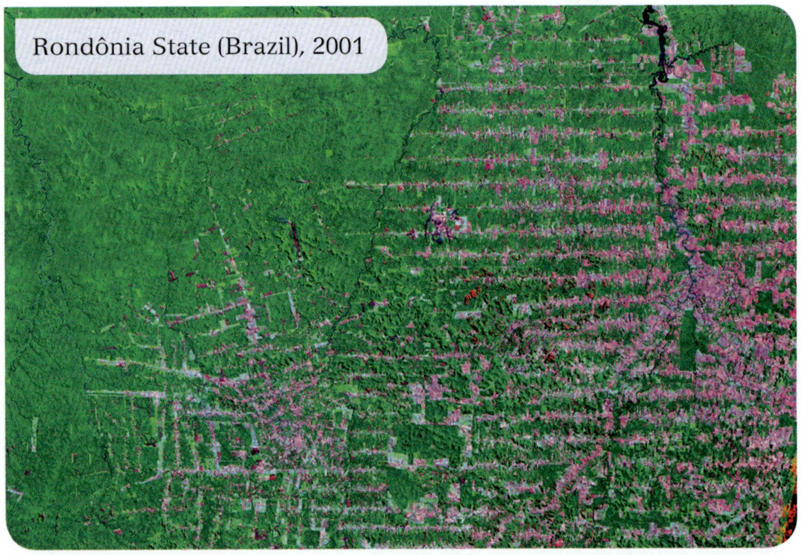

Rondônia State (Brazil), 2001

Rondônia State (Brazil), 2011

Satellite images are useful instruments for analysing the evolution of the environment. They enable us to observe large areas anywhere on the planet in very high resolution.

The **false colour technique** is applied to satellite images so that we can see some details better. This technique changes the original colours. The following colour code was used in the images on the left:

- The leafy **green vegetatio**n of the tropical rainforest is shown in bright green.
- **Water courses** are shown in dark blue.
- **Urban areas** are shown in purple.

1 Compare the images.

- List the main differences you see between the three photographs in relation to forests, water courses and urban areas.
- What happened to the forest between 1986 and 2011?
- The systematic disappearance of vegetation occurred after a road was built. Can you see the road in any of the images?

2 In groups, make a list of the consequences of the growth of urban areas in the forest.

3 Look on the Internet for more information about deserts. Find out why new deserts are being created.

PRACTICAL KNOWLEDGE. **The Sakurajima volcano**

Sakurajima in Japan is one of the most active volcanoes in the world. Explosions often occur but all that happens is that small clouds of ash and volcanic dust are emitted to the atmosphere. However, on 18 August 2013 there was a strong eruption and a large cloud of ash rose about 6 km into the air.

Volcanic clouds contain natural contaminating agents. The gases emitted, such as carbon dioxide, sulphur oxide and chlorine, pollute the atmosphere. The volcanic ash remains suspended in the air for a time. This prevents the Sun's rays from reaching the Earth in the normal way. This, in turn, increases the greenhouse effect.

1 Read the text and say if the following sentences are true or false. Correct the false sentences.

a Sakurajima is not active very often.

b Normally ash from the volcano rises high into the air.

c Volcanic clouds are harmless.

d The Sun's rays cannot get through the atmosphere easily because of the gases emitted by the volcano.

2 Look on the Internet for more information about the Sakurajima volcano.

a What is the current state of the volcano?

b Where exactly is it in Japan?

c How often does it explode?

d How long ago was the original caldera formed?

3 Working in groups, think about the following questions:

a What are the disadvantages of living near an active volcano?

b Why are people in many parts of the world prepared to live near active volcanoes?

Present your answers to the class.

9 The Age of Discovery

FIND OUT ABOUT:

- The causes of the great geographical discoveries
- Portugal and Castilla's explorations
- Pre-Columbian peoples
- Social and economic change

KNOW HOW TO:

- Analyse Columbus' journal
- Reflect on archaeological findings

How did we discover that the Earth was round?

In 1519, an expedition led by Ferdinand Magellan left southern Spain in search of the Maluku Islands in Asia.

An explorer called Antonio de Pigafetta wrote down everything he saw during the voyage. There was an especially dangerous moment when they crossed the strait between the Atlantic and Pacific Oceans.

This was the first voyage round the world. Only one of the five ships returned.

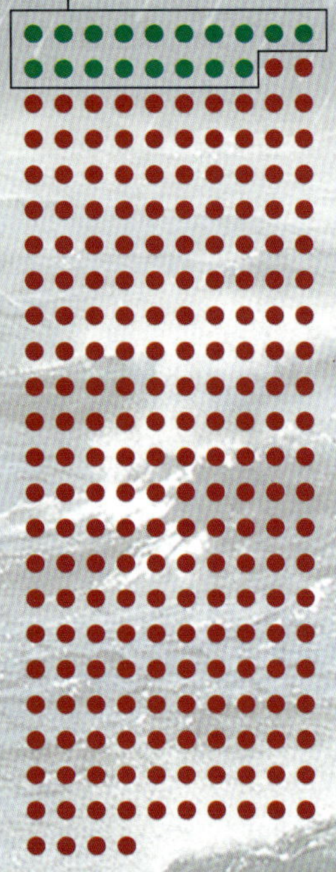

WORK WITH THE IMAGE

- What were the conditions in which Magellan and his crew travelled?
- Are these expeditions made today?
- What type of people do you think went on these voyages?
- Why was it so important to find the strait in the 16th century? Is it as important today?

Provisions for the voyage

The crew did not eat fresh food for several months. By the end of the voyage, very little food was left. They ate leather soaked in water, leftovers full of worms and even rats.

10 tons of biscuits

200 barrels of sardines

1 ton of cheese

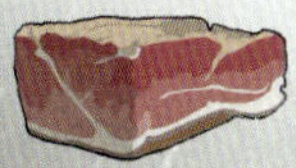

6 tons of meat

Nao Concepción

Nao Santiago

On 21 October 1520, at 52º in the southern latitude, we found a strait by miracle. It is surrounded by high mountains covered with snow. It leads to another sea, which we called the peaceful sea, or Pacific Ocean.

There was a terrible storm during the night and we were forced to pull up the anchors and let our ships drift around the bay.

We thought that two ships were lost, but after a while they returned.

Antonio de Pigafetta

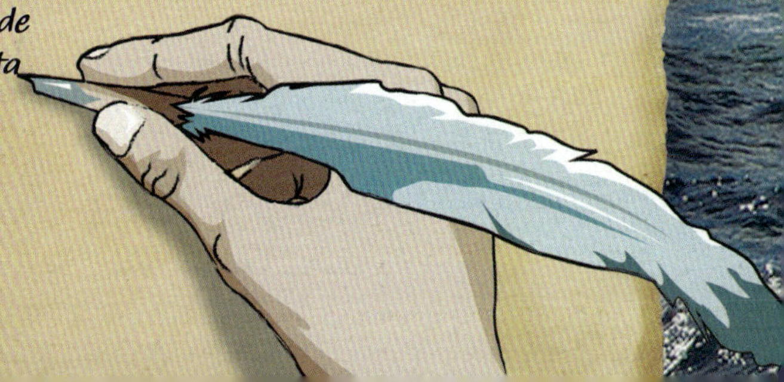

HOW DO WE KNOW?

The diaries and journals written by explorers are full of information about their experiences and the places and people that they discovered.

- Do you know any other stories told by great travellers?
- What are the advantages and drawbacks of these sources for historians?

1 The Age of Exploration

DID YOU KNOW?

Marco Polo

Marco Polo (1254-1324) was a Venetian merchant. In 1271 he set out on a long journey along the Silk Road. This took him to the court of Kublai Khan, the last ruler of the Mongol Empire in China.

After returning to Venice in 1295, he wrote *The Travels of Marco Polo*, describing his travels. This work made people more interested in the East.

WORK WITH THE MAP

1. What territories did the Silk Road go through?
2. Use an atlas. Which modern countries did it cross?

During the Early Modern period, European expeditions explored many new parts of the world. As a result, they provided detailed knowledge about the size and shape of the Earth.

The world known to Europeans

At the beginning of the 15th century, Europeans had limited geographical knowledge.

They were very familiar with all of Europe and the territories around the Mediterranean and Black Seas. They also knew about India, China and Japan because travellers and merchants like Marco Polo brought silk, spices and other products back from Asia.

But they knew very little about most of Africa. They did not know that America, Oceania and Antarctica even existed.

Trade with the East

Exotic products from the Far East, such as silk and spices, were highly appreciated in medieval Europe. The merchants carried them along the **Silk Road**, a trade route that linked Constantinople to China.

In 1453, the **Turks** conquered **Constantinople** and occupied the Byzantine Empire, disrupting the Silk Road. Consequently, European merchants wanted to find other routes to the East.

15th CENTURY TRADE ROUTES WITH THE EAST

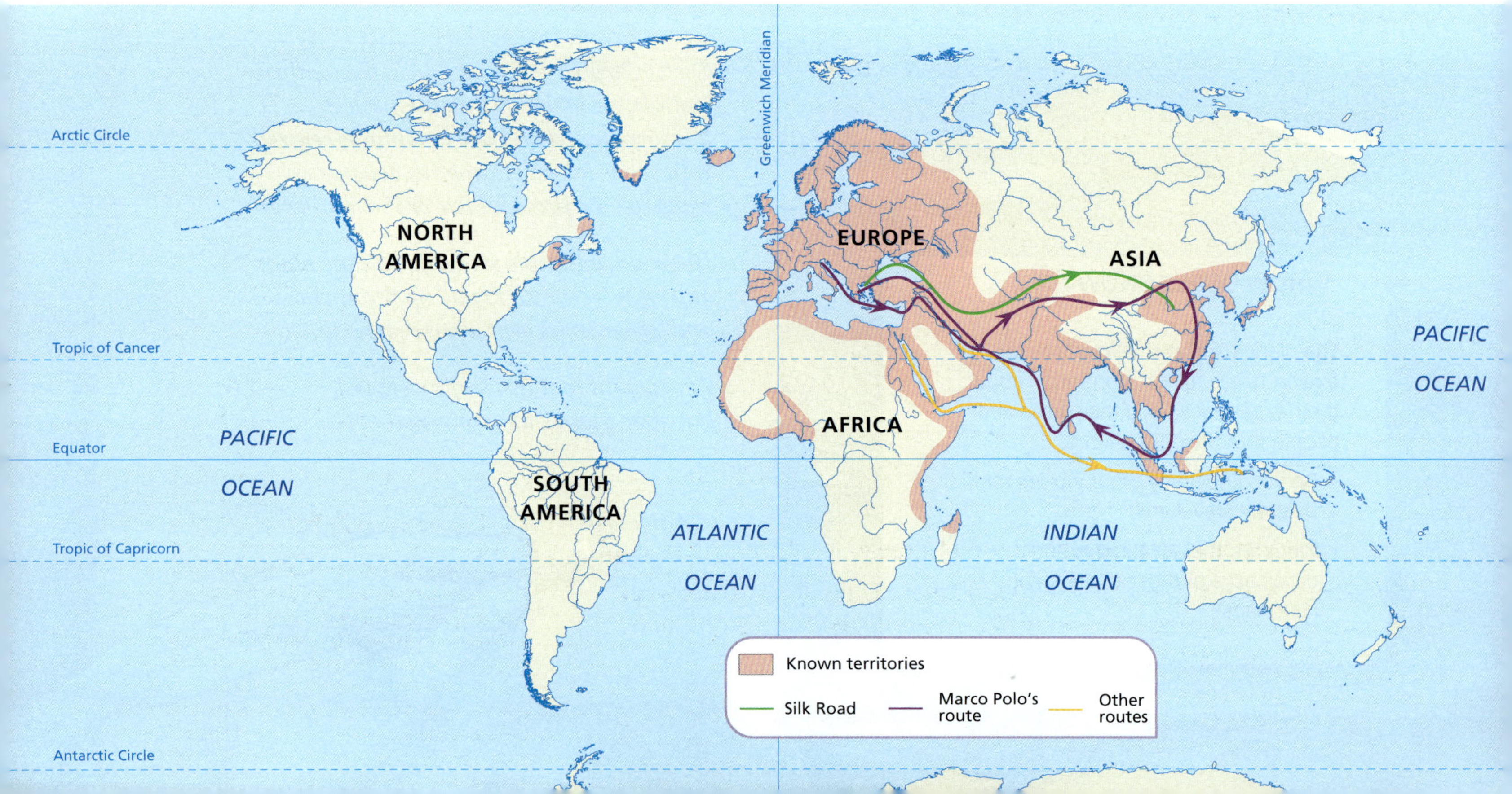

Quadrant. This established latitude by measuring the altitude of the Sun above the horizon. At night, it measured the altitude of the pole star.

A **log** (A) and **sandglass** (B). The rope was gradually released into the sea. The ship's speed was then estimated with the help of a sandglass measuring time.

A

B

Mariner's astrolabe. Like the quadrant, this instrument calculated latitude by measuring the altitude of the Sun or the pole star above the horizon.

Compass. This small box contained a magnetised needle that always pointed north.

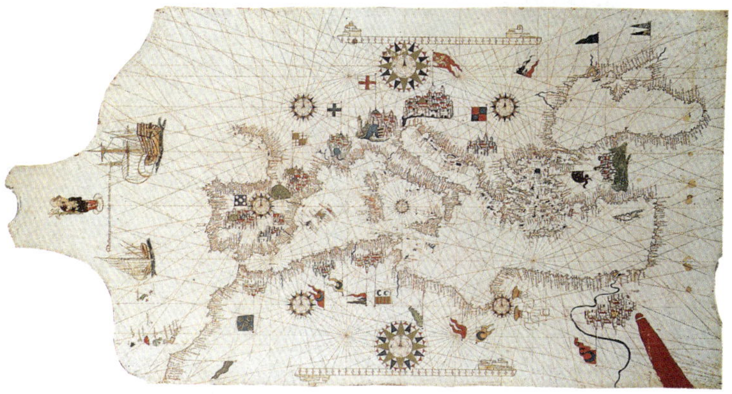

Portolan chart. This map had a compass rose that showed the cardinal points (north, south, east and west).

NAVIGATIONAL INSTRUMENTS AND CARTOGRAPHY

Advances in navigation

Important advances in navigation made it possible for European expeditions to travel long distances:

- **Cartography**. **Portolan charts** were nautical charts showing coasts, sea routes and possible obstacles. These maps only included land features (like mountains) which helped to establish orientation.
- **Navigational instruments.**
  - The **compass** was probably brought to the Mediterranean region by Arab merchants. It allowed ships to sail when land was not visible. It also made it possible to sail at night so sea voyages became quicker.
  - Sailors could calculate latitude by using the **astrolabe** and the **quadrant**.
  - The **log** and **sandglass** helped to measure the ship's speed.
- **Ship design**. New types of ships were built, like the **caravel**, which was lighter and safer. They could carry more goods than earlier ships.

ACTIVITIES

3. What territories did Europeans know in the 15th century? Which continents did they not know?
4. What influence did trade have on the voyages of discovery at the end of the 15th century?
5. What technical advances made navigation easier at that time?
6. Which technical advance do you think was most important for navigation? Discuss your opinion with your classmates.

Vasco de Gama.

2 The Portuguese expeditions

Portuguese objectives

From the early 15th century, Portugal wanted to find a **trade route** to India. This would not be the same as the trade routes across the Mediterranean Sea that were used by Italian and Arab merchants. The Portuguese believed that, in this way, they could control the silk and spice trade with the Far East.

Prince **Henry the Navigator**, the son of King John I of Portugal, promoted this ambitious project. He created a **school of navigators**, which trained crews for expeditions. Cartographers from other countries visited the school, where they drew accurate maps.

The first voyages

The Portuguese carried out a series of expeditions in the 15th century:

- After the conquest of **Ceuta** (1415), Portuguese expeditions in the **Atlantic Ocean** discovered **Madeira** (1419) and the **Azores** (1431).
- Other Portuguese expeditions explored the **coastline** of **West Africa**. They occupied the **Cape Verde** islands (1446) and sailed to the **Gulf of Guinea** (1460).
- Portuguese exploration continued after the death of Henry the Navigator in 1460. In 1487, Bartolomeu Dias reached the **Cape of Good Hope**, on the south coast of Africa. This opened a route to the **Indian Ocean**.

ACTIVITIES

1 Look for information about Vasco de Gama on the Internet. Find out where he went on his three journeys, and how he died.

DID YOU KNOW?

The caravel

The caravel was a light ship developed by the Portuguese in the 15th century. Its square sails made it possible to travel very fast, and its triangular sails were used to change direction. A caravel had a crew of about thirty men. Only the captain had a cabin to himself, and the rest of the crew slept in a single space. Caravels could travel long distances and were used on the great Portuguese and Castilian voyages of discovery in the 15th and 16th centuries.

- Imagine you are a sailor on a caravel. Find out more about the voyages of discovery and write a diary of one week aboard the ship.

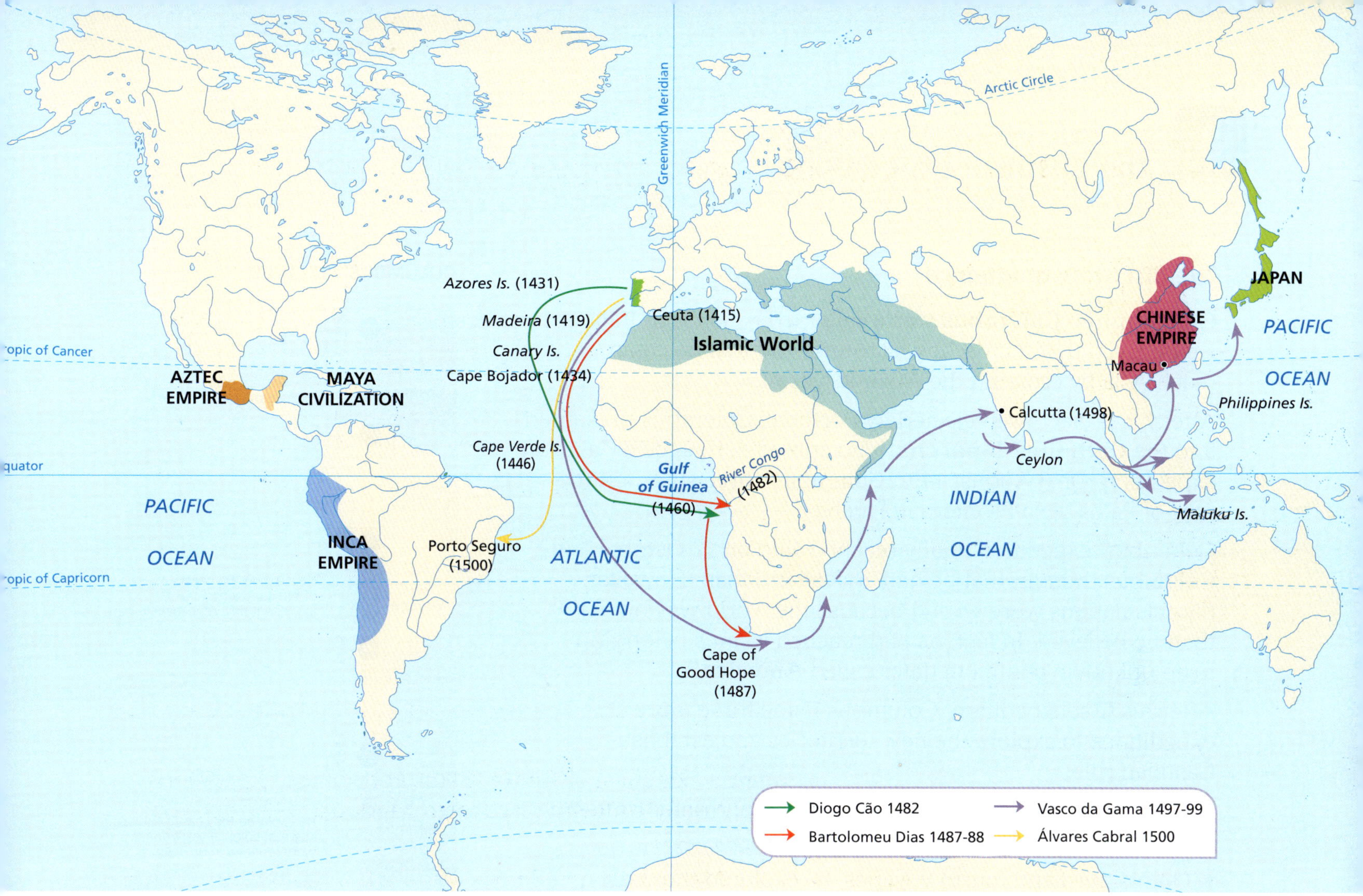

PORTUGUESE DISCOVERIES AND EXPEDITIONS

Portuguese expansion

Christopher Columbus planned to travel west across the Atlantic Ocean to establish a new route to Asia, and he requested financing from the King of Portugal. However, Columbus' project was rejected because Portugal had already discovered a new eastern route to Asia around the Cape of Good Hope.

- In 1497, **Vasco da Gama**'s expedition left Lisbon, reaching **India** in 1498. In this way, Portugal strengthened its presence in the Indian Ocean.
- In 1500, **Pedro Álvares Cabral** discovered **Brazil** after his ship sailed too far west during a voyage in the Atlantic Ocean.

As a result of these expeditions, Portugal created a great maritime empire. It was defended by fortified trading posts, called **factories**, which controlled trade and guaranteed supplies for the Portuguese fleets.

WORK WITH THE MAP

2. What territories did the Portuguese explore?
3. Which expedition first reached India by going round Africa? Describe the route.
4. Which explorer discovered Brazil?

ACTIVITIES

5. Why did the Portuguese travel east and not west to reach India?
6. Create a timeline showing Portugal's achievements.
7. Why did the Portuguese set up factories on the coast and not inland?

3 The Castilian expeditions

The discovery of America

Christopher Columbus was a sailor from Genoa in Italy. He believed that the Earth was round, and that he could reach Asia by crossing the Atlantic Ocean.

After his project was rejected by Portugal, Columbus presented it to the **Catholic Monarchs**, who agreed to finance a Castilian expedition. On 3 August 1492, Columbus' three ships left the port of Palos de la Frontera in Huelva.

Columbus' expedition finally reached land on 12 October 1492. Columbus thought that he had arrived in Japan. However, his calculations were wrong because the world was bigger than he believed. In fact, he had reached islands belonging to an unknown continent (later called **America**).

After his first expedition, Columbus made three more expeditions to explore the new territories and establish Castilian rule:

- On his second voyage, he found a faster and safer route to America.
- On his third and fourth voyages, he explored the coast of Central and South America.

Until the end of his life, Christopher Columbus was convinced that he had reached Asia. However, the Italian navigator **Amerigo Vespucci** proved that the lands were part of a new continent, which was named America in his honour.

COLUMBUS' VOYAGES

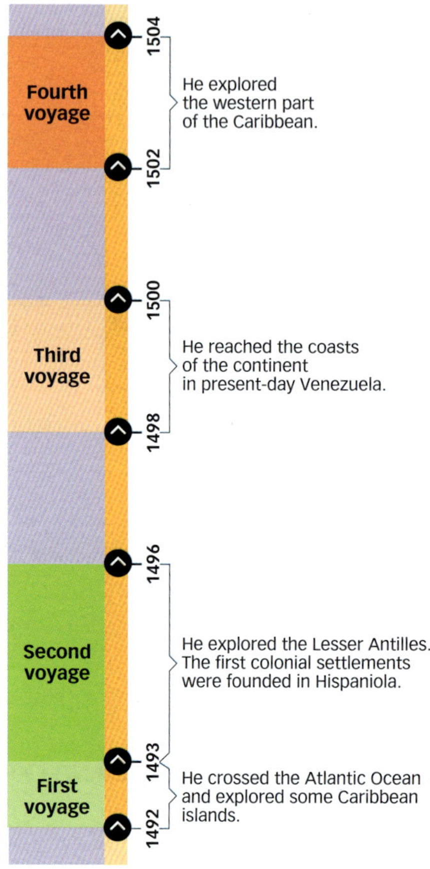

The beginning of the overseas empires

Castilla and Portugal wanted to control the territories discovered by their expeditions. To avoid conflict, they accepted mediation by the Pope. The **Treaty of Tordesillas** (1494) established that the line of division between Castilian and Portuguese territories would be along a meridian about 2,000 km west of the Azores. All the territories east of this meridian (for example, those in Africa and Asia) belonged to Portugal. All the land west of this line belonged to Castilla. Most of America was therefore Castilian.

The expeditions marked the beginning of Castilla and Portugal's great overseas empires. They were the start of European domination throughout the world.

As a result of the voyages of discovery, people learned a great deal about the new territories around the world. Many new species of plants and animals were discovered.

THE TREATY OF TORDESILLAS

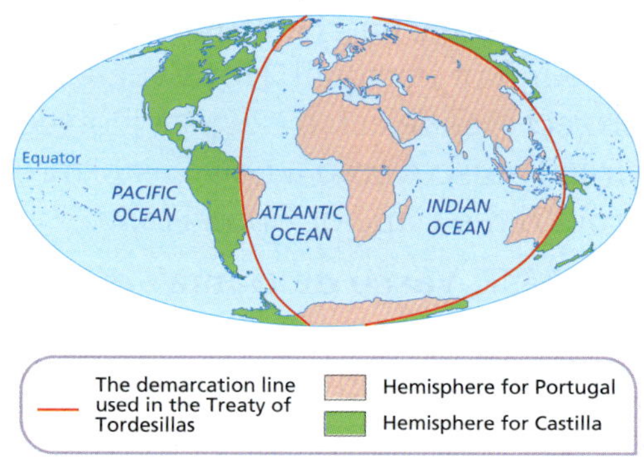

WORK WITH THE MAP

1. Brazil was discovered by a Portuguese expedition in 1500. Why did it belong to Portugal and not to Castilla?

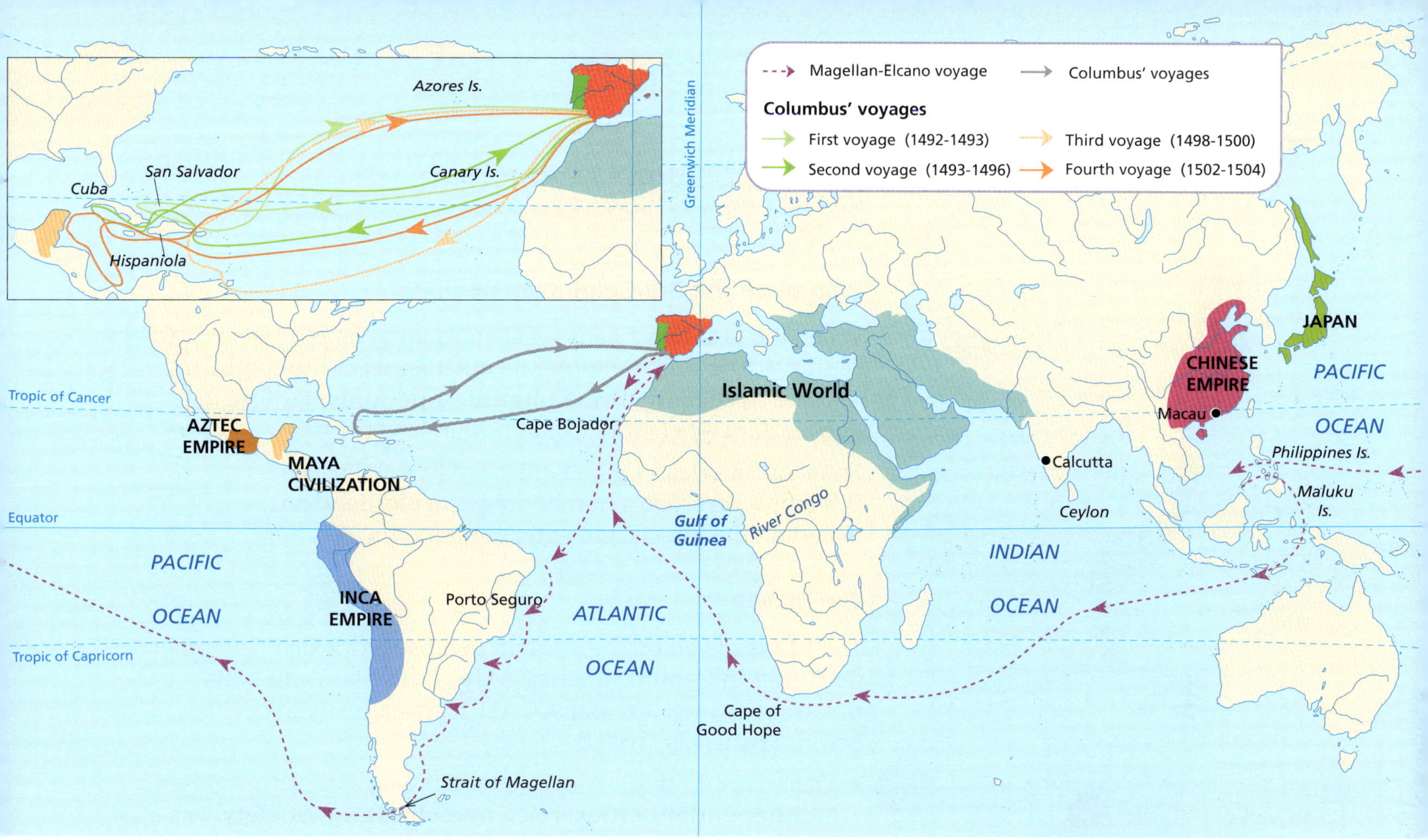

MAIN CASTILIAN EXPEDITIONS

The first voyage around the world

In 1519 an expedition of five ships left Seville under its Portuguese captain **Ferdinand Magellan**. Their goal was to find a route from the Atlantic Ocean to the Maluku Islands in the Pacific Ocean. These were known as the "spice islands".

In 1520, this expedition crossed a strait connecting the two oceans (later known as the Strait of Magellan), and sailed across the Pacific Ocean to the Philippines. Following Magellan's death, his second-in-command **Juan Sebastián Elcano** was in charge for the return journey.

This expedition completed the first journey round the world, and showed that the Earth was round.

Juan Sebastián Elcano

WORK WITH THE MAP AND THE TIMELINE

2 Which American territories did Columbus explore on each of his voyages?

3 Describe the route followed by Magellan and Elcano.

4 How did Magellan and Elcano's voyage show that the Earth is round? Discuss your explanation with your classmates.

ACTIVITIES

5 What was Columbus' goal? Did he achieve it? Explain your answer.

6 Why did Castilla and Portugal sign the Treaty of Tordesillas? What did the treaty establish?

7 Imagine that you are a 16th century scientist. Write a list of all the things that you have learned from the Castilian and Portuguese discoveries.

PRE-COLUMBIAN CIVILISATIONS

WORK WITH THE MAP

1 Compare this map to an atlas. In which modern countries were the Inca and Aztec Empires? Where were the Maya cities?

4 Pre-Columbian America

Who were the Pre-Columbian peoples?

When the first Europeans reached America in the 15th and 16th centuries, they encountered the inhabitants of the continent. These are called the **Pre-Columbian peoples** because they lived in America before Columbus arrived there.

Some of America's inhabitants lived in small tribes. But there were also a number of great **civilisations**, including the Mayas, the Aztecs and the Incas.

The Mayas

The **Mayas** lived on the Yucatán peninsula (Mexico) and other territories in Central America. Maya civilisation was at its greatest between 300 and 900 AD. It was already in decline before Spaniards arrived.

- There was no single Maya state. The Mayas were organised in **independent cities**, ruled by a **chieftain**. **Society** was made up of nobles, priests, craftsmen, peasants and slaves, who were usually prisoners of war.
- Their economy was mainly based on **agriculture**: they cultivated corn, cotton, cacao and tomatoes. There was **trade** between cities.
- The Mayas were **polytheistic**. They knew about **astronomy**, and created accurate calendars. They developed writing.
- They built palaces and temples in the form of **step pyramids**.

Ruins of the Maya city of Palenque (Mexico).

The Incas

The **Incas** were from the central Andes. They created a vast empire which included parts of present-day Peru, Bolivia, Ecuador, Argentina and Chile.

- Their emperor was the **Inca**, who was believed to be the son of the Sun. Their capital was **Cuzco**. **Society** was made up of nobles, peasants, craftsmen and slaves.
- **Terraces** were built on steep mountain sides, where crops like corn and potatoes were grown.
- **Llamas** and **alpacas** provided the Incas with food and wool. They were also used as transport.
- The Incas were **polytheistic**. Two of their main gods were Viracocha, the god of creation, and Inti, the Sun god.
- They built large temples, palaces and fortresses, as well as a network of **roads**.

The Aztecs

The **Aztecs** were warriors who settled in the Valley of Mexico in the 12th century. They collected taxes from conquered peoples and sacrificed their defeated enemies. Their empire was based on an alliance of three city-states: **Tenochtitlán** (their capital), **Texcoco** and **Tlacopán**.

- The **emperor** had political and religious power. **Society** was organised in tribes divided into groups: warriors and priests, merchants, craftsmen, peasants and slaves.
- The Aztec **economy** was mainly based on agriculture, but they were also potters, goldsmiths and silversmiths.
- The Aztecs worshipped **various gods**. Huitzilopochtli, the Sun God, was their most important god and they offered him human sacrifices.
- They built fortified cities, palaces and temples in the form of pyramids.

Saksaywaman was a huge Inca fortress high above their capital city Cuzco.

Pot representing Tláloc, the Aztec god of rain.

ACTIVITIES

2 Match the columns and write true sentences in your notebook.

• The Mayas	• built terraces to grow crops on the mountain sides.
	• offered human sacrifices to their gods.
• The Aztecs	• used llamas and alpacas for food and cloth.
	• had a federation of three cities.
• The Incas	• were organised in independent cities.
	• developed writing.

3 Form groups of three students and investigate a Pre-Columbian civilisation.

a Find at least three good websites and share your links with each other.
b Choose one civilisation, like the Olmecs or Toltecs.
c Investigate where and when it developed. Look for a map and attractive illustrations.
d Present your findings as a computer presentation or an illustrated poster.

5 Social and economic change

EVOLUTION OF THE POPULATION OF EUROPE

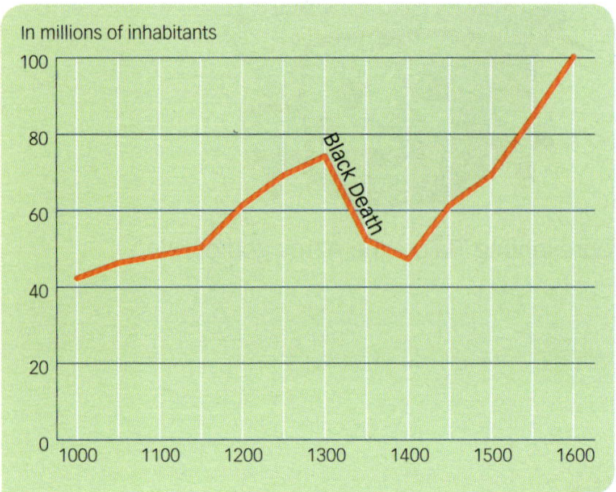

WORK WITH THE GRAPH

1. How many million people were there in Europe in 1400? In 1500? And in 1600?
2. How did European population evolve between 1000 and 1500?

The port of Hamburg. Many ports in north and north-west Europe, like Hamburg and Antwerp, expanded greatly in the 15th and 16th centuries.

Demographic recovery

- In the 14th century, there was a **demographic crisis** in Europe as a result of wars, poor harvests and the impact of a terrible epidemic called the **Black Death**.
- However, in the 15th century Europe's population began to grow again. There were fewer epidemics. Following improvements in agriculture, people had more and better food and so the death rate declined.

Society

Society was still divided into three **estates**: the nobility, the clergy and the rest of the population. However, there were significant changes:

- The **nobles** and the **clergy** still had many privileges, and they held important positions.
- The **bourgeoisie** became more prominent. Some merchants and bankers accumulated large fortunes, and their children married members of the nobility.
- Less wealthy **merchants** and **craftsmen** often came into conflict with the nobility and bourgeoisie over their role in city government.
- **Peasants** still had a hard life. However, in most parts of Europe they were **free** from their feudal lords. They were no longer serfs.

The economy

The 15th and 16th centuries were a period of economic growth.

- Population growth increased demand for food. More **land** was cultivated and agricultural production grew.
- Specialised **artisans** supplied the growing demand for a range of products (such as clothing, leather and metal work) from people living in cities.
- Both regional and international **trade** increased greatly. The voyages of discovery created new trade routes.
- Many **cities** grew rapidly, especially those with ports and craft industries. Antwerp, London, Hamburg, Venice, Lisbon and Seville were some of the greatest European cities. Many of the most important centres of economic activity were now in **north-west Europe**, and not around the Mediterranean Sea.

Carlos V and the banker Jacob Fugger, a 19th century painting by Carl Becker. Carlos V borrowed an enormous sum of money from the Fuggers to finance his campaign to become emperor.

WORK WITH THE IMAGE

3 Which social group did the Fuggers belong to? Why were they connected to the emperor Carlos V?

4 Describe the room. What clothes are the people wearing?

5 Form groups of three. Find out about one of the following:

a The Fugger family and how they made their fortune.

b Another family of bankers at the beginning of the Early Modern Period.

Present your findings to the class.

Banking and finance

Private banks played an important role in international finance. Famous banking families included the Medici in Italy and the Fuggers in Germany.

The first **public bank** was founded in Barcelona in 1401. Other public banks were created in Italian cities like Genoa and Venice.

Bankers gave merchants credit and popularised the use of **cheques** and **bills of exchange**. In this way, merchants could carry out commercial transactions without having to use cash.

Large capital investments were required for the new long distance trade routes to America and Asia. These investments were very risky as many ships sank during the voyage or were attacked by pirates. Merchants formed **trading companies** to share the profits and the losses.

DID YOU KNOW?

A **bill of exchange** was a document used as a method of payment in a commercial transaction. It was paid on a different date and at a different bank from where it came from.

A **cheque** was a payment order against a bank deposit. It could be paid at any branch of the bank where the deposit was.

ACTIVITIES

6 Why did population start to increase in the 15th century?

7 What aspects of society were the same as in the Middle Ages? What changed? Make two lists.

8 What new economic instruments were developed in this period? Why were they necessary?

ACTIVITY ROUND-UP

1 Copy and complete the table.

THE AGE OF EXPLORATION			
Geographical discoveries	Factors		...
	Portuguese discoveries		...
	Castilian discoveries		...
Pre-Columbian America	Mayas	Location	...
		Characteristics	...
	Aztecs	Location	...
		Characteristics	...
	Incas	Location	...
		Characteristics	...
Social change	...		
Economic change	...		

2 Copy the timeline and put Portugal's main discoveries on it. Then do the same for Castilla.

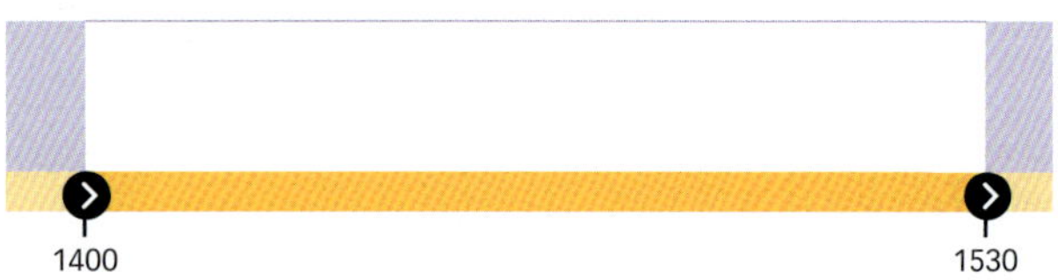

3 Briefly explain the historical role played by the following people:

- Henry the Navigator
- Amerigo Vespucci
- Bartolomeu Dias
- Ferdinand Magellan
- Juan Sebastián Elcano

4 Write a short summary of Christopher Columbus' expedition. Include:

- The aims of his first voyage.
- The source of his financing.
- The places he visited on his four voyages.

5 Draw a map of the main Pre-Columbian civilisations.

- Copy a map of South and Central America and draw the frontiers of modern-day countries.
- Colour the territories occupied by the Mayas, the Incas and the Aztecs.

6 When and why was the Treaty of Tordesillas signed? What was it? What were its consequences?

7 Describe the image.

- What are these people doing? Which social group do they belong to?
- What was banking like at the beginning of the Early Modern Period?

A 16th century tapestry showing a banker.

8 What were the implications of Magellan and Elcano's expedition for geographical knowledge?

9 How do you think the invention of cheques and bills of exchange affected trade?

10 Are the following sentences true or false? Correct the false ones in your notebook. Then listen to the recording and check your answers.

a In 1500 there were more people in Europe than before the Black Death.

b The population began to increase in the 15th century when there were fewer epidemics.

c The bourgeoisie had the same privileges as the nobles and the clergy.

d People from the bourgeoisie could rise on the social scale by marrying nobles.

e The most powerful kings gave bankers money.

f Peasants in Europe still had a hard life, but they were no longer serfs.

11 Think about the term the "Age of Discovery". Is this the best title for this unit? Can you think of a better title?

THE LEGACY OF THE AGE OF DISCOVERY

The geographical discoveries left us a very important legacy:

- **Knowledge of the shape and size of the Earth.** The expeditions of the 15th and 16th centuries proved that the Earth was round. Completely new maps were made showing previously unknown lands.
- **The art of navigation.** This improved greatly with the development of navigational instruments, improved charts and new types of ships.
- **Encounters with other cultures.** Europeans and non-Europeans learned about the existence of other cultures. Following the discovery of America, new plants like tomatoes, potatoes and cacao (used to make chocolate) reached Europe.
- **Banking and trade.** There was a great increase in trade and banking at the beginning of the Early Modern Period.

12 What does the map show? Is it like a modern map? Explain your answer.

13 Look for three good websites about native American peoples. Prepare a fact file on one people that still exists today.

14 In 1987 Unesco declared Teotihuacán a World Heritage Site. Find out more about Teotihuacán and explain this decision.

Map by Theodor de Bry (1596).

Teotihuacán (Mexico).

KNOW HOW TO... Key Competences

THINK LIKE AN HISTORIAN. **Analyse sources**

Extracts from *Christopher Columbus' Journal* (1492)

3rd August

They set out at 8 o'clock on Friday 3rd August 1492.

6th September

The Admiral stored water, firewood, meat and other things. Finally, he set sail from the island of La Gomera with his three caravels.

10th October

The sailors became impatient and complained about the length of the voyage. But the Admiral encouraged them, telling them about the profits they would receive. He added that there was no use complaining because they were going to the Indies. They would have to continue until they arrived there with the Lord's help.

11th October

The Pinta was the swiftest ship, and was ahead of the Admiral's ship. It discovered land and made the signals which had been ordered. A sailor called Rodrigo de Triana was the first person to see land.

12th October

The Admiral landed on the island in a small boat with the two captains and other sailors. They saw trees, rivers and streams, and many kinds of fruits. The Admiral told the others to witness how he took possession of the island on behalf of the king and queen.

Then many people from the island assembled together. These were the Admiral's words: "I wanted them to see us as friends, and convert them to Our Holy Religion without violence. I gave them some coloured caps and glass beads which they put round their necks, as well as other things of little value. They were very pleased with them and were wonderfully friendly with us. But I thought they were very poor people. They were all completely naked. All the people I saw were young and good-looking. Their hair was short and rough like a horse's tail, except for a small part which hung behind their heads and which they never cut.

They did not have any weapons, and they did not know anything about them. I showed them swords and they took them and cut themselves out of ignorance. They don't have any iron. Their spears are sticks with a fish's tooth at the end. I saw that some of them had wounds on their bodies and I gestured to ask what they were. They showed me how people came from other islands and attacked them and they defended themselves."

Columbus' Journal (adapted)

Christopher Columbus reaches land in 1492.

To analyse a journal, follow these steps:

- **Setting.** Situate the text in time and place. Do we know who the author is?
- **Analysis.** Describe the events and list the people and places
- **Commentary.** Check if the information matches what you have studied. Then assess its importance as a historical source.
- **Investigation.** Look for more evidence to complete your information.

1 Setting.

a When was the journal probably written? Who was the probable author?

b Who is the "Admiral"? Why does the text sometimes say "he" and sometimes "I"?

c Indicate the main characteristics of the period when it was written.

2 Analysis.

a What historical event does it narrate?

b Why did Columbus stop off in La Gomera in the Canary Islands?

c How did the crew feel before they saw land?

d What were the inhabitants of the island like?

3 Commentary.

a Did the author witness the events?

b What impression did he have of the Indians?

c Do you think the text reflects what really happened? Why? / Why not?

4 Investigation.

a Look for more information on the Internet using the keywords "Columbus Letter First Voyage".

b Was the surviving version of this document written by Columbus himself?

CULTURAL HERITAGE. **The frieze of Holmul**

A few years ago, a group of grave robbers dug into the ground at an archaeological site in Holmul in Guatemala. They found nothing of value, so they left an abandoned tunnel. Later, archaeologists went further along the tunnel and found a tomb hidden in a pyramid. Inside there was an enormous frieze, 8 metres wide and 2 metres high.

The frieze was carved in the 6th century. It has great artistic value. But it is also an extraordinary historical source because it describes a period of wars between different Maya groups, which was previously little known. It shows that there was an alliance between Holmul and other Mayan kingdoms to control Tikal and create an empire.

1 Think about what you have learned and answer the questions.

a Where did the Mayas settle?

b How were they organised politically?

c When did the Maya civilisation reach its greatest splendour? Was the frieze of Holmul made at that time?

2 Study the frieze.

a Who do you think this tomb belonged to?

b What historical information is found in the frieze?

c How are the local kings represented?

3 Draw conclusions.

a Does the frieze tell us anything about the relations between Maya groups?

b Is it true that the Mayas were always peaceful?

c Why do you think the Mayas did not create an empire like the Incas or the Aztecs?

4 Look for information about the Joya de Cerén Archaeological Site on the Internet. Prepare a fact file.

10 Renaissance and Reformation

FIND OUT ABOUT:

- Humanism
- The main characteristics of Renaissance art
- The Reformation
- The Counter-Reformation

KNOW HOW TO:

- Write a biographical profile
- Interpret a map showing religious divisions

Leonardo da Vinci, a universal genius

Leonardo da Vinci (1452–1519) was an incredibly versatile Renaissance genius. He was an extraordinary painter, engineer and inventor. He investigated in numerous fields, such as anatomy, mechanics, optics, hydraulics, architecture, meteorology and astronomy.

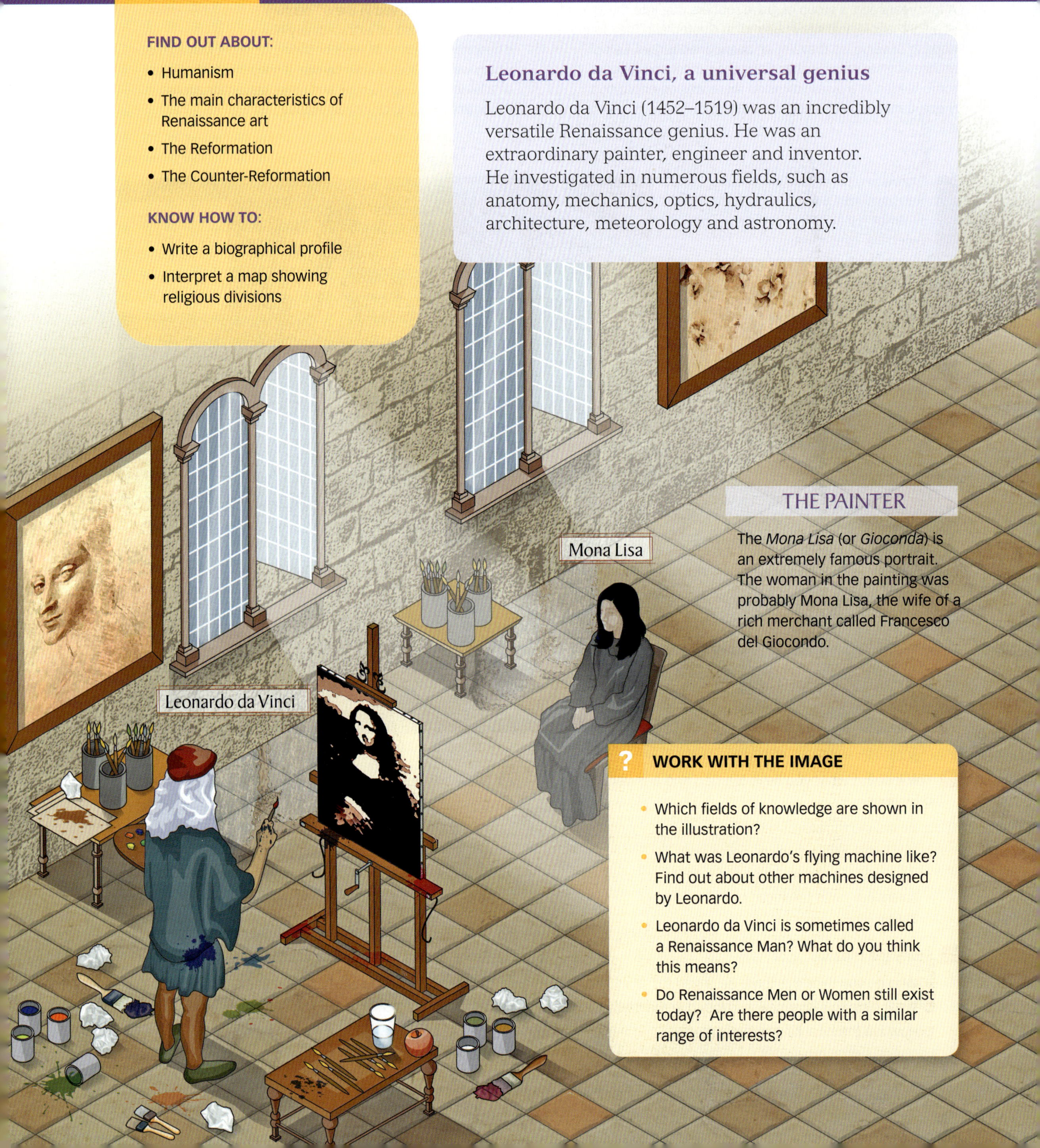

THE PAINTER

The *Mona Lisa* (or *Gioconda*) is an extremely famous portrait. The woman in the painting was probably Mona Lisa, the wife of a rich merchant called Francesco del Giocondo.

? WORK WITH THE IMAGE

- Which fields of knowledge are shown in the illustration?
- What was Leonardo's flying machine like? Find out about other machines designed by Leonardo.
- Leonardo da Vinci is sometimes called a Renaissance Man? What do you think this means?
- Do Renaissance Men or Women still exist today? Are there people with a similar range of interests?

DID YOU KNOW?

Many texts by Leonardo are written backwards. We can only read them with the help of a mirror!

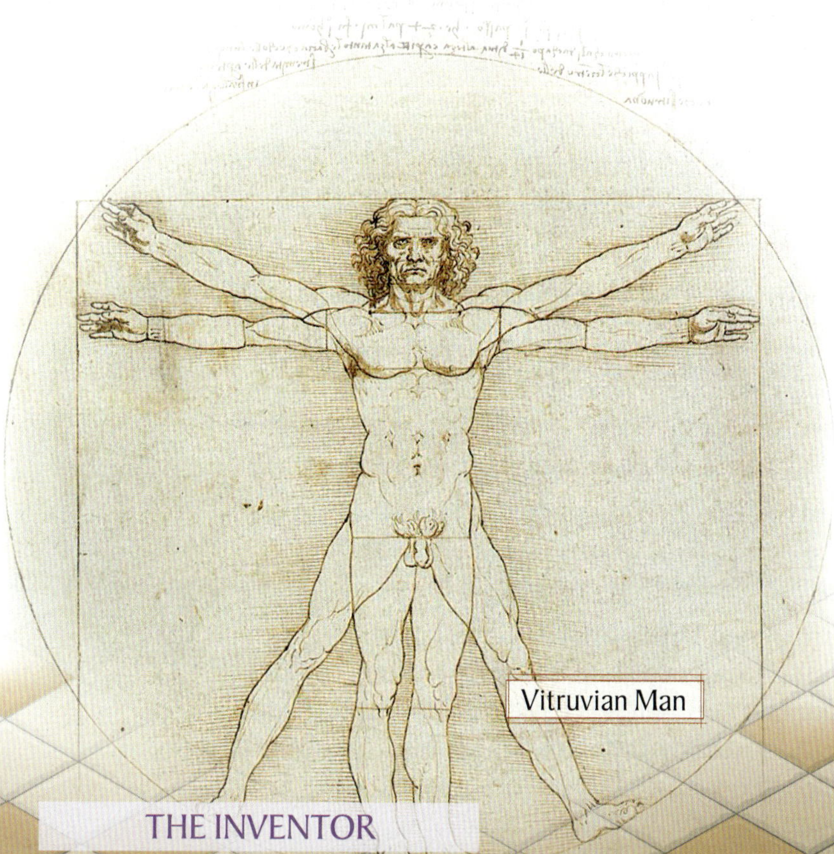

THE SCIENTIST

Leonardo was interested in anatomy and made studies of the human body. The most famous is 'Vitruvian Man'. This drawing was based on the texts of Marcus Vitruvius, a great ancient Roman architect.

THE INVENTOR

Leonardo was fascinated by the flight of birds and he designed several types of flying machine. He also designed war engines and other machines.

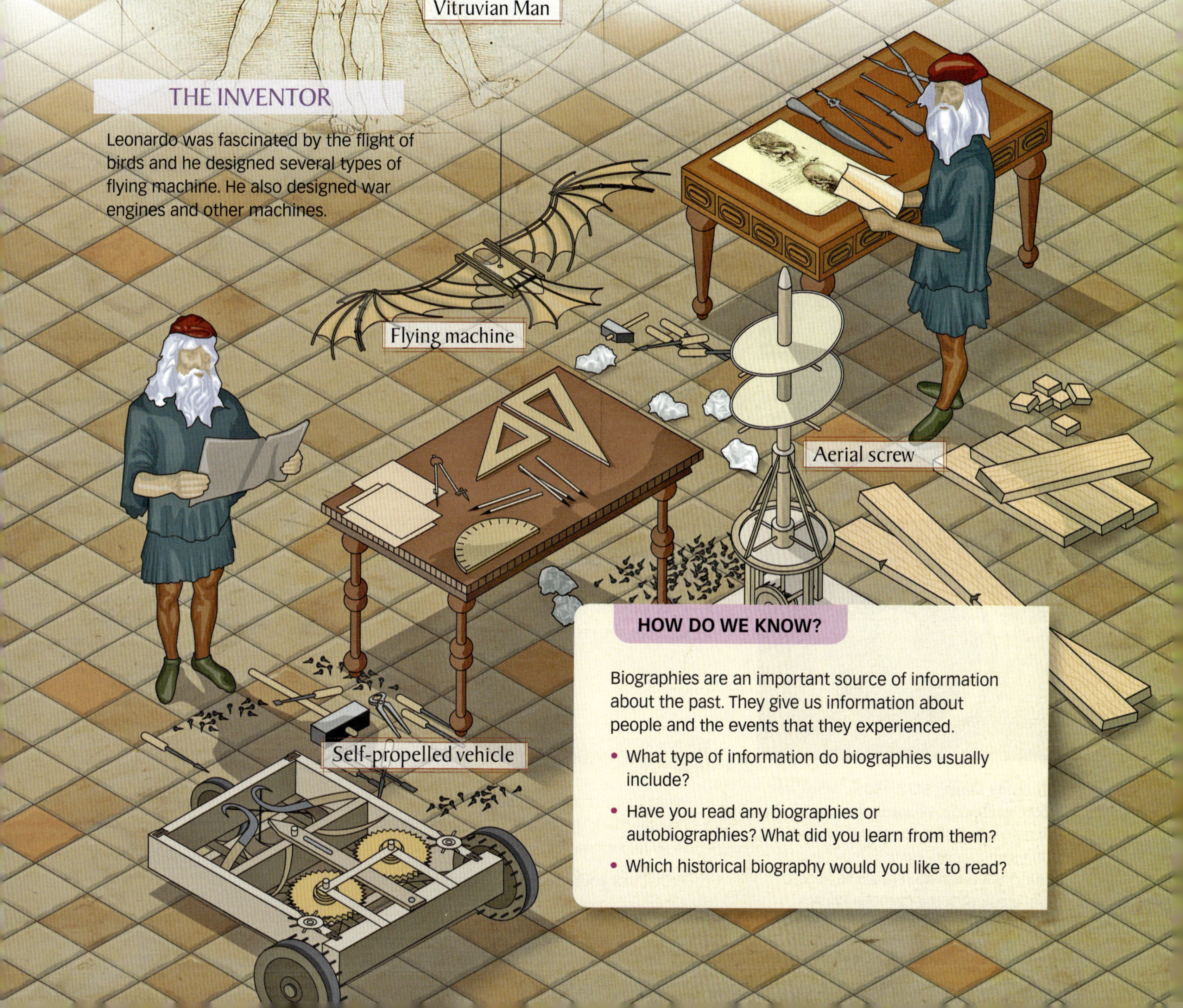

HOW DO WE KNOW?

Biographies are an important source of information about the past. They give us information about people and the events that they experienced.

- What type of information do biographies usually include?
- Have you read any biographies or autobiographies? What did you learn from them?
- Which historical biography would you like to read?

1 What was humanism?

Humanism

There were many political, social and economic changes in Europe from the 15th century onwards. These were accompanied by important changes in the way that people viewed the world.

Humanism was the philosophical and cultural movement that spread through Europe in the 15th and early 16th centuries.

- The medieval **theocentric** viewpoint was that all human activities revolved around God. Humanists still believed in God. However, they held the **anthropocentric** viewpoint that human intelligence was key to understanding and interpreting the universe.
- According to the humanists, our use of **reason** and **experience** advanced our knowledge. The humanists observed nature and carried out **experiments**.

  This approach laid the foundations for the scientific revolution of the 17th century.
- The humanists felt deep **curiosity** about the physical and spiritual world. They wanted to know about everything. The term 'Renaissance Man' refers to humanist scholars who explored all fields of human activity.
- They were inspired by the cultural inheritance of **Antiquity**. After the Turks conquered Constantinople in 1453, many Greek scholars went to live in Europe. They brought Greek and Roman texts which were of great interest to scholars.
- Most medieval texts were written in Latin. But under humanist influence many literary works were now written in modern languages.

THINK ABOUT IT

What were the key ideas of humanism in the 15th and 16th centuries? Do you think the word "humanism" is used in the same way today? How has its meaning changed?

- Discuss your ideas with your classmates.

ACTIVITIES

1. In pairs, each choose one of the following subjects. Look for information and share it with each other:
   - The life of Antonio Nebrija, and why his *Gramática* was so important.
   - The life of Luis Vives.
2. Find out about the main school of thought in the Middle Ages. What were the differences and similarities to humanism?

Erasmus of Rotterdam (1466?-1536). Erasmus was a Dutch humanist who admired ancient Greece and Rome. His famous work *The Praise of Folly* attacked ignorance and superstition and was translated into many languages.

Thomas More (1478-1535). More was born in London. He wrote *Utopia*, which imagined an ideal society where peace and justice ruled. He was executed by Henry VIII when he refused to accept the break with the Catholic Church.

A PRINTING PRESS IN THE EARLY MODERN PERIOD

The spread of humanism

- Gutenberg perfected the **printing press** in around 1440. This meant that each book no longer had to be copied by hand. Consequently, a larger number of books could be produced in less time. It became less expensive to make books so more people had access to them.
- The **academies** were cultural centres where the humanists exchanged ideas and knowledge. The Platonic Academy was founded by Cosimo de' Medici in Florence in 1459.
- **Universities,** like those of Vienna, Krakow, Bologna and Alcalá de Henares promoted the study of classical culture.
- Humanists **travelled** to meet each other in other ways. They also exchanged ideas through written correspondence. In this way humanist ideas spread throughout Europe.

WORK WITH THE ILLUSTRATION

3 Study the illustration and look for more information about Gutenberg's printing press on the Internet. With a partner, discuss what the people in the illustration are doing.

ACTIVITIES

4 Why did the printing press have such an impact in Early Modern Europe?

5 How do you think books are printed today? Do you mainly read texts on printed paper or on a screen?

2 New directions in science and art

ACTIVITIES

1 Carry out an Internet investigation using the key words "Medicine in the Middle Ages". Then carry out similar searches for "Renaissance medicine", "Vesalius" and "Miguel Servet".

Prepare a report summarising the main differences between medieval and Renaissance medicine. Do you think the care of patients really improved during the Renaissance?

A new science

During the Middle Ages, the ideas of great classical thinkers like Aristotle were accepted without discussion. But during the Renaissance, scholars had such an immense desire for knowledge that they were prepared to question everything. They carried out **experiments** to prove their theories.

- **Astronomy, medicine** and **physics** were the main branches of Renaissance science.
- Geographical discoveries led to advances in **cartography**, **botany** and **zoology**.

ASTRONOMY

Nicolaus Copernicus

In 1543, the astronomer **Nicolaus Copernicus** published *On the Revolutions of the Heavenly Spheres*, presenting his **heliocentric** theory. According to Copernicus, the Sun was the centre of the universe, and not the Earth. All the planets revolve around the Sun. However, the Church continued to insist that the Earth was the centre of the universe.

MEDICINE

Miguel Servet

In 1543, the doctor **Andreas Vesalius** published *On the Structure of the Human Body*. Based on direct observation, this book was a big step forward in the knowledge of **human anatomy**. It became the manual for students of medicine at that time. Ten years later, the doctor and theologian **Miguel Servet** discovered pulmonary circulation of the blood.

PHYSICS

Leonardo da Vinci

The most important scientist in this field was **Leonardo da Vinci**, who made numerous contributions to hydraulic and aerodynamic mechanics. He is responsible for **inventions** such as gears, propellers, windmills, the flying machine, the glider, the crane, the tank and the warship. He also carried out research in the fields of anatomy, astronomy and botany.

WORK WITH THE IMAGES

2 What were the heliocentric and geocentric theories?

3 Which movements of the Earth can you see in the picture showing the heliocentric theory?

4 Why do you think the Church rejected the heliocentric theory?

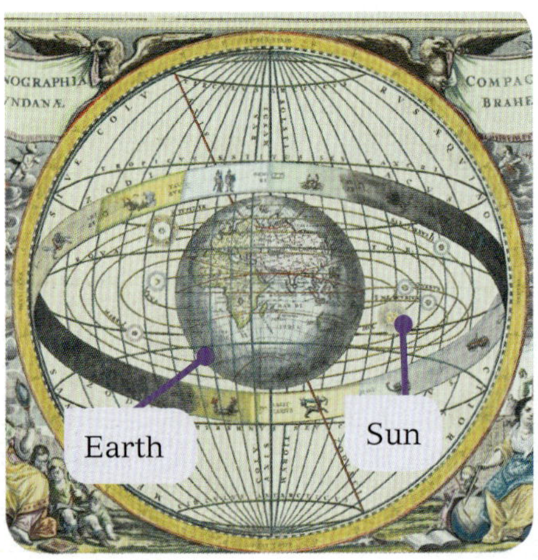

Geocentric theory.

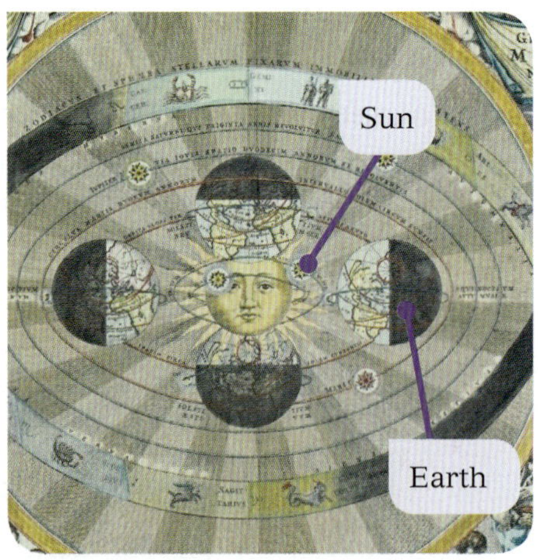

Heliocentric theory.

The Renaissance

Renaissance art broke with the traditions of the Middle Ages. Its main centre of interest was the **human being**.

Artists looked for an ideal form of beauty. They found inspiration in the **classical world** of ancient Greek and Rome. They tried to achieve harmony and proportion in their art.

The Renaissance style began in Italy and developed in two main stages:

- In the 15th century (***Quattrocento***), it centred on the city of Florence.
- During the **High Renaissance** (or *Cinquecento*) of the 16th century, it centred on Rome.

Patrons and artists

The prosperous Italian city-states were home to important **patrons**. These wealthy noblemen and burgers appreciated education and the arts. They supported artists and gave large sums of money towards building and decorating churches and palaces.

The **Medici** family of Florence were patrons of some of the greatest Renaissance artists, including Donatello and Michelangelo.

Other important patrons were the **Sforza** in Milan, the **Gonzaga** in Mantua, and **Popes Julius II** and **Leon X** in Rome.

Architects, sculptors and painters were no longer anonymous craftsmen. They were **artists** who signed their work. The conception of a work of art became as important as the technical skill required to make it.

DID YOU KNOW?

A biographical profile

A biographical profile records the main events in a person's life. It includes the following:

- Date of birth and death.
- Place of birth and death.
- Education and studies.
- Their main activities.
- The most important events in their life and their place in history.

5 Create a biographical profile of Lorenzo de' Medici.

- When and where was he born? Which family did he belong to?
- What did his family do? What part did Lorenzo play in their activities?
- What was his political role in Florence?
- Which cultural institutions did he support? For which artists did he act as patron?
- Why is he sometimes called Lorenzo the Magnificent?
- When and where did he die?

ACTIVITIES

6 Can we speak of "a new science" in the Early Modern period?

7 What was the main source of inspiration of Renaissance art? What were its main stages?

8 Explain what a patron does. Who were the main patrons of the Italian Renaissance?

9 Do you know the names of any medieval artists? How many Renaissance artists can you name? How do you explain the difference?

3 The *Quattrocento*

Basilica of Sant'Andrea in Mantua, designed by Alberti. A. Interior (central aisle) B. Exterior (main facade)

Architecture

Renaissance architects designed both religious and civil buildings, and adapted them to human needs.

Following ancient Greek and Roman models, they looked for **symmetry** and **proportion**. They made use of **classical features** such as rounded arches, pediments, barrel vaulting, domes and columns.

- **Filippo Brunelleschi** was one of the greatest architects of the *Quattrocento*. He designed the dome of Florence Cathedral and the churches of the Santo Spirito and San Lorenzo, also in Florence.
- Another important architect was **Leon Battista Alberti**, who designed the facade of the Church of Santa Maria Novella in Florence, and the Basilica of Sant'Andrea in Mantua.

WORK WITH THE PHOTOS

1 Which classical features can you see in the Basilica of Sant'Andrea in Mantua?

The doors of the Baptistery in Florence Cathedral by Lorenzo Ghiberti.

Sculpture

Sculpture, like architecture, was inspired by classical forms. It was realistic but idealised.

- It often represented the **nude human body**. Special importance was attached to proportion, beauty and anatomical accuracy. Numerous individual **portraits** and **equestrian** statues were also created.
- Important sculptors included **Lorenzo Ghiberti**, who designed the doors of the Baptistery in Florence Cathedral, and **Donatello**, whose works included *David*.

Painting

Renaissance painters made some remarkable technical discoveries.

- Their use of **linear perspective** gave great depth to their paintings.
- They achieved a sensation of volume by using **chiaroscuro** (contrasts of light and dark), and also by shading their paintings in different colour tones.

They continued to paint religious scenes, but these demonstrated the humanity of biblical figures. Portraits, landscapes and classical mythological subjects were also popular. Great painters of the *Quattrocento* included:

- **Masaccio**, who painted *The Holy Trinity*.
- **Mantegna**, who painted *The Lamentation of Christ*.
- **Botticelli**, who painted *The Birth of Venus* and *Spring*.

ACTIVITIES

2 Explain the key characteristics of the architecture, sculpture and painting of the *Quattrocento*.

3 Look on the Internet for photos of work by some of the artists, sculptors or architects mentioned on this and the previous page.

- Discuss their main characteristics in groups.

WORK WITH THE IMAGES

4 Analyse the painting by Perugino.

- Where are the most important people in this painting?
- Do you think the buildings in the background are from the time of Christ? Why / Why not?
- How does the artist use perspective in this painting?

Christ Giving the Keys to Saint Peter, by Pietro Perugino

Church of San Pietro in Montorio, by Bramante.

4 The High Renaissance

Architecture

The cultural movement of the 16th century is known as the **High Renaissance** (or ***Cinquecento***). Under the patronage of the Popes, Rome became a key centre of influence. Italy's greatest architects and artists, for example **Donato Bramante** and **Michelangelo**, worked on Saint Peter's Basilica in the Vatican.

Around Venice, the artist **Andrea Palladio** designed small palaces in the countryside, like the Villa Rotonda near Vicenza.

Sculpture

High Renaissance sculpture was large scale and monumental. **Michelangelo Buonarroti** was the outstanding sculptor. His *David*, *Moses* and the *Pietá* were among his masterpieces.

Painting

The leading Renaissance artists included the following:

- **Leonardo da Vinci's** works include *The Last Supper* and *The Mona Lisa* (*La Gioconda*).
- **Michelangelo** painted the altar wall and ceiling frescos of the Sistine Chapel in the Vatican.
- **Raphael** painted portraits and pictures with religious or philosophical subjects, such as *Madonna of the Goldfinch* and the frescos in the Vatican known as *The School of Athens*.
- **Titian** was one of the greatest representatives of the Venetian School, which was characterised by the brightness and richness of its use of colour. *The Venus of Urbino* was one of his most famous works.

WORK WITH THE PHOTO

1 Does the temple of San Pietro in Montorio remind you of a Greek or Roman building? Why?

***The Last Supper*, by Leonardo da Vinci.** Jesus is in the centre, surrounded by the Apostles.

DID YOU KNOW?

Michelangelo

Michelangelo Buonarroti (1475–1564) was one of the greatest and most versatile artists of all time. He carried out his work in Florence under the patronage of Lorenzo de' Medici, and in Rome under the Popes Julius II, Leon X, and Clemente VII.

As an architect, he was inspired by classical forms and monumental size. He designed the dome of the basilica of St Peter's in the Vatican, among other works.

Although he excelled in all the arts, he always regarded himself as a sculptor. His works show complete mastery of human proportions. He sculpted heroic characters, full of beauty and strength, like his statues of *David* and *Moses*.

As a painter, he was a master of the nude and of composition and movement. *The Last Judgement* in the Sistine Chapel in the Vatican is one of his greatest works.

The Last Judgement, by Michelangelo in the Sistine Chapel.

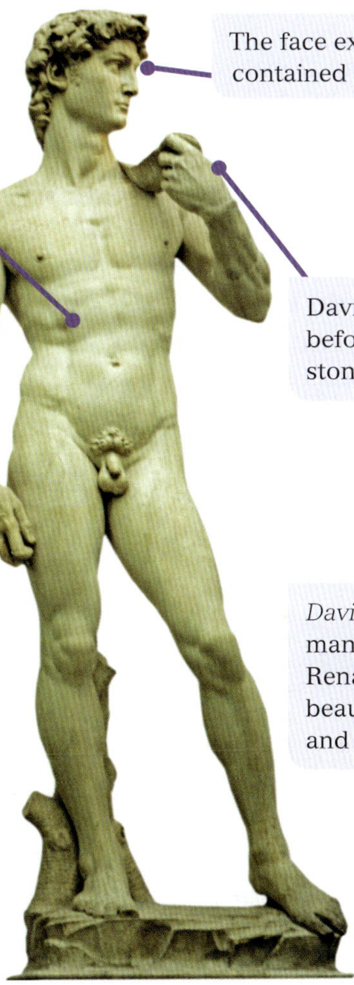

David, by Michelangelo

WORK WITH THE IMAGES

2 Which scene is represented in the *The Last Judgement*?

3 Describe the attitude and gestures of Jesus. What message does the artist convey?

4 Who are the people surrounding Jesus Christ. Who is saved and who is damned?

5 Look for information on the Internet. According to the Old Testament, who was David?

6 What is the expression on David's face? Do you think the anatomy of this statue is realistic?

Facade of Salamanca University.

Facade of the palace of Carlos V in Granada.

The Monastery of San Lorenzo de El Escorial.

5 The spread of the Renaissance

Spain

The influence of the Italian Renaissance spread beyond Italy, but it did not reach all parts of Europe immediately. In Spain, a Renaissance style developed in the 16th century.

- **Architecture.** There were three main styles:
  - The **Plateresque** style of abundant decoration developed in the early 16th century.

    The facade of Salamanca University is an outstanding example.
  - In the **Classicist** style, the harmony of a building was more important than its decoration.

    The palace of Carlos V in the Alhambra in Granada was built in the Classicist style, but this never became widespread in Spain.
  - The **Herrerian** style became influential in the second half of the 16th century. Buildings were characterised by their sobriety and almost total lack of decoration.

    The Monastery of San Lorenzo de El Escorial, designed by Juan de Herrera, was the most famous example.
- **Sculpture**. The Italian Renaissance had only limited influence on Spanish sculpture. In Spain, there was no attempt to portray ideal beauty, while sculptors like **Alonso de Berruguete** expressed intense feelings about religious subjects.
- **Painting**. **El Greco** created religious scenes full of drama and movement. El Greco's elongated figures were full of spirituality. His many great paintings include *The Disrobing of Christ* and *The Burial of the Count of Orgaz*.

ACTIVITIES

1. Where did the three styles of Spanish Renaissance architecture get their names from?

2. Which Spanish Renaissance style of architecture resembles Italian architecture?

DID YOU KNOW?

El Greco

El Greco, originally named Doménikos Theotokópoulos (1541–1614), was born in Crete. As a young man, he trained as an artist making icons in the Byzantine style. Later, he travelled to Venice, where his painting was strongly influenced by Titian. He wanted to work on the decoration of the Monastery of San Lorenzo de El Escorial. With this in mind, he presented Felipe II with the painting *The Martyrdom of Saint Maurice*, but it was rejected.

El Greco then went to Toledo where he lived until his death. In Toledo he painted masterpieces like *The Burial of the Count of Orgaz* and *The Nobleman with his Hand on his Chest*.

Detail of *The Nobleman with his Hand on his Chest*, by El Greco.

***The Burial of the Count of Orgaz*, by El Greco.** A. Saint Stephen. B. Count of Orgaz C. Saint Augustine. D. Self-portrait of El Greco. E. Saint Peter. F. Saint John the Baptist.

The Northern Renaissance

- **Dutch and Flemish art**. In north-west Europe (present-day Holland and Belgium), a highly detailed artistic style developed. Fine portraits and landscapes were created, while artists like **Pieter Brueghel the Elder** represented scenes of everyday life.
- **Germany**. **Albrecht Dürer** adopted a Renaissance style following his visits to Italy. As well as being a great painter, he was an excellent engraver and draughtsman. His works include *Adam and Eve* and the *Adoration of the Magi*.

A depiction of peasant life by Pieter Brueghel the Elder.

ACTIVITIES

3 Look on the Internet for more information about one of the following people. Write a short illustrated biography like the one on El Greco.

- Hieronymus Bosch
- Alonso de Berruguete
- Albrecht Dürer

6 The Reformation and Counter-Reformation

Martin Luther.

WORK WITH THE MAP

1 Study the map on the opposite page and answer the questions.

a What does the map show? What period in history does it represent?

b Where were there most Protestants? Where were there most Catholics?

ACTIVITIES

2 Explain the situation of the Church at the beginning of the Early Modern period.

3 What were Luther's main ideas?

4 Prepare a table with two columns summarising the main similarities and differences between Lutheran and Catholic Christianity.
Include the following headings:

- The role of the Pope and the clergy.
- Religious doctrine.
- How people worshipped.

5 What resolutions did the Catholic Church adopt at the Council of Trent?

The causes of the Reformation

The **Reformation** was a religious movement in 16th century Europe. It grew out of widespread criticism of the Catholic Church:

- The **higher clergy** usually came from noble families and lived a life of luxury, which contradicted spiritual ideals. At the same time, **parish priests** were often uneducated and did not carry out their pastoral work.
- **Indulgences** were issued by the Pope. These documents pardoned people their sins in return for money.
- **Printing** meant that more copies of the **Bible** were in circulation.

  Under humanist influence, people became more critical of spiritual authority, and wanted religion to reflect their personal reading of the Bible.

The Lutheran Reformation

In 1517, the German monk **Martin Luther** published *The 95 Theses*. This document strongly criticised the Pope and clerical corruption.

Luther was excommunicated by the Pope, and his Lutheran followers (later called **Protestants**) left the Catholic Church.

According to Luther:

- People are saved by their **faith** and not by their works.
- Christians can reach God directly through **prayer**, and can freely read and interpret the Bible. They do not require the mediation of priests.
- The only true sacraments are **Baptism** and the **Eucharist**.
- The cult of the **Virgin** and the **saints** should be suppressed.

The spread of the Reformation

The Reformation spread rapidly through central and northern Europe.

- In **Germany**, many noblemen saw an opportunity to rebel against the domination of Emperor Carlos V. They became rich by appropriating Church property.
- In Geneva (Switzerland), **John Calvin** founded **Calvinism**. According to the Calvinist idea of **predestination**, people are destined by God to be saved or damned, irrespective of their actions.

  Calvinism spread through Switzerland, the Low Countries, parts of France and Scotland.
- In **England**, King Henry VIII broke with Rome when the Pope refused to grant Henry a divorce from Catherine of Aragon. King Henry became head of the Church of England.

  Initially, **Anglicanism** was very similar to Catholicism, but it later received some Protestant influence.

The Counter-Reformation

The **Counter-Reformation** was a reform movement within the Catholic Church. It was aimed at both improving conditions in the Church and resisting Protestantism.

- **The Council of Trent** (1545–1563) adopted a series of resolutions:
  - It confirmed the authority of the Pope. It also confirmed the Church's **traditional doctrine**: the seven sacraments, the cult of the Virgin and the saints and the validity of good works in achieving salvation.
  - It also reformed some practices. It prohibited the sale of indulgences and established **seminaries** to train priests. Norms were laid down to discourage bad behaviour on the part of the clergy and encourage good morals. **Seminaries** were created to improve the clergy's education and bishops were ordered to live in their dioceses. New ways of spreading the Catholic doctrine were introduced, such as the **catechism** and the creation of **schools**.
- **The religious orders.** The **Society of Jesus** was founded by Ignacio de Loyola, and its members (the Jesuits) helped to spread the ideas of the Counter-Reformation. Other religious orders were reformed, like the **Carmelites** under Saint Teresa de Jesús.

Religious division in Europe

The Reformation and Counter-Reformation led to deep divisions in Europe, which was now divided into **two religious blocks**: Protestants and Catholics.

Some rulers used religious differences as an excuse to attack their enemies. The most important war was the **Thirty Years War** (1618-1648), which devastated many parts of Europe.

There was widespread **intolerance** in both Catholic and Protestant countries. People who practised a different form of Christianity were persecuted.

In Spain, the **Inquisition** persecuted those who did not follow Catholic doctrine, and a person with non-Christian ancestors could not hold certain posts.

RELIGIOUS DIVISIONS IN 16TH CENTURY EUROPE

ACTIVITY ROUND-UP

1 Copy and complete the table.

RENAISSANCE AND REFORMATION

- **Humanism**
  - Characteristics
    - ...
    - ...
    - ...
    - ...
    - ...
- **Science**
  - Main advances
    - In astronomy, Heliocentrism
    - ...
    - ...
- **Renaissance art**
  - Characteristics
    - ...
    - ...
    - ...
    - ...
  - Periods
    - ...
    - ...
  - Artists
    - architecture
      - ...
    - sculpture
      - ...
    - painting
      - ...
- **Religious change**
  - Reformation
    - Lutheran principles
      - ...
      - ...
      - ...
      - ...
      - ...
      - ...
    - other movements
      - ...
      - ...
  - Counter-Reformation
    - measures
      - ...
      - ...
      - ...
      - ...
    - religious orders
      - ...

2 In which fields (painting, sculpture or architecture) were these artists outstanding?

- Palladio
- Ghiberti
- Dürer
- Bramante
- Raphael
- Titian
- Mantegna
- El Greco
- Berruguete
- Alberti
- Brueghel
- Botticelli

3 Match the terms on the left with their description on the right. Then listen and check your answers.

• Heliocentrism	• a plain, unadorned architectural style
• Reformation	• a theory that places the Sun at the centre of the universe
• Counter-Reformation	• a very ornate architectural style
• Patron	• a Christian movement that opposed the Catholic Church
• Herrerian style	• someone who commissions work by an artist
• Plateresque style	• the Catholic Church's movement of renewal

4 Explain the relation between humanism and the Renaissance.

5 Copy the timeline. Investigate and write these events in the corresponding place with their dates.

- The *95 Theses*, by Luther, is published.
- The Council of Trent begins.
- The Anglican Church is created.
- The Society of Jesus is founded.

6 Answer these questions.

a What were the causes of the Lutheran Reformation?

b How did the Catholic Church react when faced with the Reformation?

c What consequences did the religious disputes between Catholics and Protestants have?

7 Look at the painting and answer the questions.

a Who is the artist and at which stage of the Renaissance did he work?

b What does the painting show?

c Name some of the people in the painting. When did they live?

d What tells you that this is a Renaissance work?

8 Explain the main differences between Italian and Spanish sculpture of the Renaissance.

The School of Athens, by Raphael.

THE LEGACY OF THE RENAISSANCE AND THE REFORMATION

The Renaissance and the Reformation have left an important legacy.

- **A new way of thinking.** Humanism developed a school of thought based on reason and experience which we still apply today.
- The development of **printing** spread knowledge to new social groups.
- Great **masterpieces** were produced. Renaissance architecture, sculpture and painting are still an important part of our civilisation.
- New **religious beliefs** emerged after the Reformation.

9 Think about the artistic legacy.

a Why is perspective important in painting? What does it achieve?

b Explain the similarities between the Villa Rotonda and the University of Virginia. When was each one built?

Protestant church in Vermont (United States).

10 What differences can you see between the Protestant church in Vermont and a Catholic church?

Villa Rotonda, designed by Andrea Palladio, 16th century.

University of Virginia, designed by Thomas Jefferson, 18th century.

ETHICS AND RELIGION. **Religious divisions in the world today**

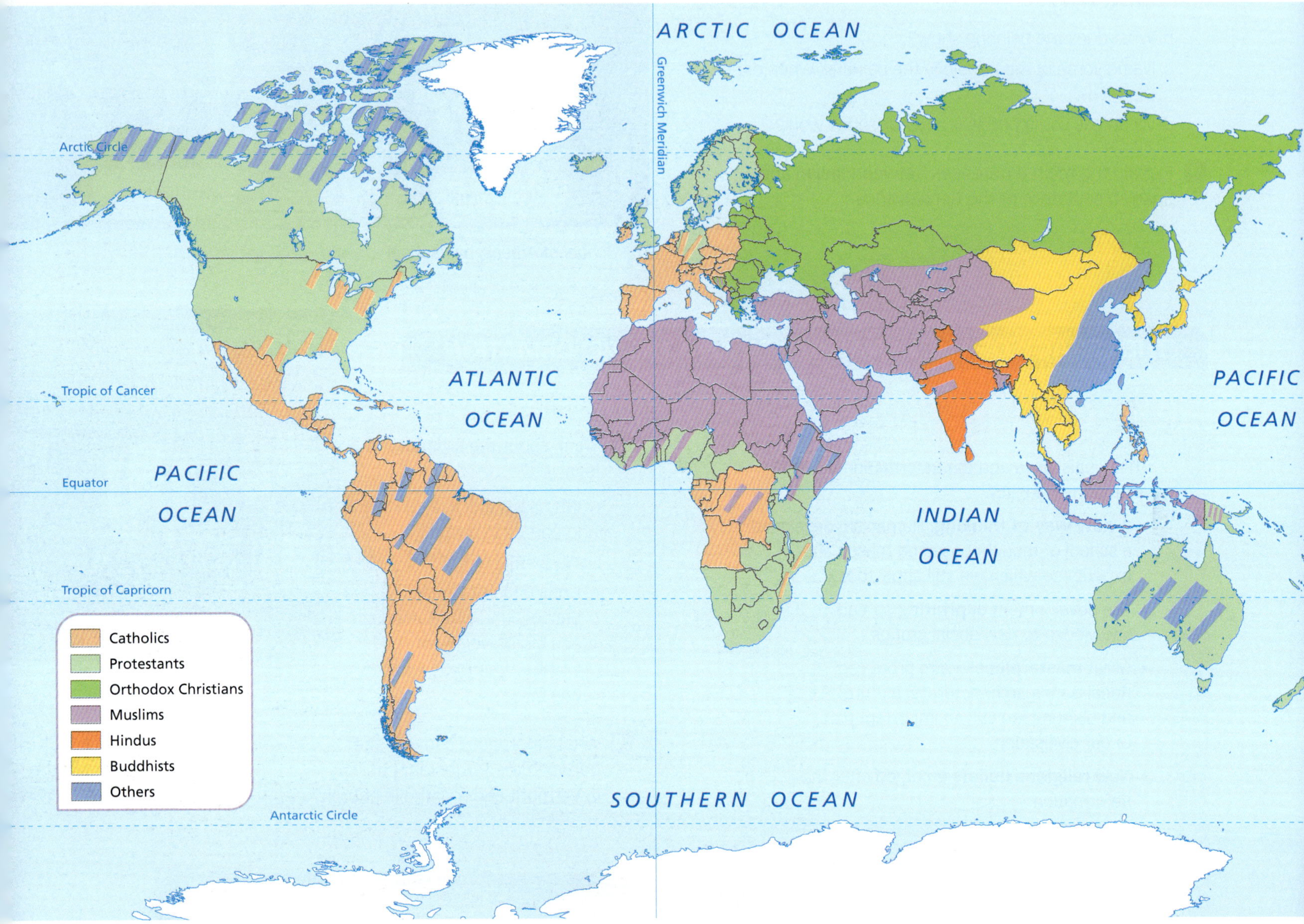

RELIGION IN THE WORLD TODAY

1 Compare the map to the map on religious division in 16th century Europe.

a Which parts of Europe have a Protestant majority today?

b Which are mainly Catholic today?

c Do these differences reflect the religious division of Europe in the 16th century?

2 Analyse the spread of Christianity in the rest of the world.

- In which parts of the world are Catholicism and Protestantism found today? Look for the likely historical explanations.

3 Think about religious issues.

a Are there still religious conflicts between Catholics and Protestants in some part of the world?

b Do you know about any regions in the world where there are religious conflicts today? Which religions are involved? What are the causes of these conflicts?

c Can religion ever justify conflict? Explain why, or why not.

d Do you think that religion is an important part of our cultural heritage? Explain your answer.

GROUP WORK. **A visit to the Uffizi Gallery**

Can you imagine a place where you can see many of the finest masterpieces of the Italian Renaissance? That place exists: it's the Uffizi Gallery in Florence. You can pay it a virtual visit by typing the keywords "Uffizi Gallery Florence" into an Internet search engine.

A. *Madonna of the Magnificat*
B. *Birth of Venus*
C. *Pallas and the Centaur*
D. *Madonna and Child and Two Angels*
E. *Allegory of Spring*
F. *Adoration of the Magi*

The Botticelli room in the Uffizi Gallery in Florence.

1 Find out more about the Uffizi Gallery using the websites you have located. For example:

- Where does the name "Uffizi Gallery" come from?
- What was the origin of the building?
- How did the works of art arrive there?
- Which Renaissance artists have paintings in the Uffizi Gallery?

2 The Uffizi Gallery is prepared to lend you ten of its masterpieces so that you can organise an exhibition in your region.

- In groups, decide which works of art you will display. For example, will the exhibition include masterpieces from different periods? Will you show a single type of painting (for example, just portraits or landscapes)?
- Together, prepare an illustrated exhibition catalogue, providing information about each work of art.

11 The Spanish Empire

FIND OUT ABOUT:

- Authoritarian monarchies
- The reign of the Catholic Monarchs
- The Spanish Empire under the Habsburgs
- The conquest and colonisation of America

KNOW HOW TO:

- Analyse maps of the Early Modern period
- Interpret a family tree
- Analyse an economic graph
- Analyse historical clothing

¿What were the *tercios*?

The *tercios* were a fighting force of the Spanish monarchy. They were created during the Italian Wars in the 16th century. The soldiers of the *tercios* were famous for being proud, professional veterans.

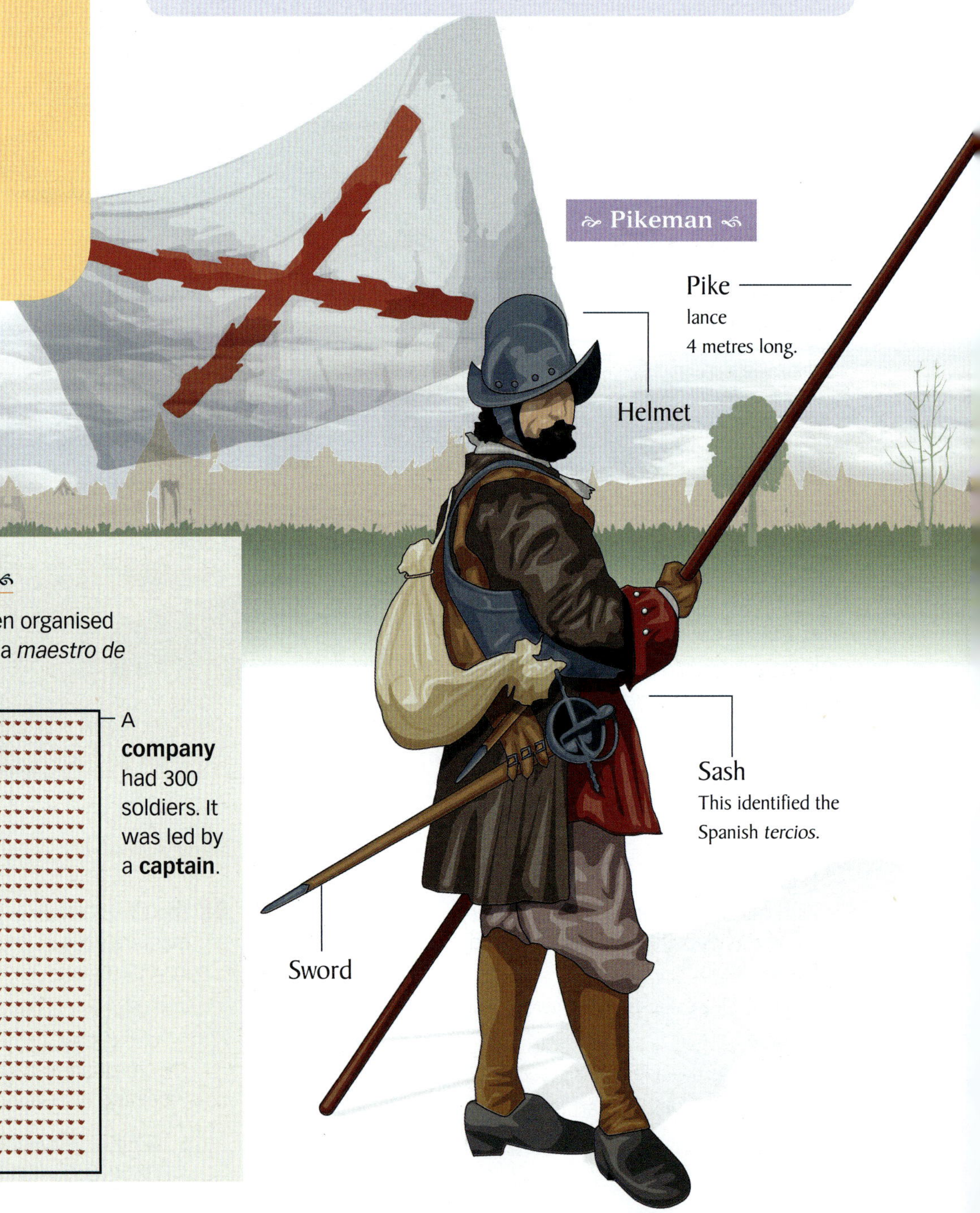

A TERCIO

A ***tercio*** was formed by 3,000 men organised into ten companies. It was led by a *maestro de campo*.

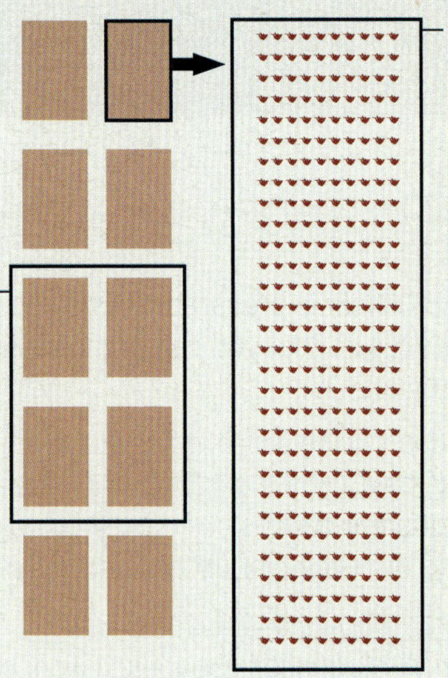

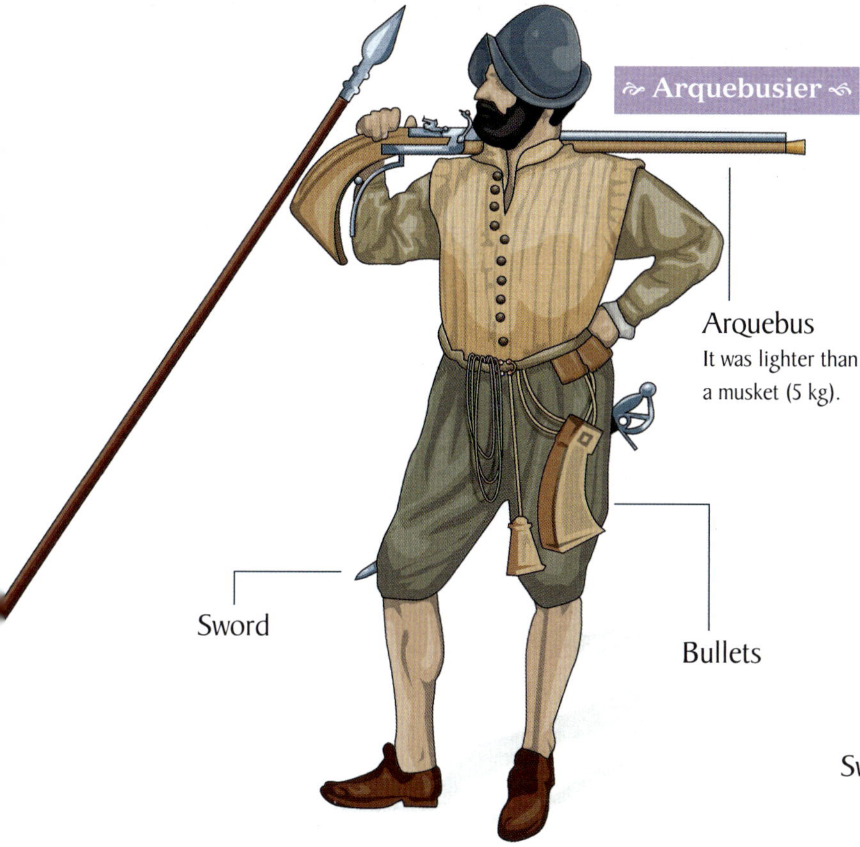

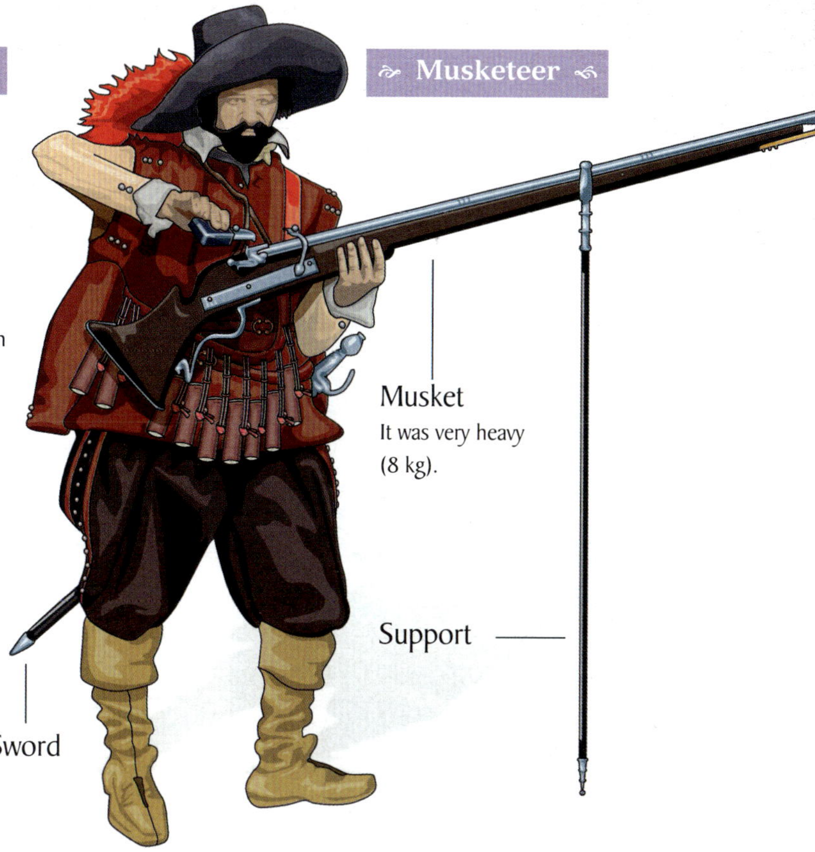

MILITARY STRATEGY

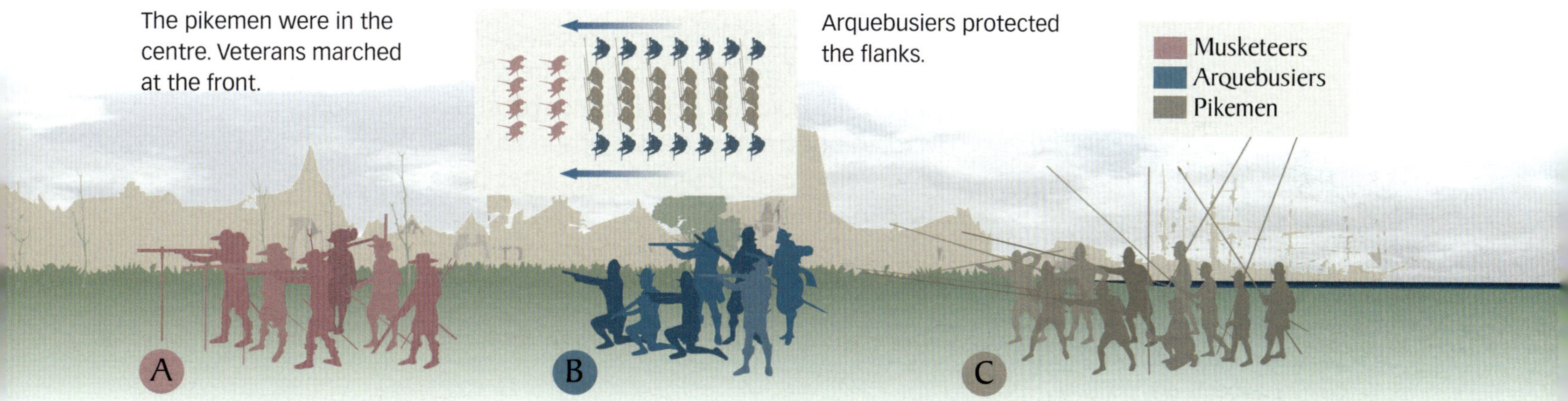

A The musketeers fired when the enemy was 100 metres away.

B Next, the arquebusiers fired.

C Finally, the pikemen used swords and daggers in hand to hand combat.

? WORK WITH THE IMAGE

- How was a *tercio* organised?
- What arms did a 16th century *tercio* use?
- Why do you think Carlos V created this military unit?

HOW DO WE KNOW?

Museums contain objects and information which are essential in studying the past.

- What type of museum would you visit to find out more about the *tercios*?
- What objects would you find in this museum?

Duke of Alba

He was one of the most famous Spanish commanders of the *tercios*.

1 The emergence of the modern state

***The Ambassadors*, Hans Holbein**. The authoritarian monarchies created diplomatic delegations in other countries.

Authoritarian monarchies

In the 15th century, European monarchs created **authoritarian monarchies**. They strengthened their authority by reducing the powers of the nobility, the Church and the municipalities.

- They created a **centralised administration**: officials were at the king's service.
- They formed **permanent royal armies**, which followed the monarch's orders.
- They limited the powers of the **representative assemblies** (the Estates-General, *Cortes* or Parliaments).
- They increased **taxes** to pay for the civil service and the army.
- They created a **diplomatic service** to defend their interests abroad.

The royal court no longer travelled around as in the Middle Ages. The king's place of residence eventually became a national **capital**. For example, Madrid became the centre of the Hispanic Monarchy under Felipe II, and later became the capital of Spain.

European monarchies

There were five great European monarchies in the early 16th century.

- **France**. After the Hundred Years War against England, French monarchs from **Charles VII** to **Francis I** unified the country and consolidated their power.
- **Spain**. The **Catholic Monarchs** unified all the Iberian Peninsula, except for Portugal, and began the conquest of other territories.
- **Portugal**. **John II** established his authority over the aristocracy and launched overseas expeditions.
- **England**. During the 15th century, a long civil war weakened the nobility. **Henry VIII** reinforced royal power in the early 16th century.
- **Russia**. **Ivan the Great** unified the country and conquered new territories. Later, **Ivan the Terrible** took the title of czar or emperor.

THINK ABOUT IT

What is the job of a diplomat? What part does diplomacy play today in the relations between different countries?

- Think about the economy, cultural aspects and political stability.
- Are diplomats different from politicians?

ACTIVITIES

1. What were the main characteristics of authoritarian monarchies? In what ways were they different from medieval monarchies?

EUROPE IN AROUND 1500

The consolidation of European states

There were many small kingdoms and principalities during the Middle Ages.

But from the 15th century onwards, several European states were joined together through **conquest** or **marriage alliances**.

In this way, a number of larger powers were formed. After a long period of evolution, some of these states grew into present-day European countries.

WORK WITH THE MAP

2 Study the map and answer the questions.

a Which countries were in the north of Europe?

b Which countries were on the Italian Peninsula?

c Which countries formed the Holy Roman Empire? Which area of Europe did this Empire occupy?

d Who ruled over the territories of the Balkan Peninsula and present-day Turkey?

e Compare this map with a modern political map. Did the Principality of Moscow occupy the same territory as modern Russia?

f Which modern countries already existed at the beginning of the 16th century?

2 The Catholic Monarchs

The monarchs Fernando and Isabel with their daughter Juana.

Dynastic union

In the first half of the 15th century, the Iberian Peninsula was divided into five large territories:

- The Crown of **Castilla**
- The Crown of **Aragón**
- The Kingdom of **Navarra**
- The Kingdom of **Portugal**
- The Islamic Kingdom of **Granada**

In 1469, the future union of the Crowns of Castilla and Aragon was promoted by a **dynastic marriage**. In that year, Isabel (Queen of Castilla from 1474) married Fernando (King of Aragón from 1479). Under Isabel and Fernando, the **Catholic Monarchs**, each kingdom maintained its own laws and institutions.

Royal authority

The Catholic Monarchs took various measures to reinforce their authority:

- They founded the Holy Brotherhood (***Santa Hermandad***), which kept order.
- They created a permanent professional **army**.
- They created courts of justice (***chancillerías***).
- They controlled the municipalities through administrators called ***corregidores***.
- They created a number of royal councils (***consejos***), which advised the monarchs on how to govern.
- They reorganised the Treasury (***Hacienda***) to increase their income.

MARRIAGE POLICY OF THE CATHOLIC MONARCHS

WORK WITH THE MAP

1. Which territories did the Catholic monarchs organise marriage alliances with?
2. What was the aim of these alliances?
3. Find out which of the children of the Catholic Monarchs married members of the royal families of other European States. Who did they marry?

POSSESSIONS OF THE CATHOLIC MONARCHS

Territorial expansion

The Catholic Monarchs had a policy of making **matrimonial alliances** with other European states. They also acquired new territories:

- In 1492, the Spanish Monarchs conquered the Kingdom of **Granada**, the last Muslim territory in the Peninsula.
- Also in 1492, they financed **Christopher Columbus'** expedition, leading to the acquisition of territory in America.
- In **North Africa**, they conquered Melilla (1497) and Oran (1509).
- They conquered the **Canary Islands** (1496).
- The Crown of Aragón acquired the Kingdom of **Naples** (1504).
- After Isabel's death, Fernando conquered the Kingdom of **Navarra** (1512).

Religious unity

In the 15th century, three religions were practised on the Iberian Peninsula: Christianity, Islam and Judaism. But the Catholic Monarchs imposed **religious unity**:

- In 1478, they established the **Inquisition**, a church tribunal which maintained Christian orthodoxy.
- In 1492, **Jews** were expelled if they did not convert to Christianity.
- In 1502, **Muslims** were also expelled if they did not convert to Christianity. This broke an agreement signed after the fall of Islamic Granada, which established tolerance.

WORK WITH THE MAP

4. Which territories did the Catholic Monarchs acquire on the Iberian Peninsula? And in other regions?
5. Make a timeline showing territorial expansion under the Catholic Monarchs.

ACTIVITIES

6. How did the Catholic Monarchs strengthen their authority.
7. Find out what each of the following terms means:
   - *Mudéjar*
   - *Morisco*
   - *Converso*

3 The beginnings of the Spanish Empire

Carlos V (in the centre) with Maximilian of Habsburg, Philip the Handsome and Mary of Burgundy.

The inheritance of Carlos V

Carlos V inherited a vast empire that made him the most powerful ruler of his time:

- In 1516, he inherited the Crowns of Castilla and Aragón and all their possessions from his maternal grandparents. As King of Spain, he was known as **Carlos I**.
- From his paternal grandparents, he inherited the title of Holy Roman Emperor and territories in Central Europe. As Emperor, he was known as **Carlos V**.

Early resistance to Carlos V

Carlos was born in Flanders. Following his arrival in Spain in 1517 he appointed Flemish advisers, which angered the Castilian nobility. He spent large amounts of Castilian money on becoming Emperor, leading to increased taxes on the urban bourgeoisie.

- The **Revolt of the *Comuneros*** (1520-1521) was a rebellion by the nobles, bourgeoisie and peasants of Castilla. The rebels protested against increased taxes, and called for greater participation in government. In 1520 cities like Toledo and Valladolid rebelled against Carlos V. The rebels had some early successes, but they were defeated in 1521 when the Castilian nobility decided to support the monarch.
- The **Revolt of the Brotherhoods** (1519-1523) was a rebellion in Valencia and Mallorca. Armed brotherhoods (***Germanías***) of artisans rebelled against the nobility. But Carlos V supported the nobles, and the rebels were defeated.

In response to this unrest, Carlos V adapted his **approach to government** in Spain:

- He spent more time in Castilla.
- He appointed Castilian nobles to the highest posts and they assisted him in his military expeditions.
- He later raised his son Felipe II at the Spanish court.

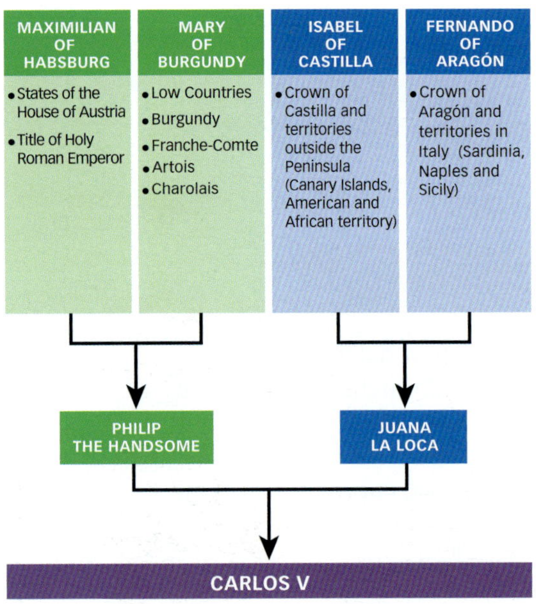

CARLOS V'S FAMILY TREE

WORK WITH THE FAMILY TREE

1. Which possessions did Carlos V inherit from a) his paternal grandparents and b) his maternal grandparents?
2. Find out more about the parents of Carlos V. Why did he inherit territories directly from his grandparents?

The Hispanic Monarchy of Felipe II

Carlos I and V abdicated in 1556. His brother Ferdinand received the title of Emperor and inherited his Austrian territories. His son **Felipe II** inherited all his other territories.

Felipe II's empire was the largest in the world:

- It consisted of **Spain**, the **Low Countries**, much of **Italy**, territories in **Central Europe**, and Castilian territories in America. It also had possessions in **North Africa** and the **Far East**.
- Felipe II was the son of a Portuguese princess. In 1580, Felipe claimed the throne when the King of **Portugal** died. This added Portuguese possessions in **Africa**, **Asia** and **Brazil**.

The monarch governed from Madrid, and he established his court there. But his attempts at centralisation met with opposition in some kingdoms. There was a revolt in Aragón in 1591.

Felipe II considered Spain to be the centre of his empire, which became known as the **Hispanic Monarchy**.

Felipe II, defender of Catholicism

Felipe II was the great defender of Catholicism against **Islam** and resisted the spread of the **Protestant Reformation**. His rule was characterised by religious intolerance. **Protestantism** was suppressed and the role of the **Inquisition** was strengthened.

The ***Moriscos*** were Muslims who had converted to Christianity. Following the defeat of an uprising in Granada (1568-1571), many *Moriscos* were dispersed through the kingdom.

The financial problems of the Empire

Both Carlos V and Felipe II had great difficulties in financing their policies. It was very expensive to maintain a large empire and fight numerous wars.

Spain received large amounts of gold and silver from America. In addition, the monarchs raised taxes, especially in Castilla. Even so, money was still short and the imperial government fell into debt.

THE POSSESSIONS OF CARLOS V AND FELIPE II

4 The organisation of the empire

The Princess of Eboli. She was a member of the high nobility during the reign of Felipe II.

ACTIVITIES

1 Look for information about the novel *Rinconete y Cortadillo* by Cervantes. Read the beginning and describe the social group to which these characters belonged.

THINK ABOUT IT

In the 16th century, Fray Antonio de Guevara wrote; "It doesn't matter how important the blood line is of *hidalgos* and gentlemen. If they own little and have no influence, it is certain that they will not be taken into account."

- What does this mean? What does it tell us about people's attitudes towards wealth and ancestry in Early Modern Spain?

Government

The **monarch** ruled the empire. He decided foreign policy, led the army and raised taxes. However, many different institutions and officials also participated in government:

- The Spanish Empire consisted of **different territories** that maintained their own laws and institutions. The creation of a new tax or the passing of a new law was approved by the **parliament** (or ***cortes***) of each territory.
- The monarch received advice from **councils** (***consejos***). There were territorial councils like the Councils of Castilla, Aragón, Flanders, the Indies, Portugal and Italy. Other councils, such as the Exchequer, dealt with specific areas. The **State Council** was the only council that took important decisions for the whole empire.
- A **Viceroy** represented the king in several territories and governed in his name.
- ***Chancillerías*** and ***Audiencias*** were supreme Courts of Justice.
- In some territories, institutions like the **Chief Justice of Aragón** guaranteed traditional rights.

Society

Spanish society continued to be divided into three **estates**.

- **Nobility**. There were big differences within the nobility.
  - The high nobility were landowners and held positions in government.
  - The low nobility was made up of ***hidalgos***, who sometimes had little money and few possessions.
- **Clergy**.
  - The high clergy included high church officials, such as cardinals and bishops, who had great influence at court.
  - The low clergy included friars, monks and parish priests.
- **The common people**. They paid taxes and carried out productive work.
  - Some merchants became wealthy through trade with America.
  - The common people also included peasants, artisans, craftsmen and servants.

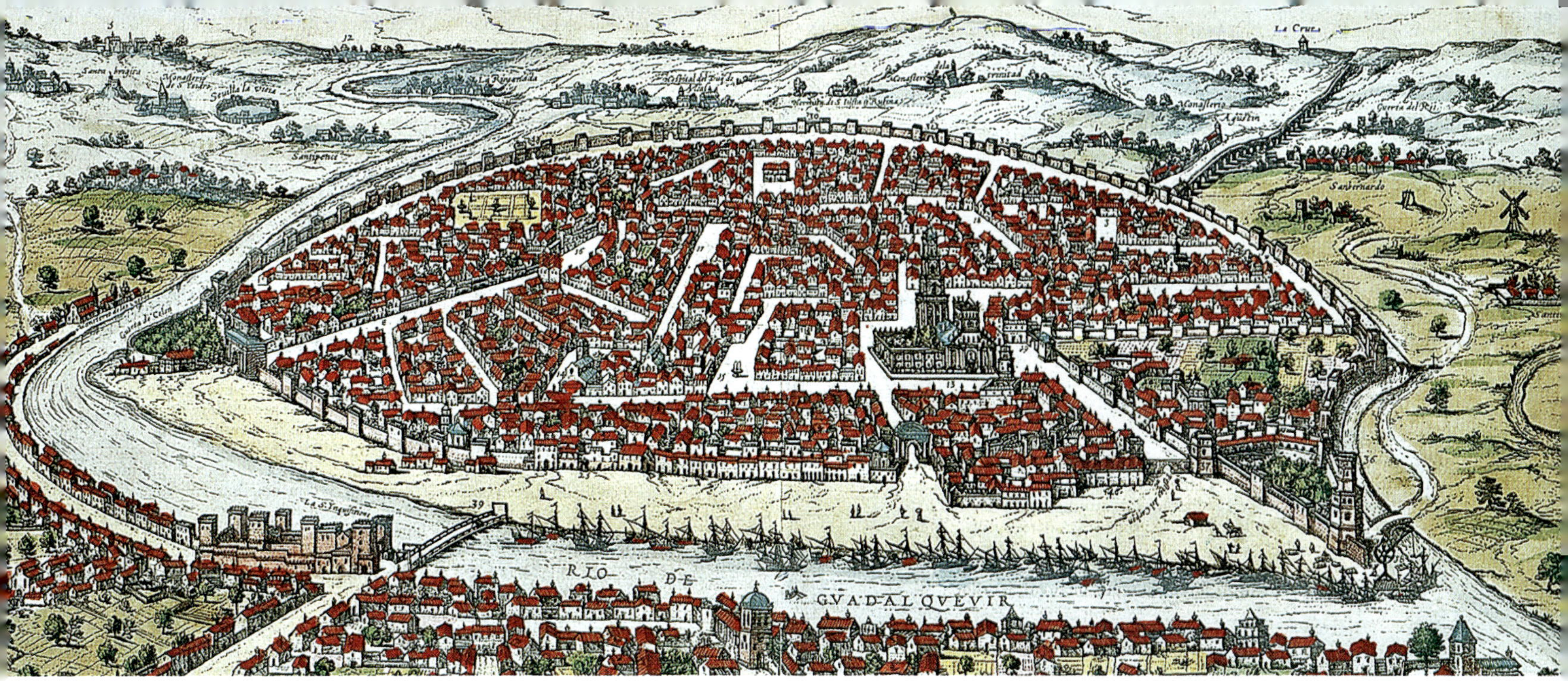

Seville in the 16th century. This city was the centre of trade with the Spanish colonies in America.

The economy

- **Agriculture and livestock breeding**. In about 1500, the Spanish economy was mainly based on agriculture in Castilla. Agricultural land could be used more intensively thanks to crop rotation. The seasonal movement of sheep across the Peninsula (called **transhumance**) continued to be important because of wool exports.
- **Craftsmanship**. Basque iron and steel manufacturing grew in importance. Woollen cloth was manufactured in Castilla.
- **Trade**. This became extremely important thanks to the **trade monopoly** with the colonial empire and the arrival of precious metals from America. All American trade went through Seville which became one of the most important cities in Europe. Cantabrian ports, like Bilbao, also grew in importance.
- **The 'price revolution'**. The Castilian economy eventually became weaker because of high taxation and inflation. The arrival of precious metals from America meant that there was too much money in circulation. Consequently, prices rose greatly. This situation has been called the **price revolution**.
- **Loans and bankruptcies**. The Spanish economy was affected by the costs of imperial policy. The wealth of Castilla was used to pay for wars in Flanders and Italy. The monarchs requested **loans** from bankers. But on several occasions, the monarchy could not repay its loans and declared bankruptcy.

WORK WITH THE ILLUSTRATION

2 Look for Seville on Google Maps. Do any of the features shown on the 16th century illustration still exist today?

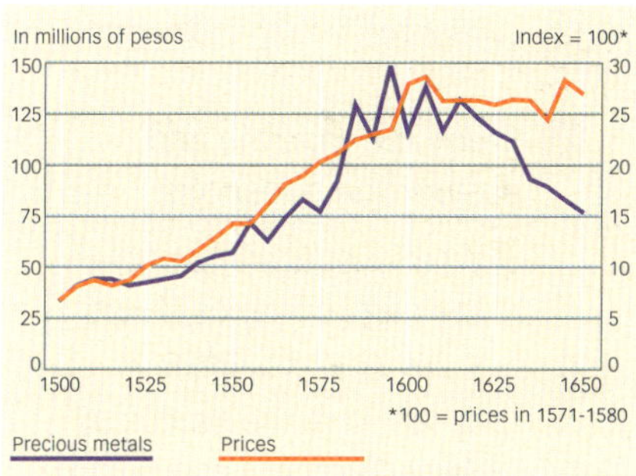

PRICES AND PRECIOUS METALS

WORK WITH THE GRAPH

3 How did the amount of precious metals coming from America evolve? What happened to prices?

4 Was there a connection between precious metals and the rise in prices? How did this affect the population?

Carlos V in Mühlberg, by Titian. This shows the Emperor after his victory against the Protestant princes of Germany.

5 Spanish foreign policy

There were three **main objectives** in the foreign policy of the Spanish Empire in the 16th century:

- The conservation of its territories.
- The strengthening of Spanish rule in Europe.
- The defence of the Catholic faith.

Rivalry with France

France was Spain's main rival:

- **Carlos V** fought Francis I of France for control of Italy after the French King occupied Milan. The Spanish *tercios* defeated the French at the **Battle of Pavia** (1525). In 1538, Carlos V incorporated the Duchy of Milan into the Hispanic Monarchy.
- **Felipe II** defeated the French at the **Battle of Saint Quentin** (1557). The **Peace of Cateau-Cambrésis** ensured Spain's domination for several decades.

The Turks

The Turkish Empire threatened Carlos V's empire in Europe and the Mediterranean:

- **Europe**. The Turks invaded parts of Europe. In 1529, they attacked Vienna but failed to take it.
- **The Mediterranean**. Carlos V organised military campaigns and conquered Tunisia. The Turks conquered Tunisia and Cyprus in 1570. Spain, Venice and the Pope defeated the Turks at the naval Battle of Lepanto in 1571.

WORK WITH THE TIMELINE

1 Study the timeline and answer the questions.

a What events stand out in the reign of Carlos I? And in the reign of Felipe II?

b When were the Battles of Saint-Quentin, Lepanto and Mühlberg? Who fought in them? What was the result?

Battle of Pavia. In this battle, the imperial troops of Carlos V took King Francis I of France prisoner.

THE SPANISH EMPIRE IN THE 16th CENTURY

CARLOS V

- 1516
- 1525 Battle of Pavia
- 1527 Sack of Rome
- 1529 The Turks lay siege to Vienna
- 1538 Annexation of the Duchy of Milan
- 1547 Battle of Mühlberg
- 1555 Peace of Augsburg

The defence of Catholicism

Protestantism spread quickly throughout Europe in the 16th century. Carlos V considered that the Spanish Monarchy was the great defender of Catholicism, and his opposition to Protestantism influenced his policies.

- Within the Holy Roman Empire, many German princes supported the Protestant Reformation. Carlos V won a military victory against the princes at the **Battle of Mühlberg** in 1547. But the Peace of Augsburg (1555) allowed each principality to decide between Catholicism and Protestantism.
- Carlos V also took part in the **French Wars of Religion** between Catholics and Calvinists.

Conflicts in north-west Europe

Calvinism spread throughout the Low Countries. Resistance to Felipe II's anti-Protestant policy, together with high taxation, led to the **Dutch Revolt** in 1568.

- The northern provinces, led by William of Orange, declared their independence under the name of **United Provinces**. This was the beginning of a very long war. The independence of the United Provinces was only recognised by Spain in the following century (1648). The rebellion interrupted trade with Flanders, one of the richest regions in Europe.
- At the same time, English support for the rebels led to a conflict with England under Elizabeth I, who became queen in 1558. Felipe II sent a powerful fleet called the **Spanish Armada** to invade England, but it was defeated in 1588.

DID YOU KNOW?

The Spanish Armada

The Great Armada, by Juan de la Corte.

In 1588, Felipe II sent his fleet to invade England. He wanted to punish Elizabeth I and the English for their support of the rebels in the United Provinces and the attacks of English pirates in America.

The Armada was made up of 130 boats and 30,000 sailors and soldiers. Its first objective was to take on board more soldiers in Flanders, but the Dutch blockade prevented this.

The English ships and bad weather pushed the Armada into the English Channel, making invasion impossible. On the return journey to Spain, storms destroyed dozens of ships and thousands of soldiers died.

ACTIVITIES

2 Which was the most important problem for the Spanish Empire in the 16th century? Discuss your ideas with a partner.

Felipe II became king after his father, Carlos V, abdicated.

Battle of Lepanto. The Christian victory ended the belief that the Turks were invincible.

FELIPE II

- … of Saint-Quentin
- **1566** Rebellion of the Netherlands
- **1571** Battle of Lepanto
- **1580** Felipe II, King of Portugal
- **1588** Spanish Armada
- **1598** Death of Felipe II

6 Spanish America

Conquests

By the early 16th century, Castilla had established its control over the West Indies. This was followed by a series of expeditions establishing Castilian rule in much of Central and South America. They were led by private individuals who received an authorisation from the King to recruit an army and conquer a particular area. The ***conquistador*** financed the expedition and took possession of the new territory on behalf of the monarch. In return, the *conquistador* was given the title of governor or captain, and received some of the land and wealth of the region.

Some of the conquered peoples had large populations and advanced civilisations. But the *conquistadors* had horses, armour and guns, unlike the Native Americans. The Europeans were also able to create alliances with local Indian populations who assisted them in their conquests.

The conquest of Mexico

In 1518, **Hernán Cortés** sailed from Cuba with 400 men to explore the Gulf of Mexico. When he reached Veracruz on the Mexican coast, he heard about the Aztec Empire. He took advantage of the confusion created by his arrival among the Indians to create alliances with some of the peoples ruled by the Aztecs.

Initially, Cortés received a friendly welcome from the emperor **Moctezuma** in Tenochtitlán, the Aztec capital. But in 1520, the Aztecs forced the Spaniards to flee from the city.

In 1521, Cortés organised the definitive conquest of the Aztec Empire. He occupied Tenochtitlán and ordered its destruction. In the following years, the rest of Mexico was conquered.

The conquest of Peru

Francisco Pizarro led the conquest of the Inca Empire. Like Cortés, he took advantage of the hostility of some of the conquered peoples to their Inca rulers. The Empire was also divided by a civil war, which broke out just before the arrival of the Spaniards.

In 1532, Pizarro and around 200 men arrived in Cajamarca, where they held the Inca emperor **Atahualpa** prisoner. Atahualpa paid a huge ransom in gold but he was eventually executed.

In 1533, the conquistadors occupied Cuzco, the Inca capital. However, it took the Spaniards many years to control the whole region because the Incas continued to resist in mountainous areas.

DID YOU KNOW?

The conquistadors

There were many expeditions in America and the Pacific in the 16th century. The conquistadors wanted to discover new lands and acquire great wealth:

- **Diego de Almagro**, one of Pizarro's companions, began the conquest of Chile in 1535. It was completed by **Pedro de Valdivia**.
- **Vasco Núñez de Balboa** discovered the Pacific Ocean.
- **Juan Ponce de León** discovered Florida.
- **Álvar Núñez Cabeza de Vaca** explored the Mississippi.
- **Francisco de Orellana** explored the Amazon River.
- **Miguel López de Legazpi** conquered the Philippines.

- Investigate and write a short biography of one of these explorers. In groups, present your findings to each other.
- Together, prepare an illustrated timeline of their discoveries.

ACTIVITIES

1. Find out about the legend of El Dorado. What was it? What connection did it have with the conquest of America?

Organisation of the New World empire

The conquered territories belonged to the Crown of Castilla, and Castilian laws and institutions were introduced into America. Carlos V created a new council to help him govern his newly acquired territories, the **Council of the Indies**.

- At first, the new territories were divided into two vast **Viceroyalties**: New Spain in Mexico and Peru. Each of these was governed by a Viceroy on the king's behalf.
- During the 16th century, a number of ***Audiencias*** were established to administer justice across a wide area.
- A governor ruled a province, or ***gobernación***. He held administrative, judicial and military authority.
- A ***corregidor*** was a judicial and administrative official who represented the Crown in a local district (***corregimiento***).

Society and the economy in America

Spanish American society was organised in the following way:

- The **Spaniards** and their descendants, called creoles, controlled trade, owned the land and held government positions.
- The ***Mestizos*** were children of a Spaniard and an Indian, and they occupied an intermediate social position.
- The **Indians** carried out manual labour and suffered harsh treatment. In 1542, the **New Laws** regulated the way that they were treated.
- **Black slaves** were brought from Africa. In very harsh conditions, they worked in regions where the native population had declined as a result of war and epidemics.

The exploitation of resources

- America was a very important source of **precious metals** (silver and gold). **Potosí** in Bolivia and **Zacatecas** in Mexico were two of the most famous mines. A fixed percentage of the value of the metals was assigned to the state. The mines were worked using the ***mita***, a labour system which required Indian communities to supply the work force.
- Under the system of ***encomienda***, a Spanish colonist was granted labour and tribute payments from a group of Indians. The ***encomendero*** was supposed to protect the native workers and instruct them in Christianity. In practice, the Indians were ill-treated and the *encomienda* became a form of servitude.
- ***Haciendas*** were large ranches. The crops cultivated in America included tobacco, maize, sugar cane, wheat, vines and olives.
- The Spanish Crown had the monopoly of **trade** with America, which was controlled by the ***Casa de Contratación*** in Seville. Precious metals and agricultural products were imported from America. For protection, these goods were transported twice a year in a naval convoy called the **Spanish treasure fleet**.

This 16th century illustration shows Hernán Cortés receiving gifts from the natives of Mexico.

ACTIVITIES

2. Which Viceroyalties were created in Spanish America in the 16th century? Which of them was in the region that had belonged to the Aztec empire? And the Inca Empire?

3. How did the Spaniards organise their labour force in America?

1 Copy and complete the table.

THE SPANISH EMPIRE

- **Authoritarian monarchies**
  - Characteristics
    - …
    - …
    - …
    - …
    - …
- **Reign of the Catholic Monarchs**
  - Union of the…
    - Domestic policy
      - …
    - Foreign policy
      - …
- **Habsburg Empire**
  - Carlos V
    - Internal problems
      - …
    - Foreign enemies
      - …
  - Felipe II
    - Internal problems
      - …
    - Foreign enemies
      - …
- **Conquest and colonisation of America**
  - Conquest of Mexico
    - …
  - Conquest of Peru
    - …
  - Territory divided into
    - …
  - Main resources
    - …

2 Copy and complete a timeline of the following events in your notebook.

- Revolt of Aragón
- Battle of Lepanto
- Conquest of Navarra by Fernando of Aragón
- Incorporation of Portugal into the Hispanic Monarchy
- Revolt of the *Comuneros*
- Battle of Pavia
- Defeat of the Spanish Armada
- Peace of Augsburg
- Abdication of Carlos V
- Rebellion of the Netherlands
- Battle of Mühlberg

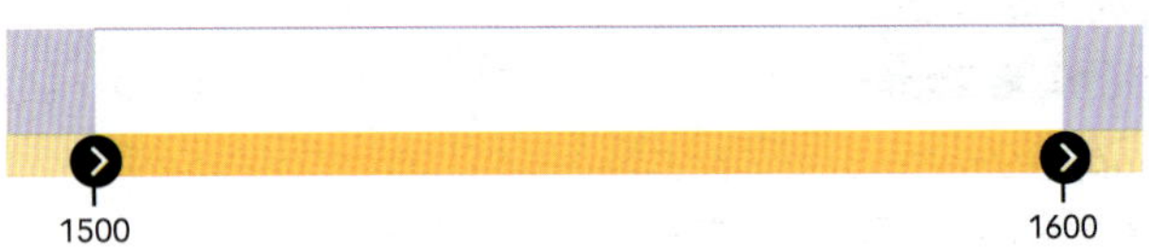

3 Make a timeline of the main events in the conquest of America.

4 Draw a map in your notebook showing the possessions of the Catholic Monarchs at the end of their reign. Write the date on which it became part of the Monarchy next to each territory.

5 What were the main objecives of Spanish foreign policy under Carlos V and Felipe II?

6 Explain the following terms.

- Authoritarian Monarchy
- *Casa de Contratación* (Institution that control commerce between Spain and America. (Seville)
- Inquisition — Kill people that were not religious. by torture, kill or expell them. (Protestants, muslims and jews)
- ~~*Encomienda*~~
- Viceroyalty — (Vireinado) Is a territory governed by a viceroyal, on the behalve the king.
- ~~*Mita*~~

7 Look for information about cochineal.

a What is cochineal?

b Where does it come from?

c Why did it become important in Europe?

8 Match the people and the descriptions.

• Carlos V	• They unified Spain.
• Francisco Pizarro	• He led the Spanish conquest of Mexico.
• Hernán Cortés	• He organised an expedition against Protestant England.
• Catholic Monarchs	• He led the Spanish conquest of Peru.
• Felipe II	• He became Holy Roman Emperor.

9 What were the similarities and differences between the Revolt of the *Comuneros* and that of the *Germanías*?

10 Draw a pyramid of Spanish society in the 16th century. Describe the different estates.

11 What were the consequences for the Spanish Empire of the following events?

- The spread of Protestantism in the German states.
- The expansion of the Turkish Empire in the Mediterranean.
- The spread of Calvinism in the Netherlands.

THE LEGACY OF THE CONQUEST OF AMERICA

The discovery and conquest of the American continent has left a profound mark on both America and Europe.

- **Language.** Today Spanish is the official language of most countries in Central and South America.
- **Catholicism.** The Spaniards spread Catholicism.
- **Cities.** The Spanish colonisers founded many American cities, which often have Spanish names.
- **Art.** The colonial style was developed in America. It blended elements of European and indigenous art.
- **New plants and animals.** The potato, tomato, cocoa, pepper and maize all came from America. The Spaniards brought wheat, olives and vines to America, as well as animals such as horses, sheep and pigs.
- **New peoples.** The Europeans and the Native Americans made contact with each other.

12 Look at a map of America in an atlas. Find three cities with the same name as a city in Spain.

13 Find out the names of three languages which were spoken in America before the Spanish Conquest.

14 Think of the food that you have eaten this week. Have you eaten anything that originally came from America?

15 In groups, prepare a report on the positive and negative results of the Spanish Conquest of America.

- Discuss your findings with the rest of the class. Did they come to similar conclusions?

This 18th century painting shows an Indian collecting cochineal.

THINK LIKE AN HISTORIAN. **Analyse historical clothing**

Fashion is an extremely interesting source of historical information. In the Early Modern period, the way that a person dressed indicated their social status.

***Market*, by Joachim Beuckelaer.**

***Prince Don Carlos*, by Alonso Sánchez Coello.** In this portrait, the prince was about 12 years old.

***Ana of Austria*, by Alonso Sánchez Coello.**

1 Analyse the images.

a Do some research. With a partner, check the meanings of the terms in the illustrations.

b Describe the clothes of Don Carlos and Ana of Austria. Do you think these clothes would be expensive? Was it possible for the people to dress themselves or would they need help?

c Was it possible to work in the fields wearing these clothes? Why? / Why not?

d Describe the clothes of the market sellers. What material are they made of? Do these clothes look new? Which colours predominate?

2 Draw conclusions.

a Which levels of society did the people in the pictures belong to? Explain your answer.

b How did the clothes of people in the Early Modern period indicate their social position?

c Think about clothes today. Are they still an indicator of social status? What else do they tell us about people?

ETHICS. **Why did the Indian population decrease?**

One terrible consequence of the conquest of America was the death of millions of Indians. The actual numbers and causes of these deaths are still argued over.

Many Indians died as a result of ill treatment and overwork. In the 16th century, people like Father Bartolomé de las Casas and Father Antonio de Montesinos spoke out against these abuses. But modern historians have shown that most Indians were victims of illnesses like smallpox which the Spaniards brought to America. The natives had no natural defences against these diseases.

Two views of the treatment of the Indians

To speak these truths, I have come here… I am the voice of Christ and you should hear what I have to say.

You are all in a state of mortal sin and you live and die in this way because of the cruelty and tyranny you practise among these innocent people.

Tell me: by what right do you hold these Indians in such cruel and horrible servitude?

On what authority have you fought such hateful wars against people who lived so peacefully in their own lands? In these wars you have killed vast numbers of them. Why have you oppressed and exhausted them? Why have you failed to feed them or heal their sicknesses? You give them far too much work and they die, or rather, you kill them in order to obtain gold every day.

Sermon,
FRAY ANTONIO DE MONTESINOS (1511)

It is not against either justice or religion to put decent, fair and prudent Spaniards in charge of these towns and villages, especially those who took an active part in the conquest. They can instruct the Indians in decent, civilised customs and initiate and educate them in the Christian religion. This should not be preached through violence but by persuasion and giving a good example. No reason of justice, humanity or Christianity prohibits domination over subject peoples or the demand for tribute. These are fair compensation for the work of princes, magistrates and soldiers.

On the just causes of war against the Indians,
JUAN INÉS DE SEPÚLVEDA (1545)

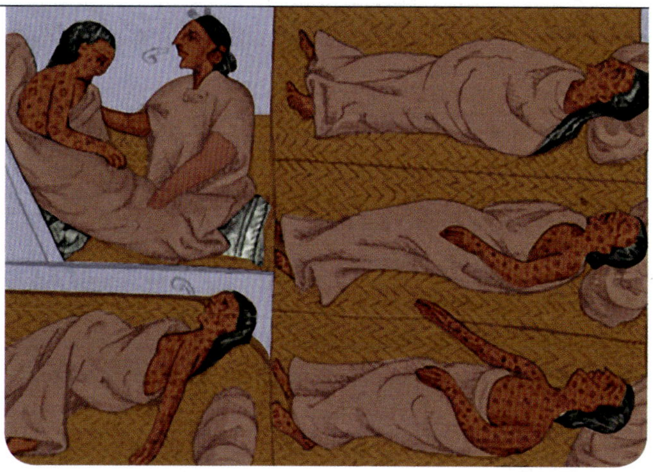

***Florentine Codex* (around 1575).** Indians suffering from smallpox.

1 Analyse the data.

- How did the number of Indians in Mexico evolve between 1519 and 1605?

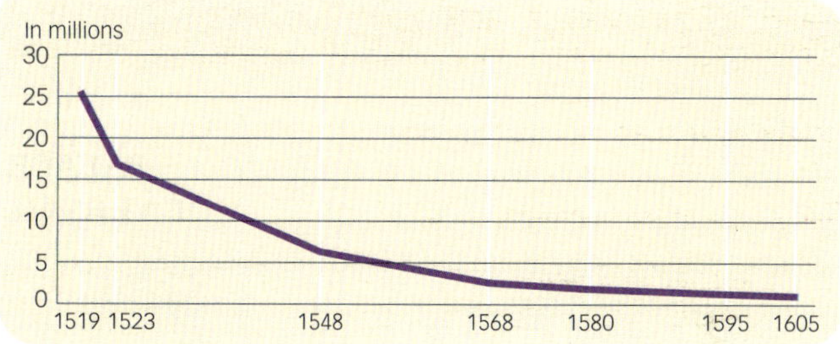

Indian population in Mexico

2 Compare the sources.

a Read the texts by Montesinos and Sepúlveda. Do they share the same opinion about the treatment of the Indians?

b Read this extract from the *Laws of the Indies* (1580). How does it suggest the Indians should be treated?

We entrust the care of the Indians to the Viceroys. We wish to compensate them for the harm they have suffered, and for them to live without annoyance or ill treatment.

3 Draw conclusions.

a Which explanation for the decrease in numbers of the Indians seems most likely?

b Do you think that the fall in the native population in America is due to one or more causes? Explain your answer.

12 Baroque Europe

FIND OUT ABOUT:

- The decline of the Hispanic Monarchy
- The rise of new powers in Europe
- The Scientific Revolution
- Baroque art and architecture
- The Spanish Golden Age

KNOW HOW TO:

- Interpret a population chart
- Interpret a historical map
- Identify the symbols of absolute power

The Palace of Versailles

King Louis XIV of France ordered his architects to convert a hunting lodge near Paris into a luxurious palace. In 1682 he moved his court there.

The Palace of Versailles was designed like a city, in which up to 20,000 people could live. Louis XIV's court was intended to display his greatness to the world.

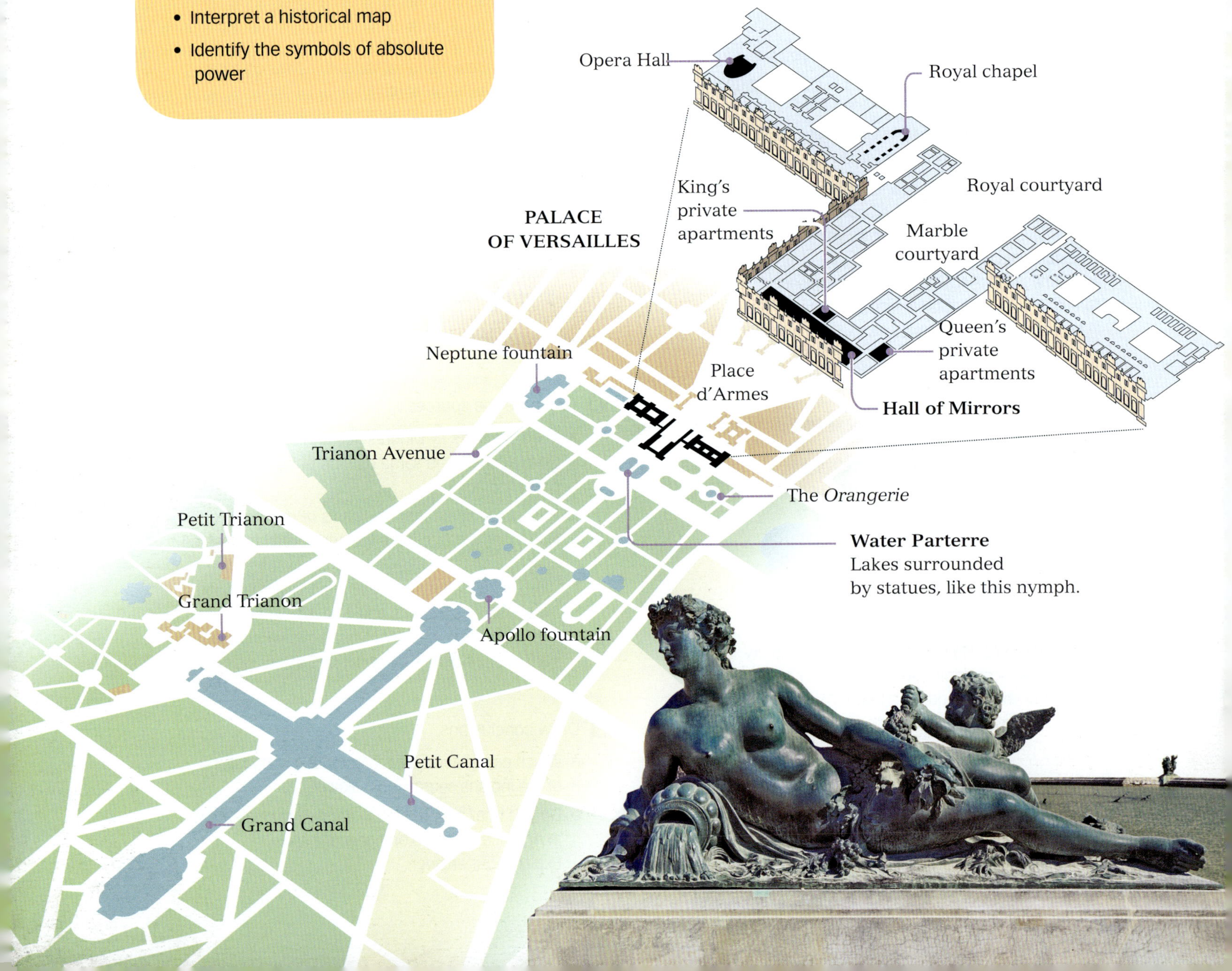

Water Parterre
Lakes surrounded by statues, like this nymph.

DID YOU KNOW?

So many people lived in the Palace of Versailles that every aspect of daily life had to be carefully planned. This included the king's day.

WORK WITH THE IMAGE

- What was the palace like?
- Would you describe this palace as luxurious or austere?
- Why was the king's day so carefullly planned?
- Look on the Internet for information about the Palace of Versailles today. Which parts of it can you visit? Make a list.

HOW DO WE KNOW?

Buildings give us an extraordinary amount of information about the period when they were built. Their size, structure, the reason they were built and their decoration tell us about people's ways of life.

- What information do the images of the Palace of Versailles give us about the people who lived there?
- Are palaces like this built today? Why? / Why not?

POPULATION GROWTH

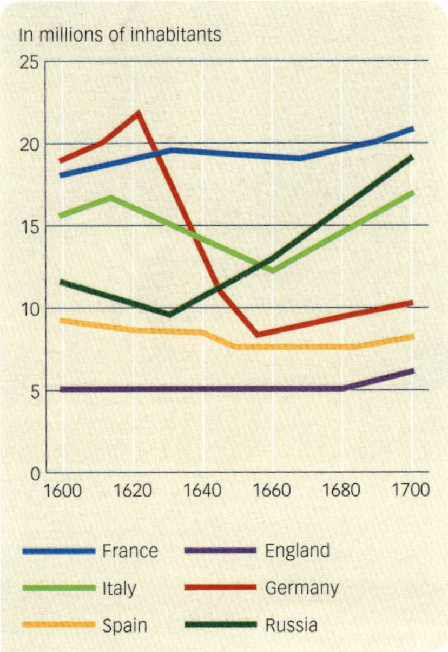

MAIN EPIDEMICS OF THE BUBONIC PLAGUE

YEAR	TERRITORY AFFECTED
1596-1603	Most of the Iberian Peninsula
1630	Italy, Central Europe, South of France and Catalonia
1647	East and southeast of the Iberian Peninsula
1652-1653	Catalonia and Aragón
1655-1656	Italy
1665	London
1676-1686	Most of Spain

WORK WITH THE CHART AND THE TABLE

1 Look at the chart and answer the questions.

a In which countries did the population vary the most in the 17th century?

b When did the population of Germany reach its highest and lowest points?

c How many epidemics of bubonic plague were there in 17th century Europe?

d Which region of Europe was most affected by epidemics?

1 A period of crisis

Historians sometimes refer to a **crisis** in 17th century Europe because many regions experienced problems in various fields.

These included demography, agricultural production, economic growth and social conflicts.

Demographic change

There was little population growth in most of Europe in the 17th century. Excluding Russia, the number of inhabitants in Europe only rose from about 70 million in 1600 to 75 million in 1700.

Population growth was low because of an increase in the **mortality rate**:

- **Famines**. These were caused by bad harvests, and the destruction and confiscation of crops in wartime.
- **Epidemics**. There were epidemics of typhus, cholera and bubonic plague. These affected the south of Europe especially.
- **Wars**. The Thirty Years' War led to the depopulation of large areas in central Europe.

There were also major **migrations**:

- Some people were forced to leave their countries. For example, 200,000 Protestants were **expelled** from France during the reign of Louis XIV.
- Other people **emigrated** voluntarily: around half a million people went to live in America.

North-west Europe grew more rapidly than the centre and south of Europe.

The rural economy

Many rural areas suffered from low productivity:

- **Agriculture** became less productive. Land was farmed too intensively using very simple techniques. Consequently, the land became less fertile.
- As a result of **climate** change, frosts and heavy rain alternated with long droughts. Many crops were lost.
- In addition, **demand** for agricultural products decreased so prices fell.

However, in England and in the Low Countries, new agricultural techniques were developed, which increased production.

HOW THE DOMESTIC SYSTEM WORKED

A. A businessman, generally a merchant, bought the raw material and the tools and delivered them to the peasant families.

B. The peasants made the product at home, when there was less work in the fields. All the family helped.

C. The businessman picked up the finished product and sold it directly. He made large profits because he sold it at the price he wanted.

The economy

- **Craft production**. In the south of Europe, craft production continued to be regulated by associations called **guilds**. However, the **domestic system** spread in north-west Europe. This was a way of obtaining cheaper products more efficiently.
- **Trade**. Before the 17th century, the main economic powers were in the Mediterranean region: Spain, Portugal and Italy. In this period, **England** and the **United Provinces** became important trading nations. They created their own colonial empires and sold their goods in Spain and Portugal's colonial territories.

Social conflicts

Economic pressures increased on many of the poorer groups:

- The states raised **taxes** from ordinary people to pay for their military expenses and the luxuries of the court.
- The **nobles** saw their income decline because of the demographic and agricultural crises. They therefore demanded higher rents and more services from the **peasants**.
- Consequently, the peasants grew poorer and had to sell their lands to pay their debts. The wealthy **bourgeoisie** benefited by buying these lands at low prices.
- People in **towns** and **cities** also grew poorer, as a result of high taxes and the crisis of the guild system.

This situation led to a number of revolts:

- **Urban revolts** were mainly caused by growing taxation and shortage of food in the cities.
- **Peasants' revolts** were in response to pressure from the nobles as well as taxation.

WORK WITH THE IMAGES

2 What are the people doing in the images?

Masaniello's Revolt in Naples, a 19th century painting by Giuseppe Mazza.

ACTIVITIES

3 Look on the Internet and find the answers to the questions.

a Who was Masaniello?

b What were the reasons for the revolt?

c How did it end?

2 The decline of the Hispanic Monarchy

Felipe III inherited the largest and most powerful empire in the world in 1598. However, he faced some extremely serious problems, which continued during the reigns of his successors **Felipe IV** and **Carlos II.**

Government by the *validos*

A 17th century king was not directly involved in government, but delegated his authority to a ***valido*** instead. In some ways, a *valido* was similar to a prime minister. However, he was also the king's personal favourite and depended entirely on his favour.

- The **Duke of Lerma** was Felipe III's *valido*. The Duke pursued a peace policy in Europe because he believed that expensive wars could ruin the monarchy financially.
- The **Count-Duke of Olivares** was Felipe IV's *valido* for over 20 years. He attempted to restore Spanish military leadership in Europe, and tried to reform the monarchy.
  - He introduced **financial reforms** to reduce expenses.
  - In 1625 he designed the **Union of Arms** to make all the monarchy's kingdoms (and not just Castilla) share the military expense of a permanent army.
  - He tried to increase **royal power** by restricting the autonomy of the kingdoms.

However, the economic difficulties of the monarchy made it impossible to apply these reforms. They also met strong resistance in the different kingdoms.

Crisis of the monarchy

At the beginning of Felipe III's reign, the position of the ***Moriscos*** weakened. They were thought to be insincere in their conversion to Christianity and were suspected of collaborating with the Turks. In 1609 the Moriscos were **expelled**, and around 275,000 people left the country. This was disastrous for many regions, which were depopulated.

In the 1640s, rebellions broke out in Catalonia and Portugal. They provoked a serious crisis for the monarchy:

- **Catalonia** rejected the Union of Arms and started a revolt, which was eventually crushed.
- The situation in **Portugal** also grew worse because of the tax pressure from the Union of Arms. Taking advantage of the rebellion in Catalonia, a revolt broke out. With the support of other European powers, Portugal became independent.

DID YOU KNOW?

The Duke of Lerma

Francisco Gómez de Sandoval Rojas, Duke of Lerma (1553-1625) was Felipe III's main *valido*.

He came from a noble family and was brought up in the court of Felipe II, where he made friends with the future king, Felipe III.

In his role as a *valido*, he was in power from 1598 to 1618. Thanks to his position, he accumulated enormous wealth and many posts for himself and his relatives and supporters. This created a system of clientelism.

The king withdrew his confidence in 1618. Lerma retired to his properties and died in Valladolid.

ACTIVITIES

1. Look for information about another *valido* of the 17th century and write a short biography.

THE HISPANIC MONARCHY IN THE 17th CENTURY

Felipe III

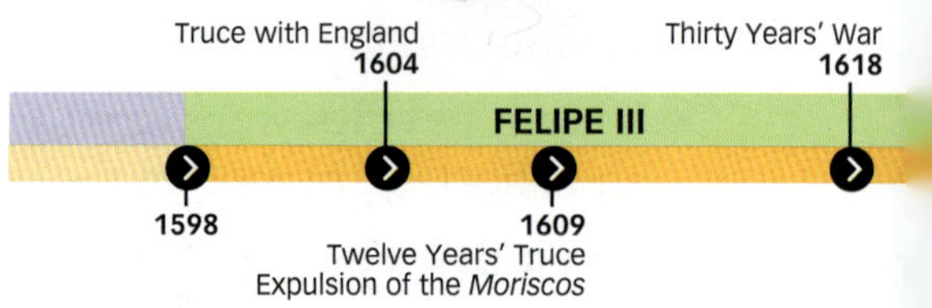

The economic crisis

Spain went through a serious economic recession in the 17th century.

- **Agricultural** and **livestock production** decreased. Population declined in large areas of the countryside and wool production fell.
- **Craft** activities decreased as a result of falling demand, higher taxes and competition from English and Flemish products.
- **Trade** with the Americas decreased as a result of competition from the English and the Dutch. The volume of precious metals coming from America also diminished.

Consequently the revenues of the Royal Treasury fell. The monarchs continued to borrow money to pay for their growing expenses. When they were unable to pay their debts, they declared **bankruptcy**.

The end of the Habsburg dynasty

Carlos II had no children. At the end of his reign, there were two candidates ready to succeed him: **Philip of Bourbon** and **Archduke Charles of Austria**. Both wanted the immense Spanish Empire, which had territories on several continents.

Carlos II died in 1700. In his will, he named **Philip of Bourbon**, the grandson of Louis XIV of France, as his heir. This led to the **War of the Spanish Succession** (1701-1713), which ended with the victory of the French Bourbon dynasty.

The water seller of Seville, by Diego Velázquez.

WORK WITH THE IMAGE

2 What social group did the people in the painting come from? What attitude do you think the painter had towards them?

ACTIVITIES

3 What role did *validos* play in the Hispanic Monarchy?

4 What were the main reasons for the economic recession in the 17th century?

5 Why did the Habsburg dynasty end?

Felipe IV

Carlos II

The Battle of Almansa was a decisive victory for the Bourbons in the War of the Spanish Succession.

Defeat by France at Rocroi 1643

Independence of Portugal 1668

War of the Spanish Sucession 1701-1713

FELIPE IV

CARLOS II

ith the Provinces

1640 Rebellion of Catalonia and Portugal

1648 Peace of Westphalia

1659 Treaty of the Pyrenees

1665

1675 Carlos II comes of age

1700 Death of Carlos II

3 The end of Spanish hegemony in Europe

Peace and war

Felipe III and his *valido*, the Duke of Lerma, followed a peace policy. In this way, they intended to strengthen the Treasury.
In 1604 they signed peace with **England**, and in 1609 the **Twelve Years' Truce** with the United Provinces.

However, Felipe IV initiated a much more aggressive policy. In 1621 Spain went back to war against the **United Provinces**.
It intervened in the **Thirty Years' War** (1618-1648) as an ally of the Holy Roman Empire against the German Protestant princes.

After a series of Catholic victories, France became concerned that the Hispanic Monarchy would become too powerful. In 1635, it decided to fight on the Protestant side. This changed the course of the war, and the Habsburgs were defeated several times.

The loss of hegemony in Europe

In 1648, the Thirty Years' War came to an end when the **Peace of Westphalia** was signed. Spain recognised the independence of the United Provinces. The war with France continued until 1659. In the **Treaty of the Pyrenees**, Spain ceded different territories in Europe to France.

These treaties ended Habsburg domination in Europe, and France became the leading power. However, Spain kept its great empire in America.

ACTIVITIES

1. Who fought in the Thirty Years' War?
2. Which faction did the Hispanic Monarchy support?
3. Why did France join in the war?
4. In your opinion, why was the Thirty Years' War costly for Spain?

THE PEACE OF WESTPHALIA (1648)

THE TREATY OF THE PYRENEES (1659)

WORK WITH THE MAPS

5 What possessions did Spain have in Europe in 1618? Which territories did it lose in 1648 and 1659?

6 Which two states became independent in 1648? Do they still exist today?

4 Absolutism in France

Louis XIV receives the Prince of Saxony, surrounded by his courtiers.

The absolute monarchy of Louis XIV

In the 17th century, the European monarchs tried to impose **absolutism**, or the divine right of kings. They believed that God had given them their power and that they were his representatives on earth. This gave them the right to exercise unlimited power. The monarch controlled all the powers of the State. He passed the laws, governed and was the supreme judge.

Louis XIV, the Sun King (1643-1715), was the absolutist monarch that other kings imitated. He consolidated his authority in various ways:

- He exercised **power** personally. His ministers could not make decisions without his approval.
- He controlled **the nobles**. He brought them to the court and gave them subsidies and posts.
- He promoted **political centralisation**. He unified the laws and taxes of the kingdom. He also appointed governors in the provinces, who were public servants with extensive powers.
- He imposed **religious unity**. He persecuted Calvinists and subjected the French church to the king's authority.

The hegemony of France

The main aim of Louis XIV's foreign policy was to consolidate France's **hegemony**. France fought wars with Spain, the United Provinces and the Holy Roman Empire. French territories were extended, mainly at the expense of the Habsburgs.

Louis XIV inherited a **colonial empire** in America. (The French had founded Quebec in the early 17th century).

Economic reforms

Louis XIV had very large expenses as a result of his aggressive foreign policy, the construction of Versailles and the maintenance of the court. To meet these expenses, the king's minister, Jean Baptiste Colbert, introduced reforms:

- **Royal factories** were protected by the State. These made luxury products like tapestries and pottery.
- **Trading companies** were given a monopoly of trade with regions in America and Asia.
- **Customs duties** were raised, increasing the price of imported products. This promoted consumption of French products.
- **Regulations** controlled the quality of products.

ACTIVITIES

1 Match the columns.

• Absolutism	• A tax on imported products
• Hegemony	• A place where luxury products were made
• Royal factory	• The domination of a single power
• Customs duties	• A form of government where a ruler has all the power

2 What did Louis XIV do to exercise his authority? List the measures that he took.

3 How did Louis XIV try to extend his territories?

5 The Dutch Republic

The rise of maritime trade

The **United Provinces** were a group of small territories which fought against Spain to obtain their independence. However, in the first half of the 17th century, they became a great economic power. This was mainly the result of the growth of their **trade**.

The Dutch became the main intermediaries in maritime trade. All kinds of products (wood and cereals from the Baltic, iron from Sweden, cloth from England, wine and wool from Spain), were brought to the ports of the United Provinces. They were then sold to other countries.

Two large trading companies were founded: the **Dutch East India Company** and the **Dutch West India Company**. These companies monopolised Asian and American trade and obtained large profits.

In the second half of the 17th century, the Dutch fought the English in several colonial wars and lost much of their power.

Two women beside a linen chest, with a child, Pieter de Hooch. Many Dutch painters represented families in wealthy bourgeois homes.

A parliamentary republic

The development of trade led to the rise of a rich **bourgeoisie**, formed by important merchants, bankers or shipyard owners. They lived in luxurious mansions in cities like Amsterdam.

Thanks to their economic power, the bourgeoisie obtained control of Parliament and acquired great political power.

The United Provinces obtained independence from Spain in 1648. It became a **republic** formed by seven provinces, which had their own Parliaments. Representatives from all the provinces met at the **States General**, where they took joint decisions.

POLITICAL SYSTEM OF THE UNITED PROVINCES

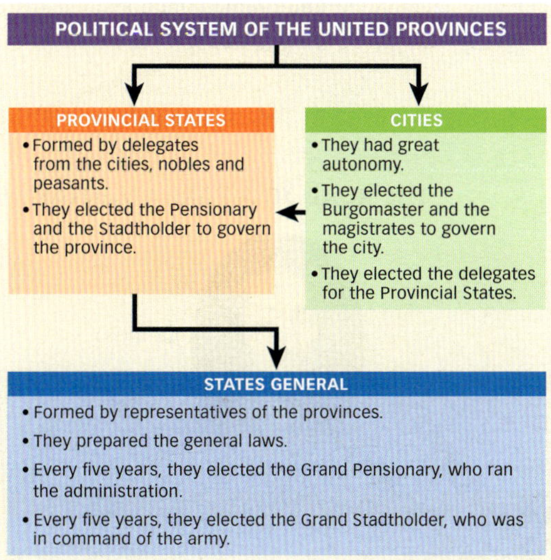

ACTIVITIES

1. What were the reasons for the commercial success of the United Provinces in the 17th century?
2. List the products that passed through Dutch ports and give their origins.
3. Why do you think the bourgeoisie became important?
4. What were the functions of the States General?

6 The rise of Parliament in England

The road to a parliamentary system

In the first half of the 17th century, the kings of the Stuart dynasty attempted to govern their territories in Great Britain and Ireland as absolute monarchs, without Parliament. However, this decision was very unpopular. The discontent, together with the revolts which broke out in Scotland over religious matters, led to a **revolution** in 1640.

The confrontations turned into a **civil war** in 1642. This ended with the fall of the Stuart dynasty: Charles I was executed and the monarchy was abolished. One of the leaders of Parliament, Oliver Cromwell, came into power and established a **dictatorship**.

After Cromwell's death, Parliament restored the monarchy. However, a second revolution, the **Glorious Revolution**, took place in 1688. This led to the deposition of King James II and, a year later, to the approval of a **Bill of Rights**. This Bill restricted the king's power and set out the rights of Parliament, laying the foundations for the separation of powers. It meant that a **parliamentary monarchy** had triumphed in England.

Social changes

The expansion of English trade in this period led to the emergence of a prosperous **bourgeoisie**. This group was mainly made up of merchants who obtained their income from trade and financial activities.

After the civil war, the prestige and importance of the **gentry** (country nobility) was consolidated. In the 17th century, the gentry owned about half of the land in Britain and used modern agricultural methods to obtain maximum profit.

DID YOU KNOW?

England's American colonies

The English founded colonies in North America. The first navigators arrived in North America in the 16th century, and the first stable settlements were founded in the 17th century.

Jamestown was the first colony, founded in 1607. In 1620, Plymouth was founded by Puritans, who were escaping religious persecution in England. In 1664, the English took New York from the Dutch, who had called it New Amsterdam.

ACTIVITIES

1. Use the Internet to find the answers to the questions:
   - a Where is Jamestown? What is the relationship between Jamestown and Pocahontas?
   - b Who were the Pilgrim Fathers? Where did they settle?
   - c What famous yearly celebration did they initiate? Why?

THE ROAD TO A PARLIAMENTARY SYSTEM

Charles I

Oliver Cromwell

England's House of Commons

- **1625** Charles I, King
- **1642-1649** Civil war
- **1649-1658** Cromwell's dictatorship
- **1660** Restoration of the monarchy
- **1688** Glorious Revolution
- **1689** Bill of

7 The Scientific Revolution

The scientific method

The importance attached to **reason** by the humanists helped to promote the development of science in the 17th century.

Thinkers of this time believed that truth could be attained through the use of reason (**rationalism**) and experimentation (**empiricism**). The Frenchman **René Descartes** was the main exponent of rationalism, while the Englishman **Francis Bacon** was the main exponent of empiricism.

The **scientific method** was developed in the 17th century and is still used today. According to this method:

- First a **hypothesis** is formulated.
- Then it is **tested** by comparing it with reality through experiments and tests.
- Finally, it is **confirmed**, rejected or revised, according to the results obtained during the testing stage.

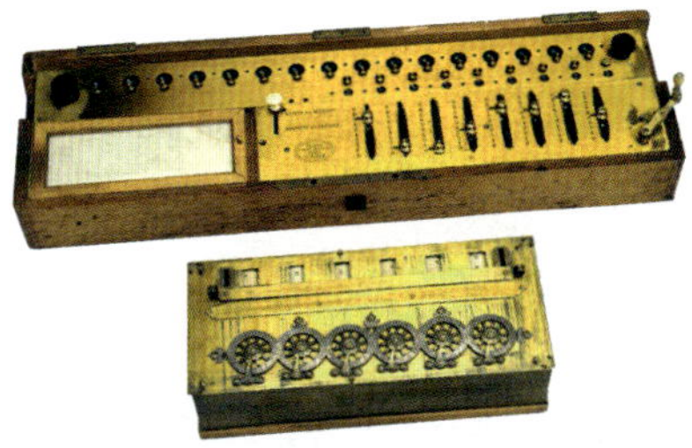

Pascal's calculator.

A 17th century microscope.

ACTIVITIES

1 **Form groups of three.** In groups of three, find out about these great scientists from the period of the Scientific Revolution. Complete the table in your notebooks.

NAME	NATIONALITY	BORN/DIED	ACHIEVEMENTS
Galileo Galilei	......	1564-1642	......
......	English	......	Law of universal gravitation
Johannes Kepler	......	1571-1630	......
......	......	1646-1742	Calculated the orbit of a comet
Robert Boyle	......	......	Boyle's law
Evangelista Torricelli	......	1608-1647	......
......	French	......	Calculator
Marcello Malpighi	......	......	Greater understanding of anatomy

2 Read these statements and match them with the achievements in the table. Then listen and check your answers.

a It explains the relationship between the pressure and the volume of a gas.

b It enabled him to discover Jupiter's satellites, sun spots and the surface of the Moon.

c It is an instrument used in meteorology to measure atmospheric pressure.

d It had eight dials and could add, subtract, multiply and divide.

e It explains the reason for the attraction of any two bodies in the universe.

f There was greater knowledge of animals, humans and plants, as the scientist was one of the first to use a microscope.

g They explain the movement of the planets around the Sun.

h The frequency of the appearance of a celestial body was predicted for the first time.

3 Choose one of the following scientists and find out more about them. Then write a short description of their life and achievements. Present your findings to the class.

- René Descartes
- William Harvey
- Gottfried Wilhelm Leibniz
- Antonie van Leeuwenhoek

V Lo que esta en el proyecto

8 Baroque culture

The Church of San Carlo alle Quattro Fontane, by Borromini. A. The facade. B. The interior.

WORK WITH THE PHOTOS

1 Describe the main characteristics of the Church of San Carlo alle Quattro Fontane.

Baroque

Baroque was a European cultural movement that developed during the 17th and 18th centuries. It was influenced by the religious tensions that divided Europe:

- In **Catholic** countries, new attitudes were expressed by an intense spirituality, which affected all areas of life.
- In **Protestant** countries, life was viewed in a more individualistic way. Economic success was highly valued.

Baroque culture

All branches of culture flourished during the Baroque period:

- **Art**. Like Renaissance art, Baroque art was inspired by classical forms. However, Baroque artists attempted to portray reality as it was, without idealising it. Their works were full of movement and contrast.
- **Literature**. Major works were written. This was the age of **Shakespeare** in England and **Molière** in France.
- **Music**. Bach, Handel and Vivaldi were among the greatest composers. The first operas were composed.

In Catholic absolutist states, Baroque art was used as **propaganda** to show the power of the monarch and the vitality of the Counter-Reformation.

Baroque architecture

Baroque architects tried to surprise the viewer. They used the following features:

- **Curved lines** and **spiral columns**.
- **Light effects** on facades by combining recesses and projections in the walls. Open pediments also created an impression of light and shade.
- **Ornamentation** on facades, combining painting, sculpture and architecture.
- **Fine materials** like coloured marble.

Rome, the centre of the Counter-Reformation, was also where the Baroque style of architecture began. Gian Lorenzo Bernini built Saint Peter's Square in the Vatican, and Francesco Borromini designed exceptional churches.

The Baroque style spread rapidly to most of Europe. But in some countries it was simpler than in Italy.

Characteristics of baroque painting

The main characteristics of baroque painting were the following:

- It represented a **range of subjects**, like religion, mythology, portraits and scenes from daily life. New subjects included the **landscape** and the **still life**, which showed objects, food or plants.
- People were represented **realistically**. They were not idealised and might have physical defects.
- Scenes showed **movement** and **energy**.
- The technique of ***chiaroscuro*** drew sharp contrasts between light and shade.
- It showed strong **emotions**, which were expressed through gestures.

Schools of baroque painting

Although Baroque painting had some common features, there was a wide range of schools and styles:

- **The Italian School**. The outstanding Italian baroque painter was **Caravaggio**. His characters were inspired by ordinary people with all their imperfections. Caravaggio's technique of using light and shade was called **tenebrism**, which was a version of *chiaroscuro*.
- **The Dutch School**. This was greatly influenced by the bourgeois taste of the merchants who commissioned and paid for the paintings. The paintings reflected their lives.

  Rembrandt was the most outstanding painter in the Dutch School. He was a master of *chiaroscuro* and his portraits were extremely realistic. **Frans Hals** was a fine portrait painter. **Vermeer** specialised in interior scenes in bourgeois homes.
- **The Flemish School**. This was greatly influenced by Italian painting in its **naturalism**, or realism, and in its attention to detail.

  Rubens was the outstanding painter in the Flemish school. His paintings stood out for their rich colour and sense of movement. **Van Dyck** was one of Rubens' disciples. He was famous for his elegant portraits and his use of colour.

Supper at Emmaus, by Caravaggio.

The Night Watch, by Rembrandt.

The Abduction of the Daughters of Leucippus, by Rubens.

ACTIVITIES

2 Analyse one of the three paintings on this page. Think about the following aspects: subject, characters and representation of light and movement.

9 The Spanish Golden Age

Facade of the Hospice of Madrid, by Pedro de Ribera.

The Spanish **Golden Age** was the period between the second half of the 16th century and the late 17th century. This was a time of extraordinary cultural achievement.

Literature

Miguel de Cervantes wrote his great novel *Don Quijote*. **Lope de Vega** and **Pedro Calderón de la Barca** wrote famous plays. **Luis de Góngora** and **Francisco de Quevedo** were the leading poets of the time.

Architecture

Architecture was characterised by the use of simple materials, such as bricks or stone, which were covered by **ornate decoration**.

Many buildings were religious, but civil buildings were also constructed. The construction of city squares (or ***plazas***) was a characteristic feature of the Spanish Baroque period.

The Churriguera brothers were the outstanding Baroque architects. They created their own style, which was called ***churrigueresco*** and is characterised by extremely ornate decoration. They built the *Plaza Mayor* in Salamanca, and the Church of San Cayetano in Madrid.

Other notable architects included Pedro de Ribera, who was responsible for the facade of the Hospice of Madrid, and Fernando Casas Novoa, who designed the western facade of the Cathedral of Santiago de Compostela.

Saint Isabel of Portugal, by Francisco de Zurbarán.

Sculpture

The sculptures of the period were mainly **religious**. Many of them were for processions and altarpieces and were made of **polychrome wood** to make them look richer. The sculptures are very expressive and dramatic.

Gregorio Fernández was the main sculptor in Castilla. He made numerous sculptures of Christ at different moments of the Passion. In **Andalusia**, the most notable sculptors were Juan Martínez Montañés, Alonso Cano and Pedro de Mena.

Christ tied to a column, by Gregorio Fernández.

DID YOU KNOW?

Diego Velázquez

Diego Velázquez (1599-1660) was born in Seville where he was educated and trained. His style was greatly influenced by tenebrism and realism.

He was a painter in Felipe IV's court and painted many portraits of the royal family and people from the court. He was expert at showing people's personalities.

During his trips to Italy, he learnt more about perspective and the nude. This is reflected in the perfection of his later works, such as *Las Meninas* and *The Spinners*.

Velázquez painted portraits of the royal family. He also painted everyday scenes (*The Water Seller*), mythological scenes (*The Triumph of Bacchus*) and historical scenes (*The Surrender of Breda*). All of his paintings showed his capacity to represent depth, and he was a master in the use of light and colour.

Self-portrait, by Diego Velázquez.

The Spinners or *The Fable of Arachne*

The painting shows a scene from mythology. Arachne told the goddess Athena that she could weave the most beautiful tapestry. But her tapestry included images that Athena did not like. The goddess was angry and turned Arachne into a spider, forcing her to weave all her life.

There are two scenes in the painting. In the foreground, Arachne is working while Athena, disguised as an old woman, looks on. The tapestry made by Arachne is in the background.

Painting

Spanish Baroque painting was used as propaganda for the Counter-Reformation. Consequently, many **religious works** were created. The Spanish school was characterised by its realism, emotion and use of symbolic messages.

In addition to **Diego Velázquez**, there were other important artists:

- **José de Ribera**, *Spagnoletto*, imitated Caravaggio's naturalism, but used more colour.
- **Francisco de Zurbarán** painted still lifes, religious scenes and portraits of saints for monasteries and convents.
- **Bartolomé Esteban Murillo**'s early works were realistic and tenebrist. His later paintings were more tranquil and intimate.

ACTIVITIES

1. What was the Golden Age? Why do you think it was given this name?
2. List the main features of Spanish Baroque architecture, sculpture and painting.
3. Choose one famous person from the Spanish Golden Age. Prepare a fact file on their life and work.

1 Copy and complete the table.

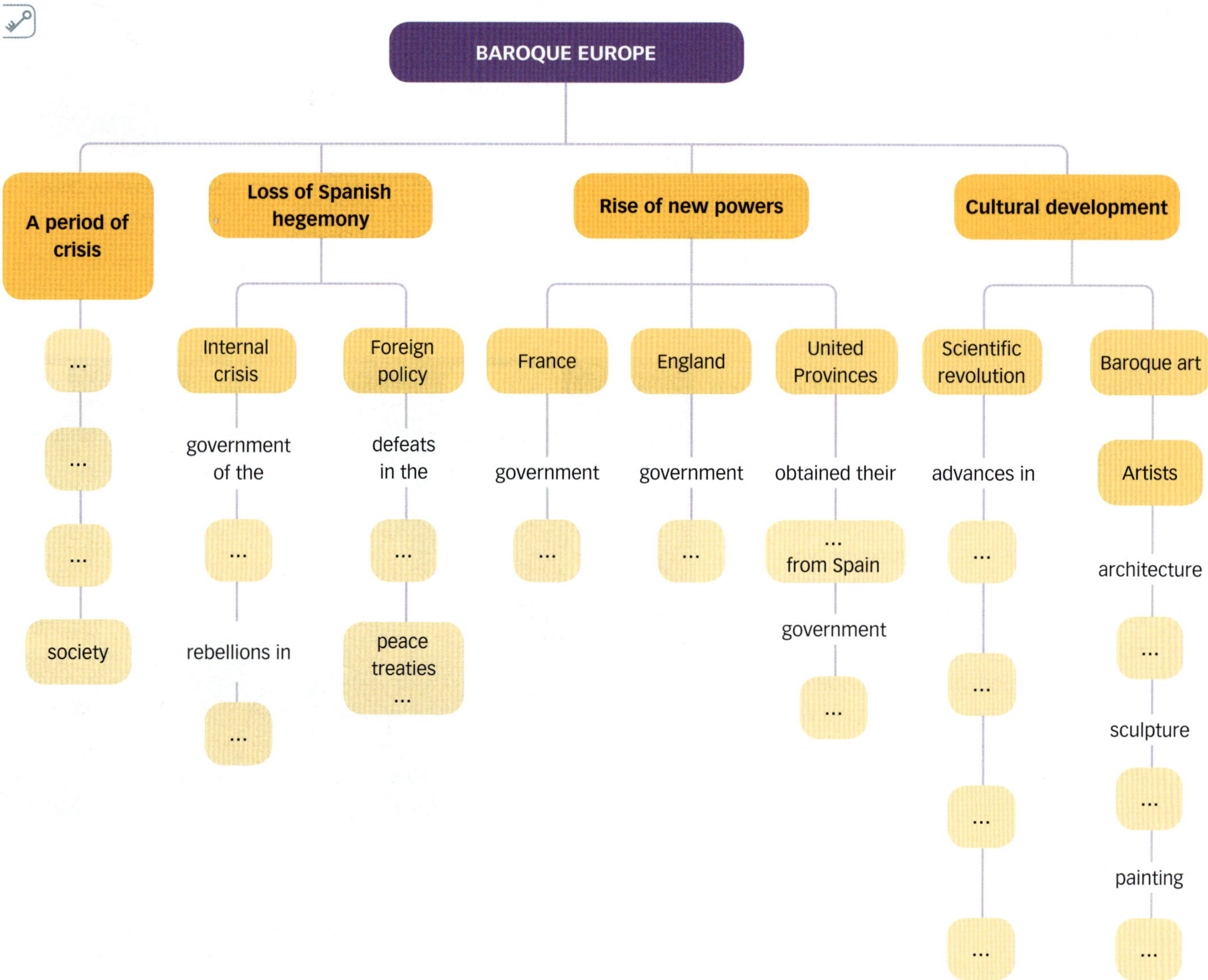

2 Write down the dates of these events and put them in chronological order. Then put them on the timeline.

- Peace of Westphalia
- Beginning of the reign of Louis XIV
- Union of Arms
- Expulsion of the *Moriscos* from Spain
- Treaty of the Pyrenees
- Twelve Years' Truce
- Bill of Rights in England
- Rebellion in Portugal

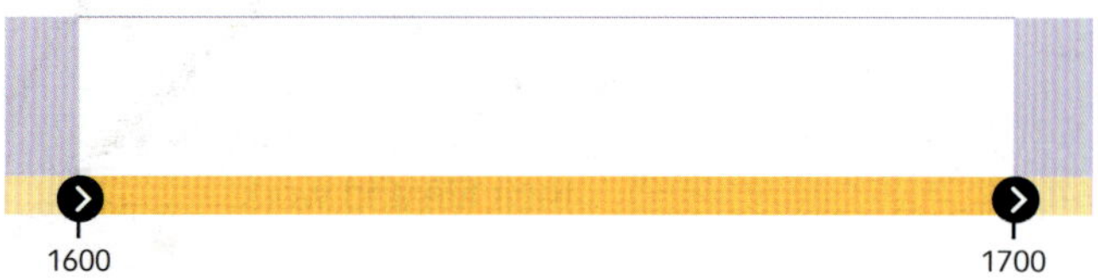

3 Explain the meaning of these terms.

- *Valido*
- Twelve Years' Truce
- States General
- Absolutism
- Baroque
- Domestic system

4 Briefly explain the historical role played by the following people in the 17th century.

- Count-Duke of Olivares
- Louis XIV
- Velázquez
- Bernini
- Felipe III
- Galileo

5 How did the Thirty Year's War affect the Hispanic Monarchy?

6 Analyse the English political system.

- What powers does the King have? Does he have power over everything?
- What is the function of Parliament?

ENGLISH POLITICAL SYSTEM

Dissolved by

KING
- He appoints the Government.
- He governs following the laws.
- He can dissolve Parliament but he has to call a new Parliament.
- He asks Parliament to pass new laws, new taxes and the recruitment of troops.

PARLIAMENT
- It establishes new laws.
- It approves the concession of new taxes to the king.
- It approves the recruitment and maintenance of troops.
- It holds regular meetings.

Controlled by

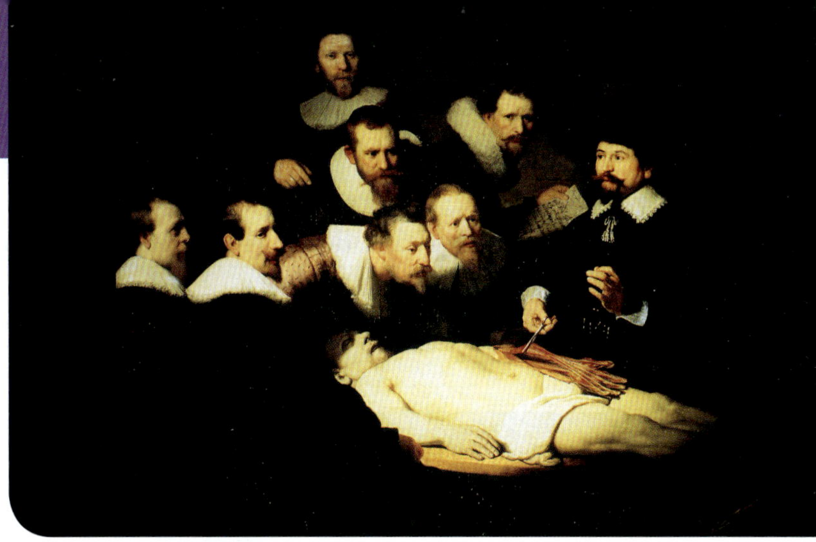

The Anatomy Lesson of Doctor Tulp, by Rembrandt.

7 Look at the painting and answer the questions.

- What is the subject of the painting? Is there a relationship between the subject and the Scientific Revolution of the 17th century? Explain your answer.

THE LEGACY OF THE BAROQUE PERIOD

The Baroque period left us a very important legacy:

- **The Parliamentary system.** Today this is the political system used in many countries in the world.
- **The scientific method.** Modern science continues to make progress thanks to experimentation.
- **Art.** Baroque artists perfected the technique of perspective.
- **Literature.** The plays written in the Golden Age are still performed today.

8 **The Parliamentary system.** Look on the Internet and answer the questions.

a What is Spain's system of government?

b What powers does the Crown have?

c What powers does Parliament have?

9 **The artistic legacy.** Look at the two paintings of *Las Meninas*. How many years' difference is there between the two paintings? Why do you think Picasso painted *Las Meninas*? How did he reinterpret Velázquez's painting?

A fragment of *Las Meninas*, by Diego Velázquez (1656).

A fragment of *Las Meninas*, by Pablo Picasso (1957).

THINK LIKE AN HISTORIAN. **Analyse symbols**

During the reign of Louis XIV, the French political system worked in the following way:

FRENCH POLITICAL SYSTEM

KING

POWERS:
- He governed.
- He passed the laws.
- He was the supreme judge.
- He established and collected taxes.
- He recruited the army.

OBLIGATIONS:
- He could not sell or transfer the kingdom.
- He had to apply God's laws.
- He had to respect his ancestors' laws and privileges.

PARLIAMENTS
- They could give their opinions on the laws proposed by the king.
- At times, they could participate in preparing some laws.

Parliament had very little power, while the king could make all the decisions because he believed he had the divine right to do so.

However, the king wanted to show that he was a modern monarch. He ordered some works of art to show this, as propaganda. In this portrait, Louis XIV is dressed in the latest fashion but he is surrounded by all the symbols of his absolute power.

1. List the symbols of power represented in Rigaud's painting and explain what they mean.

2. Describe the clothes of Louis XIV. Do you think they are expensive? Why? / Why not?

3. What image do you think Louis XIV wanted to display?

4. Work with a partner. Find videos of the King of Spain's Christmas speech on the Internet. Describe the scene and explain the symbols which surround him.

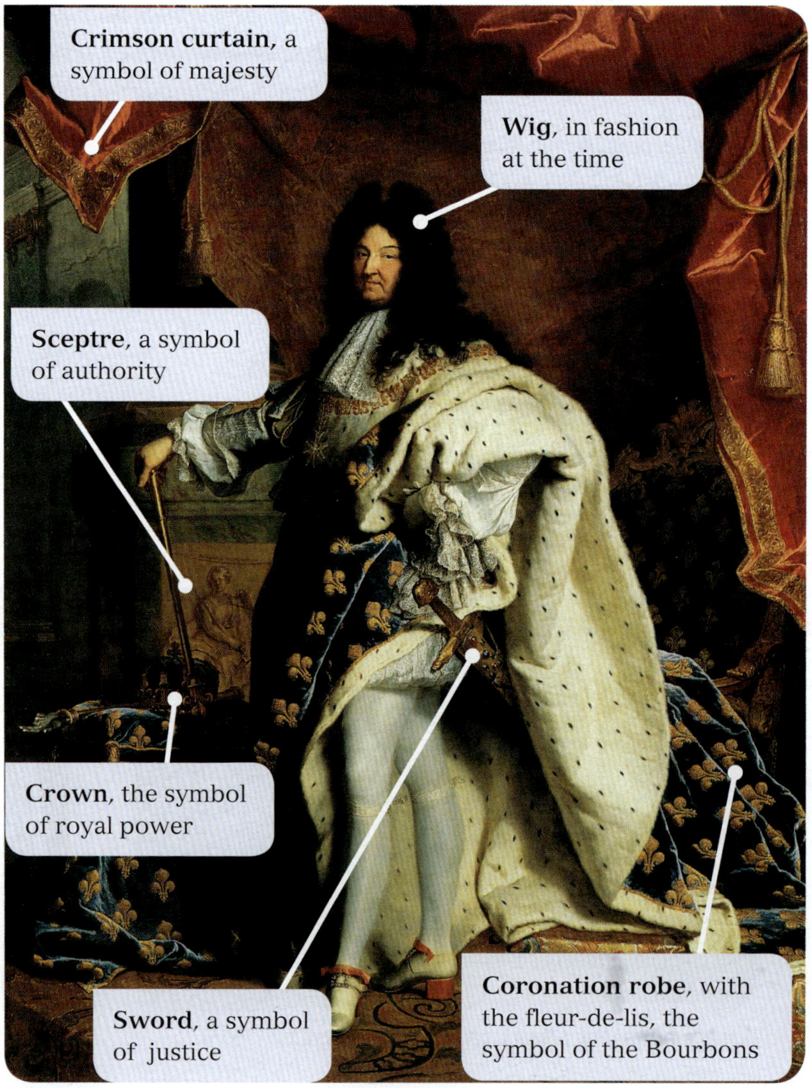

Louis XIV, by Hyacinthe Rigaud, 1701.

CULTURAL HERITAGE. **Baroque sculpture**

Investigate Baroque sculpture

Baroque sculptures were in many forms: portraits, tombs, altarpieces, monumental fountains and so on.

The main characteristics of Baroque sculpture were as follows:

- Most of the subjects were **religious**, although the sculptors also made **portraits** and sculptures of **mythological** scenes.
- **Groups** of figures became more important.
- The sculptures were very **realistic**. The sculptors tried to represent the physical features and the personality of the subject.
- The subjects showed very **strong emotions**, with exaggerated gestures.
- The figures gave a sensation of dynamic **movement**.
- The folds in the clothing and the movement of the figures accentuated the contrasts between **light** and **shade**.
- The sculptures were very **dramatic**. The figures often reached out into the surrounding space. They were made to be viewed from different angles.

Gian Lorenzo Bernini was the outstanding Baroque sculptor. He created the *Ecstasy of Saint Teresa*, *David* and a number of fountains in Rome.

The Ecstasy of Saint Teresa, by Bernini.

1 Analyse the *Ecstasy of Saint Teresa*. With a partner, think about how the sculptor represents movement and emotion. Pay special attention to Saint Teresa's clothing.

2 Think about Baroque sculpture.

a What elements of Baroque sculpture made it more realistic?

b How were Baroque sculptures different from earlier sculptures?

c In what ways was Baroque sculpture similar to Baroque art and architecture?

3 In groups, look for information on the Internet about Bernini. Prepare a project.

a Prepare a fact file on Bernini's life with the following information:
- The dates of his birth and death.
- His professions.
- His most notable works.
- Any other information you find interesting.

b Choose one of Bernini's fountains and find a photo. Write a short description, with the following information:
- Its name.
- Its location.
- What it represents.

Art director: José Crespo González.

Cover design and illustration: Carlos Cubeiro.

Design coordinator: Rosa Marín González.
Design development coordinator: Javier Tejeda de la Calle.
Design development: Raúl de Andrés González, Jorge Gómez and Ana Gómez.

Technical director: Jorge Mira Fernández.
Technical subdirector: José Luis Verdasco Romero.

Technical coordinator: Marisa Valbuena Rodríguez.
Layout: Eva Hernández Malye and Artext.
Cartography: Paul Coulbois, GEODÓS, Geotec Cartografía, Asun Jiménez Cano, Rosa López, Tania López, Alberto Solano and Marcos Testón.
Art coordination: Carlos Aguilera.
Illustrations: El Infografista, Estudio Tenllado, Carlos Fernández del Castillo and Joaquín Marín.
Photo research: Amparo Rodríguez.

Photographs: A. Real; Algar/Instituto Iberoamericano de Cooperación; C. Díez Polanco; C. Jiménez/photoAlquimia; C. Roca; C. Valderrábano e I. Hernández; F. Ontañón; F. Po; G. Aldana; I. Meléndez; J. C. Muñoz; J. Escandell.com; J. Hernández; J. Jaime; J. L. G. Grande; J. Lucas; J. M.ª Escudero; J. V. Resino; Krauel; L. Gallo; M. Blanco; Michele di Piccione; ORONOZ; Prats i Camps; R. Manent; R. Tolín; S. Padura; T. Arias; 123RF; A. G. E. FOTOSTOCK/A. Koch, SGM, Science Museum/SSPL, CORMON Francis/hemi, Albatross/Duby Tal, Zoonar/Christa Eder, Fabian von Poser, DEA/G. Cigolini, Ulrich Doering/imageBROKER, Richard Semik, Ellen Isaacs, Michael Runkel, Dave Stamboulis, Angela Prati, Ignacio Palacios, Tyler Olson, Juan Carlos Calvin, Tom Cockrem, Mark Turner, Aqua Image, Jim West, DINODIA, J. C. Calvín; ALBUM/De Agostini Picture Library, Erich Lessing, Joseph Marin, Lessing, Prisma, DeA Picture Library, Oronoz, AKG; ARCHIVO PHOTOAISA/Beba/Iberfoto, Iberfoto, Leemage; CONTIFOTO/M. Rodríguez; CORDON PRESS/ The Granger Collection, New York, SYGMA/M. Attar, SYGMA / KEYSTONE, CORBIS/Holmul Archaeological Project/Xinhua PresS, Visuals Unlimited/Dr. Richard Roscoe, firoz ahmed/Demotix, Qlai Shen, Chris Sattlberger, Danny Lehman; COVER/CORBIS/Pallava Bagla, Oronoz; EFE/Boensch, B./picture alliance/ Arco Images G/Newscom/lafototeca.com, EPA PHOTO PRESS ASOCIATION, Francis R. Malasig, EPA/Andy Rain, Rahat Dar, UPI; GARCÍA-PELAYO/Juancho/CENTRO COMERCIAL SUPERDIPLO; GETTY IMAGES SALES SPAIN/Time Life Pictures/Hank Morgan, Bloomberg/Rajan Chaughule, Chris Parker/Desing Pics, Erika Tirén/Magic Air, Stone/Zubin Shroff, AFP/Gerard Julien, Ronaldo Schemidt, Photos.com Plus, DEA/A. Gregorio, Izzet Keribar, Sean Caffrey, Gerard Sioen, Chris Mellor, DEA/G.Sosio, Thinkstock, Carini Joe, luoman, Ayan82, Gamma-Rapho/Jean-Erick Pasquier, The Image Bank/Yann Layma; HIGHRES PRESS STOCK/AbleStock.com; INDEX/Bridgeman; ISTOCKPHOTO/Getty Images Sales Spain; MUSEUM ICONOGRAFÍA/J. Martin; NASA/Jeff Schmaltz, MODIS Land Rapid Response Team; PAISAJES ESPAÑOLES; PHOTODISC; R. TOLÍN/Roberto Tolín Sommer; SCALA GROUP/De Agostini Picture Library, Andrea Jemolo; Biblioteca Medicea laurenziana, Florencia; Newberry Library, Chicago; BP; BIBLIOTECA NACIONAL DE ESPAÑA/Laboratorio Biblioteca Nacional; BlackBerry; BLOM Sistemas Geoespaciales; CONGRESO DE LOS DIPUTADOS, MADRID; Cortesía de Apple; GALERÍA DE LOS UFFIZI, FLORENCIA; HOSPITAL TAVERA, TOLEDO; J. A. Botillo; J. E. Casariego; MATTON-BILD; MONASTERIO DE LA ENCARNACIÓN, PATRIMONIO NACIONAL; MUSEO MARÍTIMO, SEVILLA; MUSEO NACIONAL DEL PRADO/Laboratorio del Museo del Prado / © MUSEO DEL PRADO - MADRID - DERECHOS RESERVADOS. PROHIBIDA LA REPRODUCCIÓN TOTAL O PARCIAL; MUSEO NAVAL, MADRID; PINACOTECA VATICANA; REAL ACADEMIA DE BELLAS ARTES DE SAN FERNANDO, MADRID; Sanz Vega; SERIDEC PHOTOIMAGENES CD; WELLINGTON MUSEUM, LONDON; ARCHIVO SANTILLANA.

Avda. de los Artesanos 6,
28760 Tres Cantos, Madrid

Richmond Publishing is an imprint
of Santillana Educación, S.L.

Printed in Spain

Richmond Publishing
58 St Aldates
Oxford OX1 1ST
United Kingdom

ISBN: 978-84-680-1978-9
CP: 509798
DL: M-25762-2015